S0-AYY-516

Image-Based Rendering

Image-Based Rendering

Heung-Yeung Shum
Microsoft Research Asia

Shing-Chow Chan
University of Hong Kong

Sing Bing Kang
Microsoft Research USA

 Springer

Heung-Yeung Shum
Microsoft Research Asia

Shing-Chow Chan
University of Hong Kong

Sing Bing Kang
Microsoft Research USA

Library of Congress Control Number: 2006924121

ISBN-10: 0-387-21113-6   e-ISBN-10: 0-387-32668-5
ISBN-13: 978-0387-21113-8   e-ISBN-13: 978-0387-32668-9

Printed in the United States of America

9 8 7 6 5 4 3 2 1

springer.com

Foreword

Ten years ago there were two worlds of research that rarely crossed. My world, the world of computer graphics, tried to solve a well-defined problem. Given the geometrical and material description of a virtual scene, a definition of the light sources in that scene, and a virtual camera, create an image that looks as close as possible to one that a real camera in a real version of the described scene would look like. In the other research world, that of computer vision, researchers were struggling with the opposite question. Given one or more real images of a real scene, get the computer to describe the geometrical, material, and lighting properties in that real scene.

The computer graphics problem was having great success, to a point. Surprising to many of us, the real problem derived from a realization that the geometric complexity of the real world overwhelmed our abilities to create it in current geometric modelers. Consider a child's fuzzy stuffed lion. To create an accurate image of such a lion could require the daunting task of describing every strand of hair and how it reflects and transmits light. This is about when we went to knock on the doors of our computer vision colleagues. We hoped, a bit prematurely, that perhaps we could just point cameras at the fuzzy lion and a computer vision algorithm could spit out the full geometric description of the lion. The state-of-the-art was (and still is) that computer vision algorithms could give us an approximate model of the scene but not the kind of detail we needed. In some ways we were lucky, for if they had given us the full model in all its detail, the computer graphics rendering algorithms most likely could not deal with such complex models.

It was this point that the ideas underlying image based rendering were born. The idea is that by using the partial success of computer vision algorithms PLUS keeping the original pixels contained in the input images, one could then leverage the partial success of computer graphics. I feel quite honored to have been a part of making some of the first realistic synthetic images of a child's stuffed lion in just this way. The results that have grown from this initial idea have been quite astounding.

The book you are about to read introduces the reader to the rich collaboration that has taken place over the past decade at the intersection of computer graphics and computer vision. The authors have lived in this exciting research world and have produced many of the seminal papers in the field. The book provides both a historical perspective and the technical background for the reader to become familiar with the major ideas and advances in the field. This can, in turn, provide the basis for the many new discoveries awaiting anyone willing to jump into the field. The results are already having a major impact on the computer game and film industry. Similar ideas are also finding their way into consumer products for digital photography and video. Hopefully the reader will find new ways to apply these ideas in yet undiscovered ways.

Michael Cohen
March 2006
Seattle, WA

Preface

When image-based rendering (IBR) first appeared in the graphics scene about ten years ago, it was greeted with a lot of enthusiasm. It was new and fresh then and it had (and still has) the potential for generating photorealistic images. Unlike traditional 3D computer graphics in which 3D geometry of the scene is known, IBR techniques render novel views directly from input images. It was this aspect of IBR that attracted much attention. Pioneering works in this area include Chen and Williams' view interpolation, Chen's QTVR, McMillan and Bishop's plenoptic modeling, Levoy and Hanrahan's light field, and Gortler et al.'s Lumigraph.

IBR is unique in graphics in that it drew significant interest not only from researchers in graphics, but researchers in computer vision as well as image and signal processing. A lot of progress has been made in this field, in terms of improving the quality of rendering and increasing its generality. For example, more representations have been proposed in order to handle more scenarios such as increased virtual motion, complicated non-rigid effects (highly non-Lambertian surfaces), and dynamic scenes. Much more is known now about the fundamental issue of sampling for IBR, which is important for the process of image acquisition. There is also a significant amount of work on compression techniques specifically geared for IBR. These techniques are critical for the practical use of IBR in conventional PCs. Interestingly, despite the large body of work accumulated over the years, there was no single book that is devoted exclusively to IBR.

This was the primary motivation for this book. Much of the material in this book is the result of years of research by the authors at Microsoft Research (in Redmond and Beijing) and through collaboration between Microsoft Research and The University of Hong Kong. The book is intended for researchers and practitioners in the fields of vision, graphics, and image processing.

Microsoft Research Asia
The University of Hong Kong
Microsoft Research

Heung-Yeung Shum
Shing-Chow Chan
Sing Bing Kang
February 2006

Acknowledgments

There are many people we would like to thank for making this book possible. The figures in this book were used with permission from the following people: Aseem Agarwala, Simon Baker, German Cheung, Michael Cohen, Xavier Décoret, Olivier Faugeras, Doron Feldman, Bastian Goldlücke, Adrian Hilton, Stefan Jeschke, Takeo Kanade, Marc Levoy, Stephane Laveau, Maxime Lhuillier, Tim Macmillan, Marcus Magnor, Wojciech Matusik, Manuel Menezes de Oliveira Neto, Shmuel Peleg, Marc Pollefeys, Alex Rav-Acha, Steve Seitz, Richard Szeliski, Dayton Taylor, Christian Theobalt, Matthew Uyttendaele, Sundar Vedula, Daphna Weinshall, Bennett Wilburn, Colin Zheng, and Assaf Zomet. Yin Li and Xin Tong wrote the first draft for Chapter 4. Sundar Vedula provided useful comments for Chapters 3 and 4.

Heung-Yeung Shum would like to thank his co-authors: Rick Szeliski, Li-Wei He, Zhouchen Lin, Jin-Xiang Chai, Xin Tong, Lifeng Wang, Tao Feng, Yin Li, Qifa Ke, King To Ng, Jian Sun, Chi-Keung Tang, Minsheng Wu, Zhunping Zhang, Baining Guo, Steve Lin, Jin Li, Yaqin Zhang, Honghui Sun, Zhengyou Zhang and Shuntaro Yamazaki. Raj Reddy, Katsushi Ikeuchi, Rick Szeliski, and Michael Cohen have been instrumental in introducing H.-Y. Shum to the area of image-based modeling and rendering. He would like to express his appreciation to Eric Chen and Ken Turkowski for introducing him to QuickTime VR at Apple. He would also like to thank Pat Hanrahan for his helpful discussion on the minimum sampling of Concentric Mosaics.

Shing-Chow Chan would specifically acknowledge the following people and organization: King To Ng (for his help in preparing the manuscript); Vannie Wing Yi Lau, Qing Wu, and Zhi Feng Gan (for their help in preparing some of the figures); James Lap Chung Koo (for his excellent technical support and the development of the plenoptic video systems); Olivia Pui Kuen Chan (for help in capturing the plenoptic videos); the University of Hong Kong, the Hong Kong Research Grant Council, and Microsoft Research (for their support).

Sing Bing Kang would like to thank his past and present collaborators: Katsushi Ikeuchi, Rick Szeliski, Larry Zitnick, Matt Uyttendaele, Simon Winder, Yingqing Xu, Steve Lin, Lifeng Wang, Xin Tong, Zhouchen Lin, Antonio Criminisi, Yasuyuki Matsushita, Yin Li, Steve Seitz, Jim Rehg, Tat-Jen Cham, Andrew Johnson, Huong

Quynh Dinh, Pavan Desikan, Yanghai Tsin, Sam Hasinoff, Noah Snavely, Eric Bennett, Ce Liu, Rahul Swaminathan, Vaibhav Vaish, Yuanzhen Li, Songhua Xu, Yuanjie Zheng, Hiroshi Kawasaki, Alejandro Troccoli, Michael Cohen, and Gerard Medioni. He is especially grateful to Katsushi Ikeuchi for having the patience to train him to be an effective researcher while he was a graduate student at CMU. He also appreciates Rick Szeliski's invaluable guidance over the past ten years.

Finally, we would like to express our gratitude to Valerie Schofield, Melissa Fearon, and Wayne Wheeler at Springer for their assistance in making this book possible. They have been particularly supportive and patient.

Dedication

To our families, with love.

Contents

1

Introduction

One of the primary goals in computer graphics is photorealistic rendering. Much progress has been made over the years in graphics in a bid to attain this goal, with significant advancements in 3D representations and model acquisition, measurement and modeling of object surface properties such as the bidirectional reflectance distribution function (BRDF) and surface subscattering, illumination modeling, natural objects such as plants, and natural phenomena such as water, fog, smoke, snow, and fire. More sophisticated graphics hardware that permit very fast rendering, programmable vertex and pixel shading, larger caches and memory footprints, and floating-point pixel formats also help in the cause. In other words, a variety of well-established approaches and systems are available for rendering models. See the surveys on physically-based rendering [232], global illumination methods [69], and photon mapping (an extension of ray tracing) [130].

Despite all the advancements in the more classical areas of computer graphics, it is still hard to compete with images of real scenes. The rendering quality of environments in animated movies such as *Shrek 2* and even games such as *Ghost Recon* for Xbox 360^{TM} is excellent, but there are hints that these environments are synthetic. Websites such as http://www.ignorancia.org/ showcase highly photorealistic images that were generated through raytracing, which is computationally expensive. The special effects in high-budget movies blend seamlessly in real environments, but they typically involved many man-hours to create and refine. The observation that full photorealism is really hard to achieve with conventional 3D and model-based graphics has led researchers to take a "short-cut" by working directly with real images. This approach is called *image-based modeling and rendering*. Some of the special effects used in the movie industry were created using image-based rendering techniques described in this book.

Image-based modeling and rendering techniques have received a lot of attention as a powerful alternative to traditional geometry-based techniques for image synthesis. These techniques use images rather than geometry as the main primitives for rendering novel views. Previous surveys related to image-based rendering (IBR) have suggested characterizing a technique based on how image-centric or geometry-

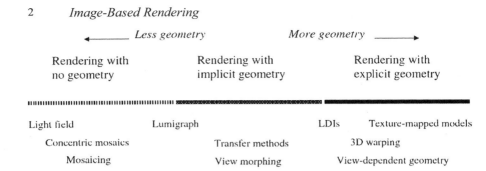

Fig. 1.1. IBR continuum. It shows the main categories used in this book, with representative members shown. Note that the Lumigraph [91] is a bit of an anomaly in this continuum, since it uses explicit geometry and a relatively dense set of source images.

centric it is. This has resulted in the image-geometry continuum (or *IBR continuum*) of image-based representations [155, 134].

1.1 Representations and Rendering

For didactic purposes, we classify the various rendering techniques (and their associated representations) into three categories, namely rendering with no geometry, rendering with implicit geometry, and rendering with explicit geometry. These categories, depicted in Figure 1.1, should actually be viewed as a continuum rather than absolute discrete ones, since there are techniques that defy strict categorization.

At one end of the IBR continuum, traditional texture mapping relies on very accurate geometric models but only a few images. In an image-based rendering system with depth maps (such as 3D warping [189], and layered-depth images (LDI) [264], and LDI tree [39]), the model consists of a set of images of a scene and their associated depth maps. The surface light field [323] is another geometry-based IBR representation which uses images and Cyberware scanned range data. When depth is available for every point in an image, the image can be rendered from any nearby point of view by projecting the pixels of the image to their proper 3D locations and re-projecting them onto a new picture. For many synthetic environments or objects, depth is available. However, obtaining depth information from real images is hard even with state-of-art vision algorithms.

Some image-based rendering systems do not require explicit geometric models. Rather, they require feature correspondence between images. For example, view interpolation techniques [40] generate novel views by interpolating optical flow between corresponding points. On the other hand, view morphing [260] results in-between camera matrices along the line of two original camera centers, based on point correspondences. Computer vision techniques are usually used to generate such correspondences.

At the other extreme, light field rendering uses many images but does not require any geometric information or correspondence. Light field rendering [160] produces a new image of a scene by appropriately filtering and interpolating a pre-acquired set of samples. The Lumigraph [91] is similar to light field rendering but it uses approximate geometry to compensate for non-uniform sampling in order to improve rendering performance. Unlike the light field and Lumigraph where cameras are placed on a two-dimensional grid, the Concentric Mosaics representation [267] reduces the amount of data by capturing a sequence of images along a circle path. In addition, it uses a very primitive form of a geometric impostor, whose radial distance is a function of the panning angle. (A geometric impostor is basically a 3D shape used in IBR techniques to improve appearance prediction by depth correction. It is also known as geometric proxy.)

Because light field rendering does not rely on any geometric impostors, it has a tendency to rely on oversampling to counter undesirable aliasing effects in output display. Oversampling means more intensive data acquisition, more storage, and higher redundancy.

1.2 Sampling

What is the minimum number of images necessary to enable anti-aliased rendering? This fundamental issue needs to be addressed so as to avoid undersampling or unnecessary sampling. Sampling analysis in image-based rendering, however, is a difficult problem because it involves unraveling the relationship among three elements: the depth and texture information of the scene, the number of sample images, and the rendering resolution. Chai *et al.* showed in their plenoptic sampling analysis [33] that the minimum sampling rate is determined by the depth variation of the scene. In addition, they showed that there is a trade-off between the number of sample images and the amount of geometry (in the form of per-pixel depth) for anti-aliased rendering.

1.3 Compression

Because image-based representations are typically image-intensive, compression becomes an important practical issue. Compression work has been traditionally carried out in the image and video communities, and many algorithms have been proposed to achieve high compression ratios. Image-based representations for static scenes tend to have more local coherence than regular video. The issues associated with dynamic scenes are similar for regular video, except that there is now the additional dimensions associated with the camera viewpoint. As a result, image-based representations have a significantly more complicated structure than regular video because the neighborhood of image samples is not just along a single time axis as for regular video. For example, the Lumigraph is 4D, and it uses a geometric impostor. Image-based representations also have special requirements of random access and selective decoding for fast rendering. As subsequent chapters will reveal, geometry has been used as a means for encoding coherency and compressing image-based representations.

1.4 Organization of book

This book is divided into four parts: representations and rendering techniques, sampling, compression, and systems and applications. Each part is relatively self-contained, but the reader is encouraged to read the Part I first to get an overall picture of IBR. In a little more detail:

Part I: Representations and Rendering Techniques

The chapters in this part survey the different representations and rendering mechanisms used in IBR. It starts with a survey of representations of static scenes. In this survey, important concepts such as the plenoptic function, classes of representations, and view-dependency are described. Systems for rendering dynamic scenes are subsequently surveyed. From this survey, it is evident that the design decisions on representation and camera layout are critical. A separate chapter is also devoted to rendering; it describes how rendering depends on the representation and what the common rendering mechanisms are.

Part II: Sampling

This part addresses the sampling issue, namely, the minimum sampling density required for anti-aliased rendering. The analysis of plenoptic sampling is described to show the connection between the depth variation of the scene and sampling density. Three different interpretations are given: using sampling theorem, geometric analysis, and optical analysis. A representation that capitalizes on the sampling analysis to optimize rendering performance (called layered Lumigraph) is also described in this part.

Part III: Compression

To make any IBR representation practical, it must be easy to generate, data-efficient, and fast to render. This part focuses on the sole issue of compression. IBR compression is different from conventional image and video compression because the non-trivial requirements of random access and selective decoding. Techniques for compressing static IBR representations such as light fields and Concentric Mosaics are described, as are those for dynamic IBR representations such as panoramic videos and dynamic light fields.

Part IV: Systems and Applications

The final part of the book showcases four different IBR systems. One system demonstrates how Concentric Mosaics can be made more compact using the simple observation about perception of continuous motion. Another system allows customized layout of representations to large scene visualization so as to minimize image capture. The layout trades off the number of images with the viewing degrees of freedom.

Segmentation and depth recovery are difficult processes—the third system was designed with this in mind, and allows the user to help correct for areas that look perceptually incorrect. This system automatically propagates changes to the user inputs to "pop-up" layers for rendering. Finally, the fourth system allows a light field to be morphed to another through user-assisted feature associations. It preserves the capability of light fields to render complicated scenes during the morphing process.

Part I

Representations and Rendering Techniques

The first part of the book is a broad survey of IBR representations and rendering techniques. While there is significant overlap between the type of representation and the rendering mechanism, we chose to highlight representational and rendering issues in separate chapters. We devote two chapters to representations: one for (mostly) static scenes, and another for dynamic scenes. (Other relevant surveys on IBR can be found in [155, 339, 345].)

Unsurprisingly, the earliest work on IBR focused on static scenes, mostly due to hardware limitations in image capture and storage. Chapter 2 describes IBR representations for static scenes. More importantly, it sets the stage for other chapters by describing fundamental issues such as the plenoptic function and how the representations are related to it, classifications of representations (no geometry, implicit geometry, explicit geometry), and the importance of view-dependency.

Chapter 3 follows up with descriptions of systems for rendering dynamic scenes. Such systems are possible with recent advancements in image acquisition hardware, higher capacity drives, and faster PCs. Virtually all these systems rely on extracted geometry for rendering due to the limit in the number of cameras. It is interesting to note their different design decisions, such as generating global 3D models on a per-timeframe basis versus view-dependent layered geometries, and freeform shapes versus model-based ones. The different design decisions result in varying rendering complexity and quality.

The type of rendering depends on the type of representation. In Chapter 4, we partition the type of rendering into point-based, layer-based, and monolithic-based rendering. (By monolithic, we mean single geometries such as 3D meshes.) We describe well-known concepts such as forward and backward mapping and ray-selection strategies. We also discuss hardware rendering issues in this chapter.

Additional Notes on Chapters

A significant part of Chapter 2 is based on the journal article "Survey of image-based representations and compression techniques," by H.-Y. Shum, S.B. Kang, and S.-C. Chan, which appeared in IEEE Trans. On Circuits and Systems for Video Technology, vol. 13, no. 11, Nov. 2003, pp. 1020-1037. ©2003 IEEE.

Parts of Chapter 3 were adapted from "High-quality video view interpolation using a layered representation," by C.L. Zitnick, S.B. Kang, M. Uyttendaele, S. Winder, and R. Szeliski, ACM SIGGRAPH and ACM Transactions on Graphics, Aug. 2004, pp. 600-608.

Xin Tong implemented the "locally reparameterized Lumigraph" (LRL) described in Section 2.4. Yin Li and Xin Tong contributed significantly to Chapter 4.

2

Static Scene Representations

In Chapter 1, we introduced the IBR continuum that spans a variety of representations (Figure 1.1). The continuum is constructed based on how geometric-centric the representation is. We structure this chapter based on this continuum: representations that rely on no geometry are described first, followed by those using implicit geometry (i.e., relationships expressed through image correspondences), and finally those with explicit 3D geometry.

2.1 Rendering with no geometry

We start with representative techniques for rendering with unknown scene geometry. These techniques typically rely on many input images; they also rely on the characterization of the plenoptic function.

2.1.1 Plenoptic modeling

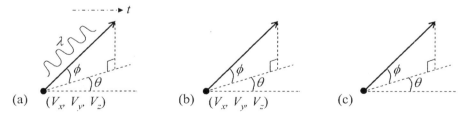

Fig. 2.1. Plenoptic functions: (a) full 7-parameter $(V_x, V_y, V_z, \theta, \phi, \lambda, t)$, (b) 5-parameter $(V_x, V_y, V_z, \theta, \phi)$, and (c) 2-parameter (θ, ϕ).

The original 7D plenoptic function [2] is defined as the intensity of light rays passing through the camera center at every 3D location (V_x, V_y, V_z) at every possible angle (θ, ϕ), for every wavelength λ, at every time t, i.e., $P_7(V_x, V_y, V_z, \theta, \phi, \lambda, t)$.

Adelson and Bergen [2] considered one of the tasks of early vision as extracting a compact and useful description of the plenoptic function's local properties (e.g., low order derivatives). It has also been shown by Wong *et al.* [322] that light source directions can be incorporated into the plenoptic function for illumination control. By removing two variables, time t (therefore static environment) and light wavelength λ, McMillan and Bishop [194] introduced the notion of plenoptic modeling with the 5D complete plenoptic function of the form $P_5(V_x, V_y, V_z, \theta, \phi)$.

The simplest plenoptic function is a 2D panorama (cylindrical [41] or spherical [291]) when the viewpoint is fixed, namely $P_2(\theta, \phi)$. A regular rectilinear image with a limited field of view can be regarded as an incomplete plenoptic sample at a fixed viewpoint.

Image-based rendering, or IBR, can be viewed as a set of techniques to reconstruct a continuous representation of the plenoptic function from observed discrete samples. The issues of sampling the plenoptic function and reconstructing a continuous function from discrete samples are important research topics in IBR. As a preview, a taxonomy of plenoptic functions is shown in Table 2.1.

Dimension	Year	View space	Name
7	1991	free	Plenoptic function
5	1995	free	Plenoptic modeling
4	1996	bounding box	Lightfield/ Lumigraph
3	1999	bounding circle	Concentric Mosaics
2	1994	fixed point	Cylindrical/Spherical panorama

Table 2.1. A taxonomy of plenoptic functions.

The cylindrical panoramas used in [194] are two-dimensional samples of the plenoptic function in two viewing directions. The two viewing directions for each panorama are panning and tilting about its center. This restriction can be relaxed if geometric information about the scene is known. In [194], stereo techniques are applied on multiple cylindrical panoramas in order to extract disparity (or inverse depth) distributions. These distributions can then be used to predict appearance (i.e., plenoptic function) at arbitrary locations. Similar work on regular stereo pairs can be found in [151], where correspondences constrained along epipolar geometry are directly used for view transfer.

2.1.2 Light field and Lumigraph

It was observed in both light field rendering [160] and Lumigraph [91] systems that as long as we stay outside the convex hull (or simply a bounding box) of an ob-

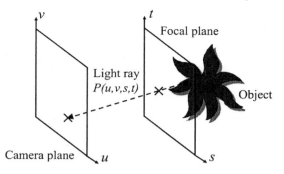

Fig. 2.2. Representation of a light field.

ject[1] and the medium is non-dispersive, we can simplify the 5D complete plenoptic function to a 4D light field plenoptic function,

$$P_4(u, v, s, t), \tag{2.1}$$

where (u, v) and (s, t) are parameters of two planes of the bounding box, as shown in Figure 2.2.

The (u, v) plane is the camera plane, where the sampling cameras are located. Figure 2.3(a) shows a visualization of the light field from the camera plane. From a point corresponding to a sampling camera location, the view is the original sampled view.

For the light field system of Levoy and Hanrahan, the (s, t) plane is the focal plane, where the scene is assumed to be located. A visualization of the light field from the focal plane is shown in Figure 2.3(b). Assuming that the surface of the scene is approximately at the focal plane, all the rays passing through a point in the focal plane are appearance samples of the same surface point from different views. This is akin to capturing the local BRDF of the scene surface for a fixed lighting condition. Rays are interpolated based on this assumption that the scene surface is close to the focal plane. Object surfaces that are located far away from the focal plane will appear blurred at interpolated views (this will be explained in the next section). On the other hand, the Lumigraph uses an approximated 3D object surface for view interpolation, which reduces the blur problem. Note that for the Lumigraph, the (u, v) plane is the focal plane while the (s, t) is the camera plane. A visualization of a subset of the full (u, v, s, t) space for the Lumigraph is shown in Figure 2.4.

In the rest of this book, we will follow the notation of Lumigraph where (s, t) is the camera plane and (u, v) is the focal plane.

Note that in general, the (u, v) and (s, t) planes need not be parallel. There is also an implicit and important assumption that the strength of a light ray does not change along its path. For a complete description of the plenoptic function for the bounding box, six sets of such two-planes would be needed. More restricted versions of Lu-

---

[1] The reverse is also true if camera views are restricted inside a convex hull.

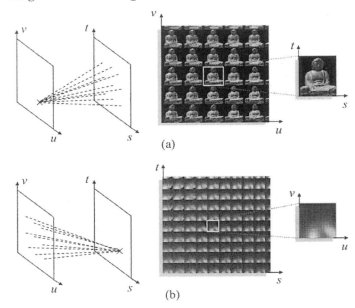

(a)

(b)

Fig. 2.3. The light field seen from (a) camera plane, and (b) focal plane. The boxed subimage is observed from a single point in the parameter plane. (Images courtesy of Marc Levoy.) ©1996 ACM, Inc. Included here by permission.

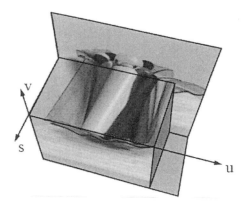

Fig. 2.4. An (s, u, v) slice of a Lumigraph. (Image courtesy of Michael Cohen.) ©1996 ACM, Inc. Included here by permission.

migraph have also been developed by Sloan *et al.* [277] and Katayama *et al* [141]. Here, the camera motion is restricted to a straight line.

The principles of light field rendering and Lumigraph are similar, except that the Lumigraph has the additional (approximate) object geometry for better compression and appearance prediction. For this reason, the Lumigraph technically belongs to the "explicit geometry" camp (Section 2.3). It is discussed here due to its strong similarity with the light field.

In the light field system, a capturing rig is designed to obtain uniformly sampled images. To reduce aliasing effect, the light field is pre-filtered before rendering. A vector quantization scheme is used to reduce the amount of data used in light field rendering, while achieving random access and selective decoding. On the other hand, the Lumigraph can be constructed from a set of images taken from arbitrarily placed viewpoints. A re-binning process (in this case, resampling to a regular grid using a hierarchical interpolation scheme) is therefore required. Geometric information is used to guide the choices of the basis functions. Because of the use of geometric information, the sampling density can be reduced.

The $P_4(u, v, s, t)$ two-plane parameterization is just one of many for light fields. Other types of light fields include spherical or isotropic light fields [113, 25], sphere-plane light fields [25], and hemispherically arranged light fields with geometry [181]. The issue of uniformly sampling the light field was investigated by Camahort [24]. He introduced an isotropic parameterization he calls the direction-and-point parameterization (DPP), and showed that while no parameterization is view-independent, only the DPP introduces a single bias.

Buehler *et al.* [22] extended the light field concept through a technique that uses geometric proxies (if available), handles unstructured input, and blends textures based on relative angular position, resolution, and field-of-view. They achieve real-time rendering by interpolating the blending field using a sparse set of locations.

2.1.3 Concentric Mosaics

Obviously, the more constraints we have on the camera location (V_x, V_y, V_z), the simpler the plenoptic function becomes. If we want to capture all viewpoints, we need a complete 5D plenoptic function. As soon as we stay in a convex hull (or conversely viewing from a convex hull) free of occluders, we have a 4D light field. If we do not translate at all, we have a 2D panorama. An interesting 3D parameterization of the plenoptic function, called Concentric Mosaics (CMs) [267], was proposed by Shum and He; here, the sampling camera motion is constrained along concentric circles on a plane.

By constraining camera motion to planar concentric circles, CMs can be created by compositing slit images taken at different locations of each circle, as shown in Figure 2.5. Two types of CMs are shown in Figure 2.6; in the first type, rays are arranged in the *tangential* direction (Figure 2.6(a)), and in the second type, rays are arranged in *normal* direction (Figure 2.6(b)). CMs define a 3D plenoptic function because they are sampled naturally by three parameters: rotation angle, radius, and vertical elevation. Clearly there is a one-to-one mapping between pixels in a CM and

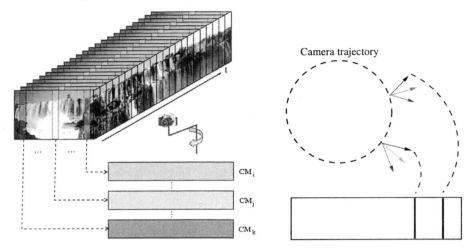

Fig. 2.5. Creation of CMs from source images. If the images are captured at regular intervals while rotated at a constant angular speed, each CM is created by just stacking the same columns from all the images in the order they are acquired. Note that the CM that consists of rays passing through the central axis of rotation is actually a (parallax-free) panorama. The left part of the figure is adapted from Figure 3 in [243].

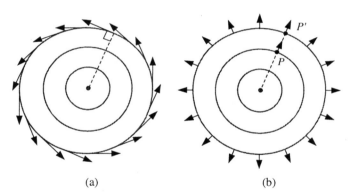

(a)   (b)

Fig. 2.6. Types of Concentric Mosaics (CMs): a plan view. A CM is assembled by unit width slit images (a) tangent to the circle; and (b) normal to the circle. We call (a) tangent CMs and (b) normal CMs. The CMs in [267] are actually tangent CMs.

their corresponding scene points. The CMs used in [267] are actually tangent CMs; unless otherwise specified, we meant tangent CMs when we mention CMs.

Novel views are rendered by combining the appropriate captured rays in an efficient manner at rendering time. Although vertical distortions exist in the rendered images, they can be alleviated by depth correction. CMs have good space and computational efficiency. Compared with a light field or Lumigraph, CMs have a much smaller file size because only a 3D plenoptic function is constructed.

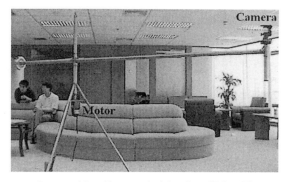

Fig. 2.7. Camera setup for acquiring Concentric Mosaics (CMs). The camera is counterbalanced by a weight; during image acquisition, it is rotated by a motor at a constant rotational speed.

Capturing CMs is almost as easy as capturing a traditional panorama except that CMs require more images. By simply spinning an off-centered camera on a rig shown in Figure 2.7, Shum and He [267] were able to construct CMs for a real scene in about 10 minutes. Like panoramas, CMs do not require the difficult modeling process of recovering geometric and photometric scene models. However, CMs provide a much richer user experience by allowing the user to move freely in a circular region and observe significant parallax and lighting changes. (Parallax refers to the apparent relative change in object location within a scene due to a change in the camera viewpoint.) The ease of capturing makes CMs very attractive for many virtual reality applications.

It has been shown [267] that a novel view inside the capturing circle can be rendered from the CMs without any knowledge about the depth of the scene. Three possible techniques for resampling CMs are shown in Figure 2.8. From densely sampled CMs, a novel view image can be rendered by linearly interpolating nearby rays from two neighboring CMs. In addition, a constant depth is assumed to find the best "nearby" rays for optimal rendering quality [33]. Figure 2.8(b) illustrates a rendering ray that is interpolated by two rays captured in nearby CMs. Despite the inevitable vertical distortion, CMs are very useful for wandering around (on a plane) in a virtual environment.

Rendered views of a lobby scene from captured CMs are shown in Figure 2.9. A rebinned CM at the rotation center is shown in Figure 2.9(a), while two rebinned CMs taken at exactly opposite directions are shown in Figure 2.9(b) and (c), respectively. It has also been shown in [225] that such two mosaics taken from a single rotating camera can simulate a stereo panorama. In Figure 2.9(d), strong parallax can be seen between the plant and the poster in the rendered images. More specifically, in the left image, the poster is partially obscured by the plant, while the poster and the plant do not visually overlap in the right image. This is a significant visual cue that the camera viewpoint has shifted.

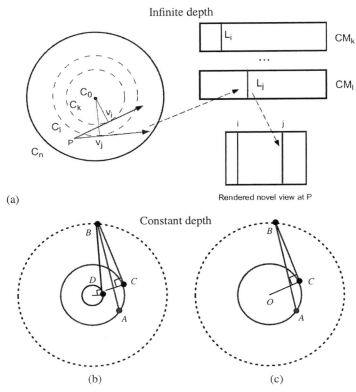

(a)

Rendered novel view at P

(b)   (c)

Fig. 2.8. Rendering CMs with (a) infinite depth, and (b,c) constant finite depth. (a) Rebinning: For a given ray in virtual view P (say v_j), the CM (CM_l) that is tangent to it is used. The column of pixels in CM_l that corresponds to the line of tangency with v_j is used to construct part of the new view. (b) View interpolation: A ray from viewpoint A is projected to the constant depth surface (represented as a dotted circle) at B, and interpolated by two rays BC and BD that are retrieved from neighboring CMs. (c) Local warping: A ray from viewpoint A is projected to the constant depth surface at B, and reprojected to the nearest CM by the ray BC.

Fig. 2.9. Rendering a lobby [267]: rebinned Concentric Mosaic (a) at the rotation center; (b) at the outermost circle; (c) at the outermost circle but looking at the opposite direction of (b); (d) parallax change between the plant and the poster.

2.1.4 Multiperspective images and manifold mosaics

A multiperspective image is assembled from rays captured from multiple viewpoints (e.g., [348]). Multiperspective images have also been called MCOP images [240], multiperspective panoramas [324], pushbroom images [97], and manifold mosaics [227], among other names. Let us consider the case of Peleg *et al.*'s notion of manifold mosaics. The manifold mosaic is created by projecting thin strips from images; the shape of these thin strips depend on the camera motion. More specifically, for each strip, the boundaries are perpendicular to the optic flow, and the width is proportional to the amount of motion. The basic idea is depicted in Figure 2.10, which also shows an example mosaic.

In this chapter, we define a *manifold mosaic* as a multiperspective image where each pixel has a one-to-one mapping with a scene point[2]. Therefore, a conventional perspective image, or a single perspective panorama, can be regarded as a degenerate manifold mosaic in which all rays are captured at the same viewpoint.

We adopt the term manifold mosaic from [226] because the viewpoints are generally taken along a continuous path or a manifold (surface or curve). For example, CMs are manifold mosaics constructed from rays taken along concentric circles [267]. Note that the concept of the manifold mosaic is widely used in Manifold hopping (Chapter 14).

---

[2] By this definition, MCOP images are *not* manifold mosaics.

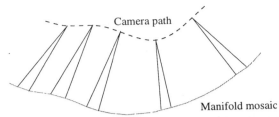

Fig. 2.10. Manifold mosaic [227]. Top: Graphical depiction of general camera motion. The wedges indicate representative parts of images used to create the manifold mosaic. Bottom: An actual manifold mosaic created using a hand-held camera. Image (courtesy of Shmuel Peleg) is from "Panoramic mosaics by manifold projection," by S. Peleg and J. Herman, IEEE Conference on Computer Vision and Pattern Recognition, June 1997, pp. 338-343. ©1997 IEEE.

Although many previous image-based rendering techniques (such as view interpolation and 3D warping) were developed for perspective images, they can be applied to manifold mosaics as well. For example, 3D warping has been used to reproject a multiple-center-of-projection (MCOP) image in [240, 216] where each pixel of an MCOP image has an associated depth.

2.1.5 Image mosaicing

A complete plenoptic function at a fixed viewpoint can be constructed from incomplete samples. Specifically, a panoramic mosaic is constructed by registering multiple regular images. For example, if the camera focal length is known and fixed, one can project each image to its cylindrical map and the relationship between the cylindrical images becomes a simple translation. For arbitrary camera rotation, one can first register the images by recovering the camera movement, before converting to a final cylindrical/spherical map.

Many systems have been built to construct cylindrical and spherical panoramas by stitching multiple images together, e.g., [187, 288, 41, 194, 291] among others. When the camera motion is very small, it is possible to put together only small stripes from registered images, i.e., slit images (e.g., [348, 226]), to form a large panoramic mosaic. Capturing panoramas is even easier if omnidirectional cameras (e.g., [207, 206]) or fisheye lens [330] are used.

Szeliski and Shum [291] presented a complete system for constructing *panoramic image mosaics* from sequences of images. Their mosaic representation associates a transformation matrix with each input image, rather than explicitly projecting all of the images onto a common surface, such as a cylinder. In particular, to construct a full view panorama, a *rotational mosaic* representation associates a rotation matrix (and optionally a focal length) with each input image. A *patch-based alignment* algorithm is developed to quickly align two images given motion models. Techniques for estimating and refining camera focal lengths are also presented.

Fig. 2.11. Tessellated spherical panorama covering the north pole (constructed from 54 images) [291].

In order to reduce accumulated registration errors, global alignment through block adjustment is applied to the whole sequence of images, which results in an optimally registered image mosaic. To compensate for small amounts of motion parallax introduced by translations of the camera and other unmodeled distortions, a local alignment (*deghosting*) technique [271] warps each image-based on the results of pairwise local image registrations. Combining both global and local alignment significantly improves the quality of image mosaics, thereby enabling the creation of full view panoramic mosaics with hand-held cameras.

A tessellated spherical map of the full view panorama is shown in Figure 2.11. Three panoramic image sequences of a building lobby were taken with the camera on a tripod tilted at three different angles. 22 images were taken for the middle sequence, 22 images for the upper sequence, and 10 images for the top sequence. The camera motion covers more than two thirds of the viewing sphere, including the top.

Apart from blending images to directly produce wider fields of view, one can use the multiple images to generate higher resolution panoramas as well (e.g., using maximum likelihood algorithms [115] or learnt image models [27]). There are also techniques to handle the exposure differences in the source image. For example, Uyttendaele *et al.* [303] perform block-based intensity adjustment to compensate for differences in exposures. More principled techniques have been used to compensate for the exposure through radiometric self-calibration (e.g., [26, 88, 186]).

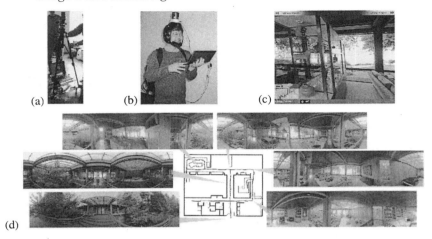

(a)   (b)   (c)

(d)

Fig. 2.12. Panoramic video [302]. (a), (b) Two versions of the capture system with Point Grey's Ladybug^{TM} 6-camera system, (c) screen shot of user interface for navigation, (d) sample panoramas along camera path network.

To produce high-quality navigation in a large environment (along a constrained set of paths), Uyttendaele *et al.*[302] capture panoramic video using Point Grey's Ladybug^{TM} 6-camera system. The resolution of each camera is 1024×768, and the capture rate was 15 fps. They mounted the Ladybug^{TM} on a tripod stand and dolly that can then be manually moved, as well as on a flattop skydiving helmet (Figure 2.12(a) and (b)). Once the images were processed to remove radial distortion and vignetting effects, they were then stitched frame by frame. The resulting panoramic video was stabilized using tracked features to provide smooth virtual navigation.

Zomet *et al.* [352] recently introduced a different way of producing mosaics called *crossed-slits projection*, or X-slits projection. What is interesting about this rendering technique is that the sampled rays passes *two* non-parallel slits, an example of which is shown at the top of Figure 2.13 (where the slits are perpendicular). The benefits are two-fold: the generated mosaics appear closer to being perspective, and interesting virtual navigation can be obtained merely by changing the location of one slit. The bottom of Figure 2.13 shows examples of visualization that can be obtained through X-slits.

2.1.6 Handling dynamic elements in panoramas

Early approaches for generating panoramas from rotated images do not compensate for exposure changes or moving elements in the scene. Once the relative transforms for the images have been computed, Davis [59] handles moving elements in the scene by segmenting the panorama into disjoint regions and sampling pixels in each region from a single input image. Uyttendaele *et al.* [303] cast the moving element problem as a graph, with nodes representing moving objects (i.e., objects that appear in one image but not in another). A vertex cover algorithm is then used to remove all but

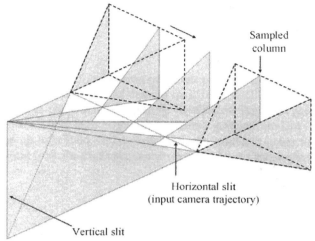

Fig. 2.13. Rendering using crossed-slits projection [352]. Top: A depiction of the idea with perpendicular slits. Bottom: The two left images are example source images of a rotating object, and the right two images are synthesized views of a virtual looming camera. Images (courtesy of Assaf Zomet, Doron Feldman, Shmuel Peleg, and Daphna Weinshall) are from "Mosaicing new views: The crossed-slits projection," by A. Zomet, D. Feldman, S. Peleg, and D. Weinshall, IEEE Transactions on Pattern Analysis and Machine Intelligence, 25(6):741-754, June 2003. ©2003 IEEE.

Fig. 2.14. Graph-based deghosting [303]. Left: Without deghosting. Right: With deghosting. Images (courtesy of Matthew Uyttendaele) are from "Eliminating ghosting and exposure artifacts in image mosaics," by M. Uyttendaele, A. Eden, and R. Szeliski, IEEE Conference on Computer Vision and Pattern Recognition, Dec. 2001, vol. 1, pp. 2-9. ©2001 IEEE.

See color plate section near center of book.

one instance of each object. A result of their technique can be seen in Figure 2.14; notice the dramatic improvement in the final panorama.

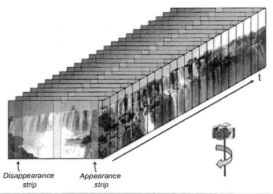

Fig. 2.15. Dynamosaic [243]. Images (courtesy of Alex Rav-Acha and Shmuel Peleg) are from "Dynamosaics: Video mosaics with non-chronological time," by A. Rav-Acha, Y. Pritch, D. Lischinski, and S. Peleg, IEEE Conference on Computer Vision and Pattern Recognition, June 2005, pp. 58-65. ©2005 IEEE.

If the image sampling is reasonably dense enough (e.g., slowly panning a camera on a scene with quasi-repetitive motion), manifold mosaics may be used (as described in Section 2.1.4). However, an interesting effect may be obtained by globally stabilizing the images in time and considering slices of the resulting space-time volume as mosaics. Rav-Acha *et al.* [243] refer to such a sequence of mosaics as a dynamosaic. As a simple example shown in Figure 2.15, the camera pans from left to right. The first mosaic is constructed by taking the "appearance strip" of every image, and the last constructed using the "disappearance strip" of every image. One such mosaic is shown at the bottom of Figure 2.15. As the mosaic is played in sequence using strips shifting from "appearance strip" to "disappearance strip", the video shows a panoramic movie of the falls, with the water flowing down. The slicing scheme can be arbitrary, creating specific effects as desired. (Given a video of a swimming meet, for example, by manipulating the spatial temporal slice shapes, a swimmer can be made to appear to swim faster or slower.)

Agarwala *et al.* [3] use a different approach to produce a video panorama from similar input. Their technique is based on video textures [259], where similar frames at different times are found and used to produce new seamless video. To produce a video panorama from just a panning video, they globally register the frames and manually tag regions as static and dynamic. The basic concept of video texture is applied to the dynamic regions to ensure that the video panorama can be played indefinitely. They construct the objective function to minimize difference between static and dynamic areas that overlap and ensure local spatial consistency for hypothesized time offsets. This function is set up as an MRF and solved.

Fig. 2.16. Video panorama [3]. Top: representative input frames. Bottom: A frame (cropped) of the extracted video panorama. (Images courtesy of Aseem Agarwala and Colin Zheng.) ©2005 ACM, Inc. Included here by permission.

2.2 Rendering with implicit geometry

The techniques described in the previous section sample directly from the source images to produce virtual views. Relative transforms between cameras or optic flow fields are computed mainly for stabilization for panorama creation. In this section, we describe a class of techniques that relies on positional correspondences (typically across a small number of images) to render new views. This class has the term *implicit* to express the fact that geometry is not directly available; 3D information is computed only using the usual projection calculations. In certain cases where the cameras are only weakly calibrated, Euclidean 3D information is not available even with correspondence information. New views are computed based on direct manipulation of these positional correspondences, which are usually point features.

The approaches under this class are view interpolation, view morphing, joint view interpolation, and transfer methods with fundamental matrices and trifocal (or trilinear) tensors. View interpolation uses general dense optic flow to directly generate intermediate views. The intermediate view may not necessarily be geometrically correct. View morphing is a specialized version of view interpolation, except that the interpolated views are always geometrically correct. The geometric correctness is ensured because of the linear camera motion. Transfer methods are also produce geometrically correct views, except that the camera viewpoints can be arbitrarily positioned.

2.2.1 View interpolation

Chen and Williams' view interpolation method [40] is capable of reconstructing arbitrary viewpoints given two input images and dense optical flow between them. This method works well when two input views are close by, so that visibility ambiguity does not pose a serious problem. Otherwise, flow fields have to be constrained so as to prevent foldovers. In addition, when two views are far apart, the overlapping parts of two images may become too small. Chen and Williams' approach works particularly well when all the input images share a common gaze direction, and the output images are restricted to have a gaze angle less than $90°$.

Establishing flow fields for view interpolation can be difficult, in particular for real images. Computer vision techniques such as feature correspondence or stereo must be employed. For synthetic images, flow fields can be obtained from the known depth values.

2.2.2 View morphing

From two input images, Seitz and Dyer's view morphing technique [260] reconstructs any viewpoint on the line linking two optical centers of the original cameras. Intermediate views are exactly linear combinations of two views only if the camera motion associated with the intermediate views are perpendicular to the camera viewing direction. To see this, let us assume the projection matrices for the two sampled viewpoints are Π_1 and Π_2. Without loss of generality, we can set $\Pi_0 = M_0(I_{3\times3}|\mathbf{0})$ and $\Pi_1 = M_1(I_{3\times3}|\mathbf{p})$. $I_{3\times3}$ is the 3×3 identity matrix, M_0 and M_1 are the intrinsic matrices, with

$$M_i = \begin{pmatrix} f_i & s_i & q_{xi} \\ 0 & a_i f_i & q_{yi} \\ 0 & 0 & 1 \end{pmatrix}.$$

f_i, a_i, s_i, (q_{xi}, q_{yi}) are the focal length, aspect ratio, skew, and principal point, respectively. $\mathbf{p} = (p_x \ p_y \ 0)^{\mathrm{T}}$ is the relative camera motion. Note that the z component of $\mathbf{p}$ is zero, which is the key to image linearity.

A given 3D point $\mathbf{x} = (X, Y, Z, 1)^{\mathrm{T}}$ is projected to $\mathbf{u}_0 = \frac{1}{Z}\Pi_0\mathbf{x}$ in view 0 and $\mathbf{u}_1 = \frac{1}{Z}\Pi_1\mathbf{x}$ in view 1. Suppose we linearly interpolate the 2D position in virtual view $\hat{I}_t$ as $\mathbf{u}_t = (1 - t)\mathbf{u}_0 + t\mathbf{u}_1$, with $0 \leq t \leq 1$. Interestingly, we have

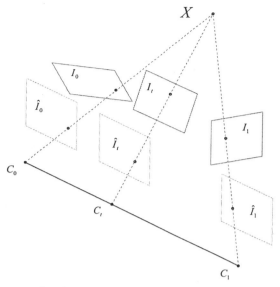

Fig. 2.17. View morphing [260]. I_0 and I_1 are the two source views at C_0 and C_1, respectively, and I_t is the synthesized view. The idea of view morphing is to rectify the source views (yielding $\hat{I}_0$ and $\hat{I}_1$), linearly interpolate (producing $\hat{I}_t$), and transform back to unrectified state (I_t). ©1996 ACM, Inc. Included here by permission.

Fig. 2.18. Two examples of view morphing. Top row: Interpolation result (middle image) for two images of the same face. Bottom row: Morphing between two different faces. In both cases, point correspondences were manually established. (Images courtesy of Steve Seitz.) ©1996 ACM, Inc. Included here by permission.

$\mathbf{u}_t = \frac{1}{2}\Pi_t\mathbf{x}$, with $\Pi_t = (1 - t)\Pi_0 + t\Pi_1$ being a valid (but virtual) intermediate intrinsic matrix. As a result, for parallel source views, it is physically correct to just linearly interpolate point positions (assuming the point correspondences are valid).

If the two source images are not parallel, a pre-warp stage can be employed to rectify two input images so that corresponding scan lines are parallel. Accordingly, a post-warp stage can be used to un-rectify the intermediate images. Note that this is possible without fully calibrating the camera. Scharstein [253] extends this framework to camera motion in a plane. He assumes, however, that the camera parameters are known.

In a more recent work, Aliaga and Carlbom [4] describe an interactive virtual walkthrough system that uses a large network of omnidirectional images taken within a 2D plane. To construct a view, the system uses the closest set of images, warps them using precomputed corresponding features, and blends the results.

2.2.3 Joint view triangulation

The biggest problems associated with view interpolation are pixel matching and visibility reasoning. Visibility reasoning is especially difficult in cases where the source images are uncalibrated; as a result, there is no relative depth information to predict occlusion in new views. Lhuillier and Quan [162] proposed the idea of *joint view triangulation* (JVT) to handle these problems.

There are two pre-processing steps to JVT: quasi-dense matching and planar patch construction. Quasi-dense matching consists of interest point extraction and matching (using zero-mean normalized cross-correlation). This step produces an initial list of correspondences sorted by the correlation score. This list is traversed in order, beginning with the best score, to search within the neighborhood of the point correspondence for more matches. The uniqueness constraint is used to ensure the final list consists of non-replicated points. The second step of planar patch construction assumes that the scene is piecewise smooth. It is also performed to remove possible mismatches. One of the images is subdivided into a regular patch grid; RANSAC (Random Sample Consensus) [74] is then applied to each patch to extract its homography.

Quasi-dense matching and planar patch construction are followed by the actual JVT algorithm. The basic idea of JVT is to generate Delaunay triangulations on both source images such that there is one-to-one correspondence in vertices and edges. The vertices and edges correspond to those of the precomputed patches. The patches are added raster style; they are labeled as matched or unmatched as appropriate. Patches that have no matches are given hypothesized transforms to preserve continuity with those that have matches. View interpolation is then done by rendering unmatched patches, followed by matched patches.

Lhuillier and Quan [163] later extended their work to using epipolar geometry for more robust correspondence extraction. To overcome the restrictions of using coarse preset patches, they added edge-based partitions to better fit object boundaries. Results for an outdoor scene using their JVT algorithm can be seen in Figure 2.19.

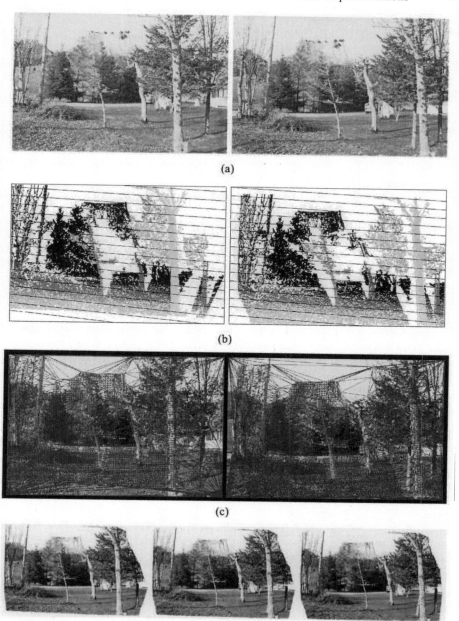

Fig. 2.19. Joint view triangulation. (a) Source views with extracted point matches. (b) Computed disparity at points with correspondence. The darker the pixel, the smaller the disparity. White pixels represent those without any correspondence. Epipolar lines (shown as dark lines) are superimposed. (c) Resulting meshes with constraint edges (in red). (d) Interpolated views. Images courtesy of Maxime Lhuillier.

See color plate section near center of book.

2.2.4 Transfer methods

Transfer methods (a term used within the photogrammetric community) are characterized by the use of a relatively small number of images with the application of geometric constraints (either recovered at some stage or known *a priori*) to reproject image pixels appropriately at a given virtual camera viewpoint. The geometric constraints can be of the form of known depth values at each pixel, *epipolar constraints* between pairs of images, or *trifocal/trilinear tensors* that link correspondences between triplets of images. The view interpolation and view morphing methods above are actually specific instances of transfer methods.

Using fundamental matrix

Laveau and Faugeras [151] use a collection of images called reference views and the principle of the *fundamental matrix* [72] to produce virtual views. The fundamental matrix F is a 3×3 matrix of rank 2. More specifically, if $\mathbf{u}_1$ and $\mathbf{u}_2$ are two corresponding points in views 1 and 2 respectively, we have $\mathbf{u}_2^T F_{12} \mathbf{u}_1 = 0$. Also, $\mathbf{u}_2$ lies in the *epipolar line* given by $F_{12}\mathbf{u}_1$. Another important concept is the *epipole*: All epipolar lines intersect at the epipole, and the epipole is the projection of the other camera projection center. The epipole $\mathbf{e}_{12}$ in view 2 is the null space of F_{12}, i.e., $F_{12}\mathbf{e}_{12} = 0$.

Suppose the point correspondences and fundamental matrix for a pair of images have been extracted. The virtual camera viewpoint (view 3) is specified by the user choosing two points $\mathbf{e}_{13}$ (in image 1) and $\mathbf{e}_{23}$ (in image 2) such that $\mathbf{e}_{23}^T F_{12} \mathbf{e}_{13} = 0$. The image plane associated with view 3 is then interactively choosen by specifying three pairs of corresponding points plus one point in one of the images. It is not necessary to manually pick the last corresponding point in the other image because it can be automatically obtained using the collinearity and epipolar constraints.

To avoid holes, the new view is computed using a reverse mapping or raytracing process, as shown at the top of Figure 2.20. For every pixel $\mathbf{m}_3$ in the new target image, a search is performed to locate the pair of image correspondences in two reference views. More specifically, for the ith pixel $\mathbf{m}_{1i}$ along the epipolar line in view 1 (given by $F_{31}\mathbf{m}_3$), we check if its corresponding point $\mathbf{m}_{2i}$ (part of cor($F_{31}\mathbf{m}_3$)) in view 2 satisfies the epipolar constraint $\mathbf{m}_{2i}^T F_{32}\mathbf{m}_3 = 0$. In other words, we search along $F_{31}\mathbf{m}_3$ until the curve cor($F_{31}\mathbf{m}_3$) and epipolar line $F_{32}\mathbf{m}_3$ intersects. The pixel is transferred if such a point of intersection is found. Cases where no such point exists or multiple locations exist are discussed in [151]. An example result of using this technique is shown in Figure 2.20.

Note that if the camera is only weakly calibrated, the recovered viewpoint will be that of a projective structure (see [72] for more details). This is because there is a class of 3D projections and structures that will result in exactly the same source images. Since angles and areas are not preserved, the resulting viewpoint may appear warped. Knowing the internal parameters of the camera removes this problem. In a later work, Faugeras *et al.* [73] use geometric information of the scene (such as line orthogonality) to recover Euclidean structure from uncalibrated images.

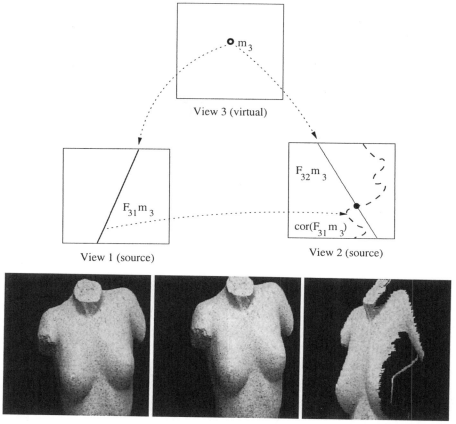

Fig. 2.20. View synthesis using the fundamental matrix [151]. Top: Process of finding the corresponding points in source views 1 and 2 that projects to point m_3 in virtual view 3. Bottom, from left to right: two source images, and novel oblique view. Images (courtesy of Stephane Laveau and Olivier Faugeras) are from "3-D scene representation as a collection of images," by S. Laveau and O.D. Faugeras, International Conference on Pattern Recognition, Oct. 1994, vol. A, pp. 689-691. ©1994 IEEE.

Using trifocal tensor

While the fundamental matrix establishes projective relationship between two rectilinear views without any knowledge of scene structure, the trifocal (or trilinear) tensor establishes a projective relationship across *three* views. The trifocal tensor is a $3 \times 3 \times 3$ matrix with a number of properties related to relationships of points and lines across three views and extraction of fundamental and projection matrices [101].

One particularly interesting property is that given a pair of point correspondences in two source images, the trifocal tensor can be used to compute the corresponding point in the third image without resorting to explicit 3D computation. Note that the 2-image epipolar search technique of [151] fails when the two epipolar lines in the virtual image are coincident (or becomes numerically unstable and sensitive to noise nearing this condition). Fortunately, the trifocal tensor avoids this degenerate case because of the flexible nature of relationships between points and lines across the three views. For example, suppose we wish to compute point $\mathbf{m}_3$ in the third view given corresponding points $\mathbf{m}_1$ and $\mathbf{m}_2$ in the first and second views, respectively, and trifocal tensor $\mathcal{T}$, with the (i, j, k)th element indexed as $\mathcal{T}_i^{jk}$ (using the terminology in [101]). We can find line $\mathbf{l}_2$ perpendicular to the epipolar line given by $F_{21}\mathbf{m}_2$, after which $\mathbf{m}_3$ can be determined from the relationship $(\mathbf{m}_3)^k = (\mathbf{m}_1)^i (\mathbf{l}_2)_j \mathcal{T}_i^{jk}$. $(\mathbf{m})^k$ refers to the kth element of point $\mathbf{m}$ and $(\mathbf{l})_j$ refers to the jth element of line $\mathbf{l}$.

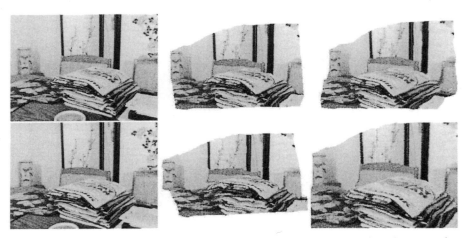

Fig. 2.21. Example of visualizing using the trilinear tensor. The left-most column are the source images, with the rest synthesized at arbitrary viewpoints.

The point transfer property of the trifocal tensor has been used to generate novel views from either two or three source images [6]. Here, the idea of generating novel views from two or three source images is rather straightforward. First, the "reference" trilinear tensor is computed from the point correspondences between the source images. In the case of only two source images, one of the images is replicated and regarded as the "third" image. In [6], the camera's intrinsic parameters are assumed known, which simplifies the specification of the new view. The trifocal tensor associated with the new view can be computed from the known pose change (i.e., changes in rotation and translation) with respect to the third camera location. With the trifocal tensor and correspondences across two source views known, points can then be transferred through forward mapping (i.e., transferring pixels from source to virtual views). It is not clear how visibility is handled in [6], but the modified painter's algorithm can be used without explicit depth reasoning. In addition, splatting, where a pixel in the source image is mapped to multiple pixels, can be used to remove holes in the new view. (The issues of forward mapping, modified painter's algorithm, and splatting are discussed in Chapter 4, Section 4.3.1.) A set of novel views created using this approach can be seen in Figure 2.21.

2.3 Representations with explicit geometry

Representations that do not rely on geometry typically require a lot of images for rendering, and representations that rely on implicit geometry require accurate image registration for high-quality view synthesis. In this section, we describe IBR representations that use explicit geometry. Such representations have direct 3D information encoded in them, either in the form of depth along known lines-of-sight, or 3D coordinates. The more traditional 3D model with a single texture map is a special case in this category (not described here, since its rendering directly uses the conventional graphics pipeline).

Representations with explicit geometry include billboards, sprites, relief textures, Layered Depth Images (LDIs), and view-dependent textures and geometry. Sprites can be planar or have arbitrary depth distributions; new views are generated through 3D warping. LDIs are extensions of depth per-pixel representations, since they can encode multiple depths along a given ray. View-dependent texture mapping refers to mapping multiple texture maps to the same 3D surface, with their colors averaged using weights based on proximity of the virtual viewpoint relative to the source viewpoints.

2.3.1 Billboards

In games, *billboards* are often used to represent complex objects such as trees. They are either single texture-mapped rectangles that are kept fronto-parallel with respect to the viewing camera (i.e., view aligned), or sets of two rectangles arranged in a cross. Their popularity stems from the low footprint and ease of rendering (directly using the traditional graphics pipeline), but they typically work well only when viewed at a distance. The flat appearance is very pronounced when seen close up; very complex objects may appear unsatisfactory even at a reasonable distance.

To reduce these problems, Decoret *et al.* [63] proposed the use of *billboard clouds* (see Figure 2.22). A billboard cloud is just a set of textured, partially transparent billboards, with each billboard having an independent size, orientation, and

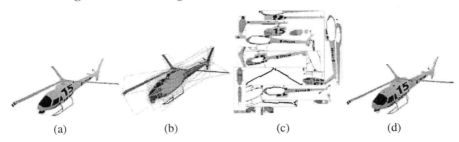

(a)   (b)   (c)   (d)

Fig. 2.22. Example of billboard cloud [63]. (a) Original 3D model, (b) locations of billboards, (c) textured billboards, (d) view of combined billboards. (Images courtesy of Xavier Décoret.) ©2003 ACM, Inc. Included here by permission.

texture resolution. Because a billboard cloud does not require topological information such as polygon connectivity, its format is easy to create, store, and read. Starting with a 3D model, Decoret *et al.* use an optimization approach to produce a set of representative textured planes that produce geometric errors within a specified threshold. To simplify the search, plane parameters are discretized into bins; planes are extracted sequentially by iteratively picking the bin with the minimum error. (There is the subsequent adaptive refinement in plane space—details can be found in [63].) Despite the improvements over regular billboards, billboard clouds are not intended for extreme close-ups.

2.3.2 3D warping

When the depth information is available for every point in one or more images, 3D warping techniques (e.g., [192]) can be used to render nearly all viewpoints. An image can be rendered from any nearby point of view by projecting the pixels of the original image to their proper 3D locations and re-projecting them onto the new picture. The most significant problem in 3D warping is how to deal with holes generated in the warped image. Holes are due to the difference of sampling resolution between the input and output images, and the disocclusion where part of the scene is seen by the output image but not by the input images. To fill in holes, splatting is used. Chapter 4 has a more detailed description of this process.

To improve the rendering speed of 3D warping, the warping process can be factored into a relatively simple pre-warping step and a traditional texture mapping step. The texture mapping step can be performed by standard graphics hardware. This is the idea behind relief texture, a rendering technique proposed by Oliveira and Bishop [216]. A similar factoring approach has been proposed by Shade *et al.* in a two-step algorithm [264], where the depth is first forward warped before the pixel is backward mapped onto the output image.

The 3D warping techniques can be applied not only to the traditional perspective images, but also multi-perspective images as well. For example, Rademacher and Bishop [240] proposed to render novel views by warping multiple-center-of-projection images, or MCOP images.

2.3.3 Layered Depth Images

To deal with the disocclusion artifacts in 3D warping, Shade *et al.* proposed Layered Depth Images, or LDIs [264], to store not only what is visible in the input image, but also what is behind the visible surface. In their paper, the LDI is constructed either using stereo on a sequence of images with known camera motion (to extract multiple overlapping layers, see Figure 2.23) or directly from synthetic environments with known geometries. In an LDI, each pixel in the input image contains a list of depth and color values where the ray from the pixel intersects with the environment.

Fig. 2.23. Layered depth image example [264]. Five source images were used to generate the layered representation of the scene using the technique in [9]. Top: Extracted layers. Bottom: Reconstructed views. Images (courtesy of Richard Szeliski) from "A layered approach to stereo reconstruction," by S. Baker, R. Szeliski, and P. Anandan, IEEE Computer Society Conference on Computer Vision and Pattern Recognition, June 1998, pp. 434-441. ©1998 IEEE.

Though an LDI has the simplicity of warping a single image, it does not consider the issue of sampling density. Chang *et al.* [39] proposed LDI trees so that the sampling rates of the source images are preserved by adaptively selecting an LDI in the LDI tree for each pixel. While rendering the LDI tree, only the level of LDI tree that is comparable to the sampling rate of the output image need to be traversed.

2.3.4 View-dependent texture mapping

Texture maps are widely used in computer graphics for generating photo-realistic environments. Texture-mapped models can be created using a CAD modeler for a synthetic environment. For real environments, these models can be generated using a 3D scanner or applying computer vision techniques to captured images. Unfortunately, vision techniques are not robust enough to recover accurate 3D models. In addition, it is difficult to capture visual effects such as highlights, reflections, and transparency using a single texture-mapped model.

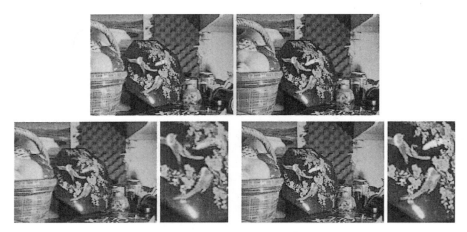

Fig. 2.24. Importance of view-dependent texture and geometry. Depth maps were extracted with the source images as reference views using the multi-view stereo technique described in [140]. Top: Source images. Notice the significant changes in the highlights. Bottom: Interpolated (left) vs. actual (right) views, with close-ups of the highlights. The highlights are a little blurred in the virtual view but resemble the actual version.

To obtain these visual effects of a reconstructed architectural environment, Debevec *et al.* in their Façade [61] work, used view-dependent texture mapping to render new views by warping and compositing several input images of an environment. This is the same as conventional texture mapping, except that multiple textures from different sampled viewpoints are warped to the same surface and averaged, with weights computed based on proximity of the current viewpoint to the sampled viewpoints. A three-step view-dependent texture mapping method was also proposed later by Debevec *et al.* [60] to further reduce the computational cost and to have smoother blending. This method employs visibility preprocessing, polygon-view maps, and projective texture mapping. For the unstructured Lumigraph work, Buehler *et al.* [22] apply a more principled way of blending textures based on relative angular position, resolution, and field-of-view. Kang and Szeliski [139] use not just view-dependent textures, but *view-dependent geometries* as well. This is to account for the fact that stereo is only locally valid for scenes with non-Lambertian properties. They blend warped depth images (depth maps and textures) to produce new views, as shown in Figure 2.24.

There are other approaches designed to handle non-rigid effects for IBR (mostly for synthetic scenes). For example, Heidrich *et al.* [105] handle reflections and refractions by decoupling geometry and illumination. This is accomplished by replacing the usual ray-color mapping with ray-ray mapping. Rendering is done by constructing this geometry light field and using it to look up the illumination from an environment map. On the other hand, Lischinski and Rappoport's [171] idea for handling non-diffuse scenes is based on layered depth images (LDIs) [9, 264]. They partition the scene into diffuse (view-independent) and non-diffuse parts. The view-independent parts are represented as three orthogonal high-resolution LDIs while the non-diffuse parts are represented as view-dependent lower-resolution LDIs. Rendering is done by warping the appropriate LDIs.

Another representation that accounts for non-rigid effects is the surface light field [323], which handles complex reflections in real-world data. However, they also require detailed geometry (obtained with a laser scanner) and a very large number of input images to capture all the effects. We now describe an IBR representation which was designed to handle some non-rigid effects without the use of detailed geometry. Issues associated with the difficulty of modeling non-rigid effects are also discussed.

2.4 Handling non-rigid effects

In this section, we describe an IBR representation to handle non-rigid effects compactly, called *locally reparameterized Lumigraph* (LRL)[3] [299]. The LRL is based on the use of local and separate *diffuse* and *non-diffuse geometries*. The diffuse geometry is associated with true or approximately true depth while the non-diffuse geometry has virtual depth that provide local photoconsistency with respect to its neighbors. This is similar to [171] in that there is the notion of using layers. In contrast to [171], however: (1) all local geometries are view-dependent, and (2) the rendering mechanism is different. The local geometries are used for depth correction, and do not contain radiance information. In [171], rendering is accomplished by warping LDIs. The LRL was designed to handle two common non-rigid effects: planar reflection and transparency, and specularity off low-curvature surfaces.

The concept of the LRL can be explained by first analyzing the diffuse and non-diffuse effects using the Epipolar Plane Image (EPI) [17] as a visualization tool. An EPI is basically a 3D representation (u, v, t) of a stacked sequence of camera images taken along a path, with (u, v) being the image coordinates and t being the frame index, and is therefore a 3D slice through a 4D Light Field. It has been used for stereo and multiview rendering (e.g., [100]).

---

[3] This name is actually inspired by the term "dynamically reparameterized Light Field" used in [116].

2.4.1 Analysis using the EPI

For the case of a laterally translated camera (along the x-direction) as shown in Figure 2.25(a), a typical EPI slice parallel to the $u - t$ plane is shown in Figure 2.25(b). In the EPI slice, multiple trails correspond to points similar in color and brightness moving across the EPI image. Trails that correspond to diffuse parts of the scene track the same points, and thus are straight. In fact, the slope of a diffuse trail k is proportional to the depth of its corresponding point. On the other hand, a specular trail does not necessarily track the same scene point; as a result, it is often curved.

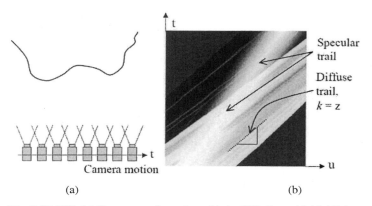

(a)   (b)

Fig. 2.25. EPI. (a) Camera configuration, (b) An EPI slice with highlights.

Figure 2.26 shows a different common phenomenon, that of planar reflection. The flower painting is being reflected off the glass covering the Mona Lisa painting. The corresponding EPI slice shows two overlapping sets of trails. The first set corresponds to those in the Mona Lisa painting, while the second corresponds to the flower painting. The slopes for the Mona Lisa trails are linked to their depths. The slopes for the flower trails are linked to their *virtual* depths, by considering the principles of optical reflection. Note that the flower trail slopes are less steep than those of the Mona Lisa counterparts, indicating that the flower painting is (at least virtually) *behind* the Mona Lisa painting.

It is thus easy to see why using a single global or even multiple view-dependent local geometries may not be adequate to handle such effects. The easiest recourse would be to sample the images more densely. However, there is a much better way: diffuse and non-diffuse scene components are modeled separately using what are called local diffuse and non-diffuse geometries.

2.4.2 Local diffuse and non-diffuse geometries

In an EPI slice, two local slopes can be observed within the vicinity of where a highlight or reflection occurs: one corresponding to the diffuse component, the other the

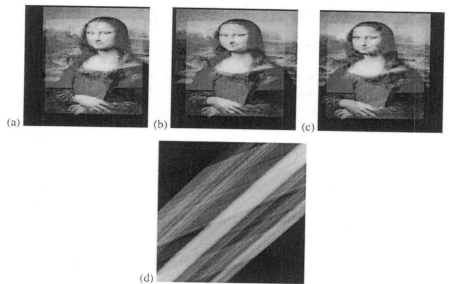

Fig. 2.26. Visualizing the planar reflection effect for an object moving from right to left. (a-c) Three snapshots of the sequence, (d) EPI of a highlight constructed by stacking middle rows of images.

non-diffuse component. Both of them are represented with the appropriately named local geometries to provide depth compensation for Light Field rendering. (In general, the non-diffuse trail may not stay within an EPI slice, but rather jump from slice to slice. It is more proper to say that *3D EPI trails* are tracked, and the argument of local geometries still applies. This simple scenario is used for illustrative purposes.)

Local geometry is defined to be view-dependent geometry used for depth compensation only within the neighborhood of the viewing camera location. The tighter the sampling camera configuration, the smaller this neighborhood is. This is in the same spirit as [237], for example. Note that the local geometry associated with a non-diffuse area is *virtual*, i.e., there may be no physical entity in the scene that corresponds to that area, as shown by the reflection phenomenon. *The function of local geometry, real or virtual, is to approximate the EPI trail as much as possible.* For the LRL, a fronto-parallel plane is used to represent local geometry. Stereo data of real scenes were not used because such data tend to be less reliable in the presence of occlusions, non-rigid effects, and untextured surfaces, all of which are prevalent in image sets of real scenes. In addition, the analysis detailed in [33] showed that it is not necessary to use exact geometry for antialiased rendering.

In synthetic environments where the geometry, diffuse, and non-diffuse parts are known, using a single global diffuse geometry is adequate. However, for images of real scenes, it is very likely that multiple local diffuse geometries will be needed. This is to compensate for errors in camera parameters and shape, or incorrect separation of diffuse and non-diffuse components. However, it is expected that the local diffuse geometry would change much more slowly than its non-diffuse counterpart.

The implications for using two different "layers" in the form of these geometries can be seen in Figure 2.27. The analysis shown by Chai *et al.* [33] indicates that it is the depth variation and not absolute depth in the scene that dictates the sampling rate. The bigger the depth variation, the larger the sampling rate required for antialiased rendering. The presence of the non-diffuse component has the effect of expanding the perceived depth variation, as can be seen in Figure 2.27. If just a single geometry (be it view-dependent or global) is used, the sampling rate has to be high to accommodate the non-diffuse effect. However, if the non-diffuse can be separated from the diffuse, both can be depth compensated separately, yielding a tighter perceived depth variation, as seen to the right of Figure 2.27(b). As a result, antialiased rendering is possible using a lower sampling rate.

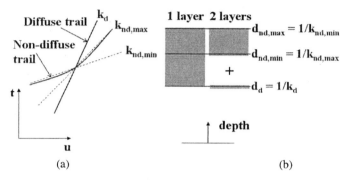

Fig. 2.27. Benefit of using separate layers. (a) Closeup view of vicinity of non-diffuse trail, (b) Depth variation (shaded regions) required to represent both diffuse and non-diffuse component using 1 and 2 layers. The dotted lines in (a) are the bounding slopes for the non-diffuse trail at the point of intersection between the diffuse and non-diffuse trails.

2.4.3 Implementation

To separate the non-diffuse from the diffuse in the real scene experiment with planar reflection, the following was done:

1. Choose the image with little or no reflection as the reference.
2. Perform dominant motion estimation between the reference and the others. In principle, a 2D perspective or homography motion should be used. The affine transform was used because it was adequate and had fewer parameters to estimate.
3. Compute a *min-composite* of the registered images, since this in principle optimally removes the reflection [289]. A min-composite is extracted by taking the color associated with the *minimum* luminance (across all registered images) at each pixel.

4. For each image, perform image difference between it and the motion-compensated min-composite to estimate the reflective components of the scene.

A big assumption here is that the reflective components are additive and that the surface is uniformly reflective. This is reasonable as long as there is no pixel intensity saturation. Portions of the reflected image do, of course, get "trimmed" (e.g., by the borders of the picture frame shown in Figure 2.28). This is one reason why a single global view of the reflection layer is inadequate, and a local view-based representation is preferable. For details on a more accurate means for separating the reflection component from the diffuse component, see the work of [289].

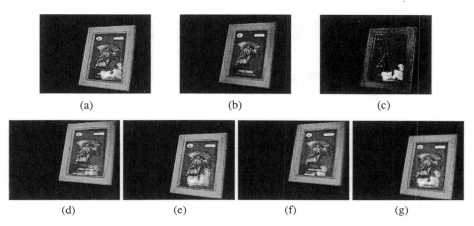

(a)   (b)   (c)

(d)   (e)   (f)   (g)

Fig. 2.28. Results for a real scene with reflection: (a) An original image, (b) Same image with only diffuse component, (c) Same image with only non-diffuse (reflection) component, (d,e) Rendering with single local depth at two different virtual camera poses, (f,g) Rendering with two local depths at the same virtual camera poses.

The LRL renderer is similar to the one described for the Lumigraph [91], i.e., it uses the two-slab, 4D parameterization of light rays. As with the Lumigraph, each rendering ray is computed based on quadrilinear interpolation of rays from the nearest four sampling cameras using the local geometry for depth compensation. After rendering each layer separately, the results are then directly added to produce the output view.

2.4.4 Results with two real scenes

This section describes the results of two experiments involving real scenes: the first with strong reflection components and the second with highlights. Both sets were acquired using a camera attached to a vertical precision X-Y table that can accurately translate the camera to programmed positions. The first scene consists of a picture frame with a toy dog placed at an angle to it on the same side as the camera. A grid

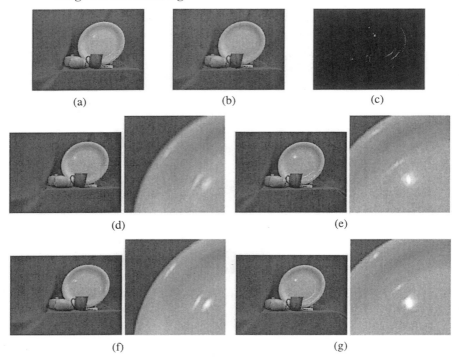

Fig. 2.29. Results for a real scene with highlights: (a) An original image, (b) Same image with only diffuse component, (c) Same image with only non-diffuse (specular) component, (d,e) Rendering with single local depth at two different virtual camera poses, (f,g) Rendering with two local depths at the same virtual camera poses. The right subimages in (d-g) are closeups of one of the highlight areas.

of 9×9 images, each of resolution 384×288, were captured. Figure 2.28 shows the rendering results for this image set. The rendering resolution is also 384×288. The layers are separated using the dominant motion estimation technique as descibed earlier. The rendering mechanism is exactly the same as in the previous synthetic experiment, with local fronto-parallel planes as local geometries. These planes are prespecified in the experiments.

As can be seen from Figure 2.28(a-c), the layers have mostly been separated, with some residual errors. Despite these errors, the rendering results using the LRL representation look markedly better than using just a single geometry. The rendered reflections shown in Figure 2.28(f,g), which are the result of using two local geometries (LRL), are much sharper than those shown in Figure 2.28(d,e), which are the result of using only one local geometry. The slightly blurred frame and picture in Figure 2.28(f,g) are caused by errors in separating the layers.

In the second set, the scene captured was that of a collection of household articles, including a cup and a plate. The images were taken at positions in a 65×5 grid, and both the original and rendering resolutions are 768×576. In this case, we took two sets of images at the same camera locations; one set with the lamp switched off and another with the lamp on. The highlights are computed based on the difference between these two sets. Even with this crude means of extracting layers, we are able to generate good rendering results, as shown in Figure 2.29. The highlights as shown in Figure 2.29(f,g), which are the result of using two local geometries (LRL), are visibly crisper than those shown in Figure 2.29(d,e), which are the result of using only one local geometry.

2.4.5 Issues with LRL

In addition to handling the diffuse and non-diffuse components separately, another important feature of the LRL representation is that it can handle *negative depths* to account for possible negative slopes in the EPI trail. On the other hand, the system of Lischinski and Rappoport [171], which is based on warping of LDIs, cannot accommodate negative depths explicitly.

The LRL uses only two local "layers" or geometries. It would be reasonably straightforward to extend this representation to accommodate multiple local geometries (diffuse and multiple non-diffuse). This would be useful in handling cases where multiple non-diffuse components overlap within the captured images.

In practice, it is difficult to separate the diffuse and non-diffuse components completely for images of real scenes. However, partial separation appears to be better than none at all. This notion was verified to a certain extent by the experiments involving real scenes, since diffuse/non-diffuse separation was not perfect.

One limitation of the LRL is that it is not able to adequately account for complex BRDF behavior, such as rapidly changing BRDF over the scene surface. An example would be the shimmering surface of satin. In addition, occlusions cannot be handled perfectly, as geometry within the vicinity of scene discontinuity cannot be extracted and hence represented exactly for real scenes. Furthermore, human eyes are very sensitive to edges and can detect anomalies very easily, which compounds the problem.

2.5 Which representation to choose?

There are many factors influencing the choice of the representation to use: ease of data capture, ease of processing, rendering speed, memory footprint, database size, degrees of freedom and spatial extents of navigation, and quality of reconstruction. We discuss a subset of these factors here.

By definition, for IBR representations, only images are required. As we have seen, most IBR representations require additional data, be it image correspondence or geometry. In IBR approaches such as those described in [237, 323], geometry captured using a 3D scanner is used. In others (e.g., [106, 139]), stereo is used to automatically extract depth maps. Manual assistance has also been used to produce the desired morphing results (e.g., [260]) or geometry for rendering (e.g., [22, 61]).

The image capture process varies substantially from one representation to another. Custom equipment is required for light fields [160] and Concentric Mosaics (CMs) [267]. Such a requirement is relaxed for the Lumigraph [91] and the plenoptic modeling work of [106], where a hand-held camera is used. Similarly, Chai *et al.* [31] demonstrated visualizations with similar quality to CMs using images taken with only approximate circular trajectories.

View-dependent textures on global geometry are used to account for view-dependent appearance changes such as highlights and non-Lambertian behavior (e.g., [61]). However, this assumes that such a consistent global geometry can be extracted easily. This is not true for a general real scene with unknown surface properties and lighting conditions. To reduce the severity of this problem, Pulli *et al.* [237] and Kang and Szeliski [139] use view-dependent geometry as well.

For representations that are based on implicit geometry, correspondence between the source images has to be somehow obtained for view transfer. Ideally, establishing correspondence should be automatic, and indeed, techniques for this do exist [101] (though for sparse correspondence). In addition, to facilitate novel viewpoint specification, full camera calibration is required to ensure Euclidean view reconstruction. It is much less intuitive to specify a new viewpoint if cameras are only weakly calibrated; in addition, view reconstruction is only up to a projective transform. This may result in an unnatural-looking skewed scene. However, if the centers of the source cameras lie in a line (or in general within a plane), it can be shown that the disparities are linear with camera motion once the images have been rectified [260]. Regardless, *full* frame correspondence is in general a very difficult problem, especially in the presence of occlusion and non-linear effects such as highlights and transparency.

For a representation to be compelling, it has to allow a reasonably wide range of viewpoints to be selected. While most representations allow a wide selection of viewpoints, they require a substantial amount of data to be captured—this is not attractive from a practical standpoint. In addition, a certain amount of specialized knowledge about cameras is required to minimize the number of images captured. Content creators need to have some measure of understanding of complex issues such as trade-offs between field of view and resolution, type of scenes to avoid, relative placement of the camera to the scene (to avoid degenerate camera motions), and density of sampling. It is thus not surprising that despite IBR as a field having been around for some number of years, most of its representations have not been adopted for widespread commercial use. The notable exceptions are the simplest ones such as panoramas.

2.6 Challenges

This chapter shows that representations and rendering techniques can differ radically, depending on design decisions related to ease of capture, use of geometry, accuracy of geometry (if used), number and distribution of source images, degrees of freedom for virtual navigation, and expected scene complexity. IBR as an area of research has been around for about ten years, and substantial progress has been achieved in effectively capturing, representing, and rendering scenes. However, many challenges remain.

The ability to handle general complex scenes remains a big issue for IBR. The easiest scenes to render remain those with mostly Lambertian surfaces. While there are techniques that can handle reflection or translucency and highlights to a certain extent (e.g., [139, 166, 287, 301] and the LRL described in Section 2.4), a substantial amount of work is still required to ensure robustness. The surface light field [323] handles such effects, but it requires accurate geometry and many source images. What if the scene is highly complicated, like a bush or a very cluttered office? How can we capture the surface subscattering or inter-reflection effect of an object with just images? How many images are enough? Should the new representation be view-dependent and multi-layered to account for depth, matting, and non-linear effects?

Since IBR, by definition, uses source images for rendering, interacting with IBR representations remains a challenging issue. Chapter 17 describes a technique for morphing from light field to another, and while it is interesting, it is also a very specific operation. What about typical operations such as object removal and insertion? How can we properly relight real scenes?

IBR techniques that use transfer methods for generating virtual views tend to use a small number of source images. The issues associated with the standard computer vision problems of feature selection and correspondence, occlusion handling, and structure from motion apply. Again, most techniques assume Lambertian surfaces.

Most of the IBR techniques described in this chapter are designed for static scenes. While photorealistic visualization of static scenes can be compelling, there is a limit on the amount of information that can be conveyed from an appearance frozen in time. The ability to rendering dynamic scenes is substantially more appealing. The next chapter surveys work done on capturing and rendering dynamic scenes.

3

Rendering Dynamic Scenes

Almost all of the early work on IBR involves the capture and rendering of static scenes. While rendering photorealistic static scenes is intriguing in its own right, its appeal and use is limited. When the "bullet time" effect (the illusion of stopping time and changing the camera viewpoint) appeared in the action film *The Matrix* in 1999, the effect was fresh and looked spectacular. It was not long before this effect, also known as the "freeze frame" effect, was emulated (and spoofed) in many productions, especially commercials. While there is some debate about who actually invented this effect, this effect has actually been used in film before *The Matrix*.

Fig. 3.1. Two versions of Time-Slice camera. Left: The "Slade" camera in action, the moment of exposure of "Splash." Photo from British Artist's Film and Video Collection at the Tate Britain. ©1984 Tim Macmillan. Right: The "Josephine" camera in action (in Bahamas) for the BBC documentary series "Supernatural." ©1997 Time-Slice Films Ltd.

There is documentation that the "freeze frame" effect was first demonstrated by Tim Macmillan in the early 1980's [217]; he calls his camera system the "Time-Slice" camera[1]. Two versions of his film-based Time-Slice camera are shown in Fig-

---
[1] http://www.timeslicefilms.com/

ure 3.1. An earlier version (called "Slade," left photo of Figure 3.1) consists of 360 pinhole film cameras arranged in a circle looking towards the center of the circle. Filming was done in the dark using a flash. Film stock was fed into the camera via a magazine and advanced 360 frames after each take. A later version (called "Josephine," right photo of Figure 3.1) has 120 glass lenses covering 90°. It is capable of very fast sequential exposures and can create a 5-second clip of "freeze frame" footage[2].

Beginning 1995, Macmillan used the "freeze frame" effect in his broadcast TV work. Contemporaneously, Dayton Taylor used his film-based Timetrack system[3] to produce commercials. In both cases, the "freeze frame" effect were produced by rapidly jumping between different still cameras arranged along a path to give the illusion of moving through a frozen slice of time. Michel Gondry's "Like a Rolling Stone" music video clip for the Rolling Stones in 1995 has also been credited with using the "bullet-time" effect as well. In his case, morphing between two simultaneous camera shots was done to produce this effect [276].

Fig. 3.2. Movia digital camera system (photo taken in April 2003). The 36 cameras are synchronized and saved uncompressed directly to digital disk recorders. The frame resolution is 1024×768 and the capture rate is 15 fps. Photo courtesy of Digital Air.

The huge interest in the "freeze frame"/"bullet time" special effect has probably helped spur research in capturing and rendering dynamic scenes. The "bullet time" effect in *The Matrix* was a one-time, pre-planned affair. The viewpoint trajectory was planned ahead of time, and many man hours were spent to produce the desired interpolated views. Newer systems such as Digital Air's Movia are based on video camera arrays (see Figure 3.2), which initially relied on having many cameras to avoid software view interpolation. Digital Air has been using view interpolation based on

---

[2] Personal communication with Tim Macmillan.
[3] http://www.timetrack.com/

optical flow since 1996, and is currently working on complete 3D reconstruction for dynamic view interpolation (using up to 80 high-definition cameras at 30 fps)[4].

Extending IBR to dynamic scenes beyond just a few seconds and with arbitrary viewpoint selection while the scene is changing (or "bullet time" on demand), is not trivial. Some of the problems are associated with the difficulty and cost of synchronizing so many cameras as well as acquiring and storing the images. However, decreasing costs of hardware (such as PCs and cameras), cameras that can be easily synchronized, faster PCs, and higher capacity drives have helped make the capture of dynamic scenes and their subsequent processing more practical. Another problem is the difficulty of automatically generating seamless interpolation between views for arbitrary scenes. In this chapter, we review IBR approaches that handle dynamic scenes and highlight several of them.

3.1 Video-based rendering

IBR approaches designed to handle dynamic scenes may be considered as "video-based rendering," with view synthesis accomplished in both along the space and time dimensions. The term "video-based rendering" was used in both [176] and [259], though [259] refers to synthesizing views only in the time dimension. A repository of related work can be found in http://www.video-based-rendering.org/.

One of the earliest attempts at capturing and rendering dynamic scenes was Kanade *et al.*'s Virtualized Reality work [133]. The first version of their system uses 51 cameras distributed on a 5-meter geodesic dome. More details of this work are given in a later section. Moezzi *et al.* have similar goals for rendering dynamic scenes. They call their system Immersive Video [200], and use three to six synchronized cameras to capture different viewpoints of a scene. The static portion of the scene (background) is first manually built; dynamic objects are extracted as time-varying voxel representations extracted through volume intersection, from which isosurface objects are created and subsequently rendered. All model construction is done of-fline.

Matusik *et al.* [190] use images from four calibrated FireWire cameras (each 256×256) to compute and shade visual hulls. The computation is distributed across five PCs, which can render 8000 pixels of the visual hull at about 8 fps. Yang *et al.* [335] designed an 8×8 grid of 320×320 cameras for capturing dynamic scenes. Instead of storing and rendering the video data, they transmit only the rays necessary to compose the desired virtual view. In their system, the cameras are not genlocked; instead, they rely on internal clocks across six PCs. The camera capture rate is 15 fps, and the interactive viewing rate is 18 fps.

Using the Lumigraph structure with per-pixel depth values, Schirmacher *et al.* [258] were able to render interpolated views at rates ranging from 2 to 9 fps (depending on image size, number of input cameras, and whether depth data has to be computed on-the-fly).

---

[4] Personal communication with Dayton Taylor.

Yang *et al.* [336] use graphics hardware to compute stereo data through plane sweeping and subsequently render new views. They are able to achieve a rendering rate of 15 fps with five 320 × 240 cameras. However, the matching window used is only one pixel, and occlusions are not handled.

As proof of concept for storing dynamic light fields, Wilburn *et al.* [317] demonstrated that it is possible to synchronize six cameras (640 × 480 at 30 fps), and compress and store all the video data in real time. They have since increased the size of the system to 128 cameras [305]. The resulting system, called the Stanford Light Field Camera, is described later.

The MPEG community has also been investigating the issue of visualizing dynamic scenes, which it terms "free viewpoint video." The first ad hoc group (AHG) on 3D audio and video (3DAV) of MPEG was established at the 58th meeting in December 2001 in Pattaya, Thailand. A good overview of this MPEG activity is presented by Smolić and Kimata [280].

3.2 Stereo with dynamic scenes

Many images are required to perform image-based rendering if the scene geometry is either unknown or known to only a rough approximation. If geometry is known accurately, it is possible to reduce the requirement for images substantially [91]. One practical way of extracting the scene geometry is through stereo. Within the past 20 years, many stereo algorithms have been proposed for static scenes (see, for example [254] for a review of stereo techniques). However, stereo data can be more reliably extracted from multiple synchronized video streams by taking advantage of temporal coherency, assuming that objects do not move too fast.

As part of the Virtualized Reality^{TM} work, Vedula *et al.* [309] proposed an algorithm for extracting 3D motion (i.e., correspondence between scene shape across time) using 2D optical flow and 3D scene shape at each time step. In their approach, they use a voting scheme similar to voxel coloring [261], where the measure used is how well a hypothesized voxel location fits the (linearized) 3D flow equation.

Zhang and Kambhamettu [344] also integrated 3D scene flow and structure in their framework. A 3D affine motion model is used locally, with spatial regularization, and discontinuities are preserved using color segmentation. Tao *et al.* [293] assume the scene is piecewise planar. They also assume constant velocity for each planar patch in order to constrain the dynamic depth map estimation.

In a more ambitious effort, Carceroni and Kutulakos [28] recover piecewise continuous geometry and reflectance (Phong model) under non-rigid motion with known lighting positions. They discretize the space into surface elements ("surfels"), and perform a search over location, orientation, and reflectance parameter to maximize agreement with the observed images.

In an interesting twist to conventional local window matching, Zhang *et al.* [343] use matching windows that straddle space and time. The advantage of this method is that there is less dependence on brightness constancy over time. Their best results are for experiments involving structured lighting.

Active rangefinding techniques have also been applied to moving scenes. Hall-Holt and Rusinkiewicz [98] use projected boundary-coded stripe patterns that vary over time. Depth is computed by comparing temporally adjacent stripe patterns. Because of the temporal dependence in extracting stripe patterns, this technique is not designed for fast dynamic scenes.

There is also a commercial system on the market called ZCam^{TM}, which is a range sensing video camera add-on used in conjunction with a broadcast video camera (NTSC and PAL formats). [5] Its resolution is about 1 cm. However, it is an expensive system, and provides single viewpoint depth only, which makes it less suitable for free viewpoint video.

3.3 Virtualized Reality^{TM}

One of the first attempts at capturing dynamic scenes and rendering them at arbitrary viewpoints was Kanade *et al.*'s Virtualized Reality^{TM} work [133].

3.3.1 Video acquisition system

Their first system involved 51 cameras arranged around a 5-meter geodesic dome (see Figure 3.3(a)). The resolution of each camera is 640×480 and the capture rate 30 fps. Later improved versions of their dynamic capture system (with larger workspace, better lighting control, better quality cameras, and longer recording time) can be seen in Figure 3.3(b) and (c). For example, in the second and later versions, video from multiple cameras was recorded directly to hard disks.

To handle the high bandwidth, Kanade *et al.* acquire their video in two steps: real-time recording and off-line digitization. Real-time recording involves standard CCD cameras and off-the-shelf VCRs. The cameras are synchronized using a common external synch signal. The timing information is stored as time stamps in the video tapes. In the second step, the video tapes are digitized.

3.3.2 Camera calibration and model extraction

The intrinsic and extrinsic camera parameters are calibrated using Tsai's technique [300]. Because of the wide distribution of the cameras, the calibration process is done in two steps. In the first step, the intrinsic parameters of each camera are extracted individually using a movable planar calibration pattern. With the cameras in their positions for acquisition, their relative poses are extracted using a known set of dots on the floor (ensuring that the pattern is visible to all cameras).

To extract the shape of the scene being modeled, Kanade *et al.* compute multiple stereo depth maps, each from a different vantage point. Each stereo depth map is computed at a given vantage point using immediate neighboring cameras and applying the multibaseline algorithm developed by Okutomi and Kanade [215].

---

[5] http://www.3dvsystems.com/products/zcam.html

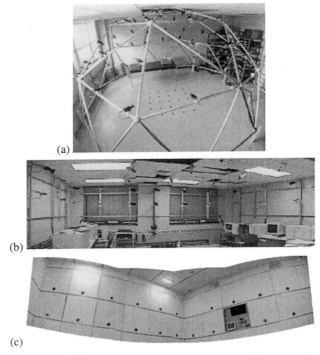

(a)

(b)

(c)

Fig. 3.3. Virtualized Reality^{TM} [133] arrangement of cameras: (a) First generation; (b) Second generation; (c) Third generation. Images courtesy of Takeo Kanade and colleagues.

These stereo depth maps can be used as view-dependent geometries for new view synthesis. In one version, the closest reference view is used, with two other nearest views being used to fill holes. In another version, all the depth maps are merged into a single surface model using a version of Curless and Levoy's volumetric integration technique [58]. The second version is used because of the ease of programming for hardware rendering. In an earlier version of the work [133], the models are computed independently of time. Results of this work are shown in Figure 3.4. They subsequently improve model extraction by capitalizing on both spatial and temporal coherence.

3.3.3 Spatial-temporal view interpolation

A multiple video-based acquisition system that captures a dynamic scene essentially discretely samples the scene in time and viewpoint space. The goal of rendering the dynamic scene is to smoothly interpolate the samples using the closest views in time and viewpoint. This concept is illustrated in Figure 3.5.

To achieve spatial-temporal view interpolation, Vedula *et al.* explicitly recover the 3D scene shape at every time frame as well as 3D scene flow [306]. "3D scene

Fig. 3.4. Results for an earlier version of Virtualized Reality^{TM}: (a) two of 51 input images, (b) extracted 3D geometry over time, and (c) rendered texture-mapped 3D geometries in a synthetic gym. Images courtesy of Peter Rander, Sundar Vedula, and Takeo Kanade.

Fig. 3.5. Spatial-temporal view interpolation. Images courtesy of Sundar Vedula, Simon Baker, and Takeo Kanade.

flow" refers to the local instantaneous 3D non-rigid temporal deformation of the dynamic scene. It connects points on the 3D scene surface at time t to scene points at time $t - 1$ or time $t + 1$. Vedula *et al.* apply the voxel coloring algorithm [261] to extract the 3D scene shapes at every time instant independently, from which 3D scene flow is extracted at every voxel on the surface of the scene. (Other formulations for 3D scene flow are detailed in [307, 308]. The cases covered are: when geometry is known at every time instant, when image correspondences are given, and when there is no knowledge of the 3D surface.) The inputs used for spatial-temporal view interpolation are shown in Figure 3.6.

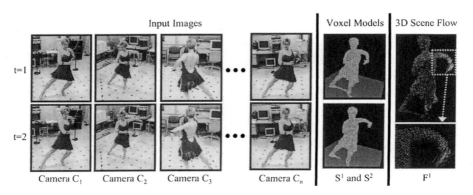

Fig. 3.6. View interpolation through 3D scene flow. Figure courtesy of Sundar Vedula, Simon Baker, and Takeo Kanade.

Figure 3.7 shows the process of constructing the novel view at time t^* (between time t and time $t + 1$). First, the 3D scene flow information is used to interpolate the (smoothed) shape S^* between the scene shapes at time t and time $t + 1$. Using the process of ray-casting, the point $X_i^{t^*}$ on the surface of shape S^* projects to a given pixel (u, v) in the novel view can be computed.

Using the 3D scene flow information again, points X_i^t (in shape S^t at time t) and X_i^{t+1} (in shape S^{t+1} at time $t + 1$) corresponding to $X_i^{t^*}$ can be estimated. The colors of pixels that X_i^t and X_i^{t+1} are projected onto their respective images are then blended to produce the color at (u, v) for the novel view. The weights used in blending the different contributing pixels depend on a combination of temporal proximity and spatial proximity measured by the angle subtended between the virtual ray at time t^* and the actual rays at time t and time $t + 1$.

To reduce the pixelation effect due to discretization of the 3D scene surface, two steps are taken. First, smooth surfaces are fitted to the voxel centers and used in the ray-casting process. Second, "duplicate voxels" are used. This step is necessary because scene flow in one direction can be potentially many-to-one and the number of voxels arbitrary. As a result, a 3D point in a given time instance may map to multiple 3D points or none at all at a different time instance. The one-to-one property

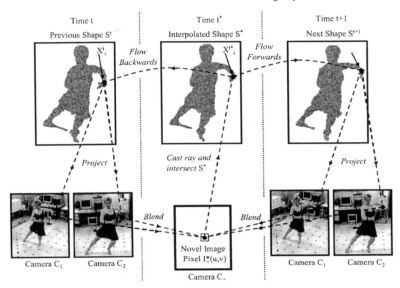

Fig. 3.7. Computing the virtual view by ray-casting across space and time. Figure courtesy of Sundar Vedula, Simon Baker, and Takeo Kanade.

is enforced by replicating voxels if necessary. Several view interpolation results at different virtual times and viewpoints are shown in Figure 3.8.

Fig. 3.8. Sample spatio-temporal view interpolation results. Images courtesy of Sundar Vedula, Simon Baker, and Takeo Kanade.

3.4 Image-based visual hulls

Matusik *et al.* [190] use the visual hull, which is an approximate geometric representation, to represent a 3D scene. A visual hull is constructed by casting the visible
silhouette information from a collection of input images to 3D space and intersecting the cast volumes. The creation of the visual hull of a teapot from three image
silhouettes is illustrated in Figure 3.9.

Fig. 3.9. Three extruded silhouettes of a teapot (left), and the resulting visual hull (right).
(Figure courtesy of Wojciech Matusik.)

While the visual hull has been used in computer vision in the past (mostly in the
context of object recognition), Matusik *et al.* demonstrated fast dynamic scene rendering based on the visual hull without explicit geometric or volumetric construction.
As such, they refer to their representation as image-based visual hulls (IBVHs). The
color information from the input images are used to provide texture for the IBVHs.

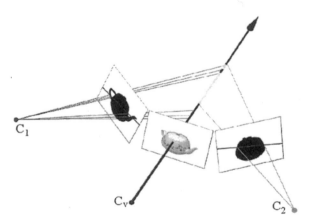

Fig. 3.10. Reduction of polyhedron-line intersections (for visual hull computation) to 2D intersections. C_v is the location of the virtual view while C_1 and C_2 are two input views. (Figure
courtesy of Wojciech Matusik.) ©2000 ACM, Inc. Included here by permission.

3.4.1 Computing the IBVH

Their technique for computing the IBVH is similar to finding CSG (constructive solid geometry) intersections using ray-casting. Instead of constructing the 3D visual hull and finding the intersection of the 3D viewing ray with the 3D geometry, they perform 2D ray intersections instead, which is faster. The idea is shown in Figure 3.10. Suppose the desired view is at C_v, with the two inputs being at C_1 and C_2. All the input cameras are fully calibrated so that the epipolar geometry (Euclidean projection) is known. For a given point **p** in the virtual image, the corresponding 3D ray can be constructed and projected onto the input views. These projections are the epipolar lines corresponding to **p**.

Next, intervals where the epipolar lines crosses the silhouettes are determined and projected back ("lifted") to the 3D ray. The intersection of these lifted intervals yields the intersection between the ray for **p** and the visual hull. The 3D point on the visual hull surface is thus the closest part of the intersection of the lifted intervals to the virtual camera. Speedups can be achieved by capitalizing on incremental computations enabled by epipolar geometry.

3.4.2 Texture-mapping the IBVH

The IBVH uses the input images as view-dependent textures. At each pixel of the virtual view, the input images are ranked based on the angle between the desired viewing ray and the rays from the surface of the visual hull to the input images. The smaller the angle, the better.

To avoid using occluded texture, visibility is checked using the visual hull. This check is performed pairwise, each time between the desired view and one of the input views. To compute the visibility of an IBVH sample with respect to an input image, a series of IBVH intervals are projected back onto the input image in an occlusion-compatible order (i.e., scanning from front to back). The front-most point of the interval is visible if it lies outside of the unions of all preceding intervals. A slightly more complicated algorithm is used to conservatively account for the discretization of the image.

3.4.3 System implementation

Matusik *et al.* use four calibrated Sony DFW500 Firewire video cameras, with each camera attached to a separate 600 MHz PC. Each PC captures video, corrects for radial distortion using a look-up table, segments out the foreground, and sends out the silhouette and texture over to a separate PC (which acts as a server). The server, which is a quad-processor 550 MHz PC, is used to compute and render the IBVH.

A typical IBVH covers about 8000 pixels in a 640×480 image, and rendering speed is slightly higher than 8 fps. Sample rendering results can be seen in Figure 3.11. The polyhedral-like shape is attributed to the small number of cameras used. As with all visual hulls, concavities that cannot be observed as silhouettes cannot be handled.

Fig. 3.11. Rendered IBVH at different views. The back of the person was unseen by any camera, which accounts for the missing texture (right image in middle row). Used with permission (courtesy of Wojciech Matusik).

3.5 Stanford Light Field Camera

The Stanford Light Field Camera started as a proof-of-concept 6-camera system [317], which was later expanded to a system of 128 CMOS cameras [305]. Two possible set ups are shown in Figure 3.12. The light field acquisition system is a modular design based on the IEEE 1394 high speed serial bus (Firewire). Each module is custom-made; it has an image sensor and is capable of MPEG compression. The image sensor has a resolution of 640×480 and is capable of capturing at 30 fps. Even though CMOS sensors generally have worse noise characteristics than CCD sensors, Wilburn *et al.* [317] chose CMOS sensors because they are easier to digitally interface with and control (for example, dealing with exposure time, gain, and gamma correction). The system is designed to return live, synchronized, slightly compressed (8:1 MPEG) video from all 128 cameras at once, and to record these video streams through four PCs to a striped disk array.

Goldlücke *et al.* [87] used a subset of the Stanford Light Field Camera for acquiring and displaying dynamic scenes. They first calibrate the cameras to extract their intrinsic and extrinsic parameters. The cameras are also calibrated to reduce radial

Fig. 3.12. Two configurations of the Stanford light field camera array. Top: Arrangement of 8 rows of 16 cameras each. Camera spacing is about 2 inches. Bottom: Arrangement of 5 panels in an arc, with each panel containing 25 cameras. Camera spacing is about 9 inches. Used with permission (courtesy of the Stanford Multi-Camera Array Project).

distortion as well as color and brightness variation across cameras. The camera color characteristics are calibrated using matrices obtained from MacBeth color chart images. The brightness variation across cameras is reduced through image equalization.

3.5.1 Depth map extraction

Depth maps are used to warp sample views to the new view. Suppose we are to compute the depth map for input camera view C_j and that there are k input cameras $(C_i, i = 1, ..., k)$. For each pair of cameras C_j and $C_i(i \neq j)$, the chosen disparity λ_{ij} is computed by searching along the epipolar line for best matches in a manner that attempts to minimize spatial depth changes while preserving depth discontinuities at image edges. Legitimate disparities are chosen to be those that fall within a prescribed range; the "optimal" disparity is then computed to be the mean of these disparities. The discrepancy in the disparities is most likely caused by image noise

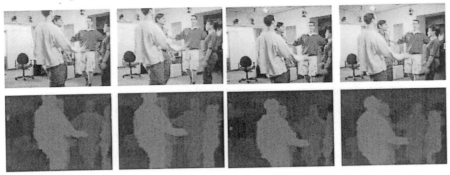

Fig. 3.13. Four sample images from the Stanford Light Field Camera (top row), and estimated disparity images (bottom row). Used with permission (courtesy of Bastian Goldlücke, Marcus Magnor, and Bennett Wilburn).

and blocking artifacts from the MPEG-encoded input images. As a result, it is difficult to produce depth maps that are correct to within a pixel at the boundaries. Examples of computed depth maps are shown in Figure 3.13.

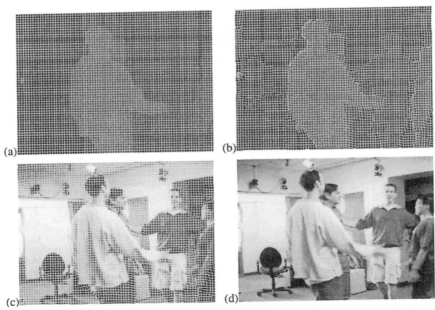

Fig. 3.14. Generating the virtual view: (a) triangle mesh superimposed on disparity image, (b) mesh warped to the new view, (c) texture warped to the new view, and (d) final result of blending the warped views of the four nearest cameras. Used with permission (courtesy of Bastian Goldlücke, Marcus Magnor, and Bennett Wilburn).

3.5.2 Interactive rendering

The interactive rendering step assumes that the dense disparity maps for all the images and timeframes have been precomputed. For a chosen novel view, the four nearest input depth images (color and depth maps) are used to generate the novel view. To exploit the polygon processing capabilities of OpenGL as well as hardware texturing and blending, Goldlücke *et al.*create a regular (downscaled) triangle mesh covering each of the input depth images. The disparity at each vertex of the downscaled triangle mesh is computed to be the average of disparities at the surrounding pixels.

A vertex program is used to warp each contributing depth image to the novel view. The warped color images are combined using weights assigned based on proximity to the novel view (more specifically, the closer the input image is to the novel view, the higher the weight). Backfacing triangles are culled during rendering since they are obscured by a closer object.

The warping and blending results are shown in Figure 3.14. With a mesh resolution of 160 × 120 (block size of 2 × 2 pixels), the frame rate obtained using a 1.7 GHz PC with an nVidia GeForce 4 graphics card is about 11 fps. Notice that because the triangular mesh is continuous, the appearance in the vicinity of significant depth discontinuities will in general be incorrect (the main artifact being blurring).

3.6 Model-based rendering

We have seen that the Virtualized Reality^{TM} work uses a single global geometry for rendering. Meanwhile, the work associated with the Stanford Light Field Camera uses view-dependent local geometries to produce novel views. In both cases, the geometries were extracted without any knowledge of the scene (that is, no shape priors were assumed).

In this section, we highlight three approaches that fit a human model to the dynamic scene and subsequently render the model using captured textures. Having prior knowledge of the scene and using an appropriate model to represent the scene reduces the ambiguities associated with multi-view reconstruction. The three approaches described in this section are those of Carranza *et al.* [29], Starck and Hilton [282], and Cheung *et al.* [43].

3.6.1 Free viewpoint video of human actors

In Carranza *et al.*'s work [29], an actor's movement is captured using multiple video cameras at 15 fps at a resolution of 320 × 240. The cameras are synchronized using an external trigger, and each pair of cameras is controlled by a 1 GHz PC that streams the video data directly to disk. The cameras are set up in a convergent configuration around the actor. They are pre-calibrated so that all the intrinsic and extrinsic parameters are known. In addition, the cameras are white-balanced to reduce color variation across the different cameras.

OFFLINE

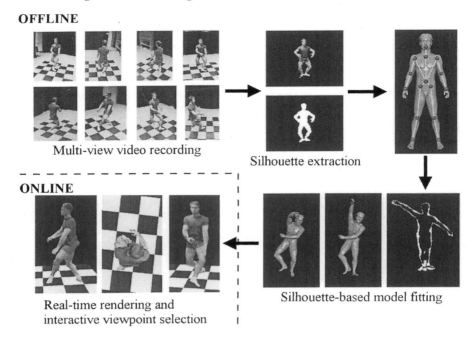

Multi-view video recording

Silhouette extraction

ONLINE

Real-time rendering and
interactive viewpoint selection

Silhouette-based model fitting

Fig. 3.15. Overview of free-viewpoint video system for rendering human movement [29]. Used with permission (courtesy of Christian Theobalt).

The graphical overview of their method is shown in Figure 3.15 (eight cameras are used here). Each image is processed to produce a binary image of the silhouettes of the human actor; this is done by making use of the color statistics of background pixels. From the video stream of the background, the mean and standard deviation of each background pixel in each color channel are computed. Significant deviation from any of the color channel tags the pixel as a foreground pixel. Morphological dilation and erosion are then applied to remove isolated noisy pixels.

The key component of their method is the fitting of their human model to the collection of silhouettes. Their human model consists of 16 articulated body parts, each of which is represented by a closed triangular mesh. There are 17 joints in the model, and 35 parameters are required to fully define the body pose.

Prior to motion capture, the scales of body parts are initialized by capturing a specific pose in which arms and legs are bent. The process of computing the pose and body part scales is the same as for motion capture, except that the scales are fixed for motion capture.

The error metric used to extract body pose parameters is the sum of differences between the recovered silhouettes and projected silhouettes of the currently estimated body pose. Their implementation is based on pixel-wise XOR using the OpenGL stencil buffer. Because the silhouette image is binary, a bit-plane of the stencil buffer is sufficient to render the result of the overlap between the actual and predicted silhouette. Using an 8-bit stencil buffer and the depth buffer, the error metric for 8 camera views can be efficiently evaluated in graphics hardware with only a single frame buffer read and write.

The model pose is automatically initialized using a grid sampling of the parameter space. For subsequent time frames, the model parameters are computed using a non-linear minimization approach. For robustness, Carranza *et al.* split the parameter estimation problem into a sequence of optimizations (Powell's method) on subparts of the body. First, the global translation and rotation is estimated. This is followed by estimation of head and hip joint rotations, poses of arms and legs, and finally, hand and foot orientation.

In the rendering step, view-dependent texturing [60] is *not* used for the reason that the geometry extracted is inexact and thus causes blurring. Instead, a view-independent, consistent texture map is created at every time step. Vertex weights are calculated based on the angle between the visible surface normal and viewing vector towards the input camera. A weighting function which has characteristics close to "winner-take-all" is used to preserve as much detail as possible.

The renderer takes as input the precomputed model parameters, per-vextex texture weights, and color images from the cameras. It is implemented on NVidia's GeForce3 rendering architecture, and is capable of rendering at 15 fps at 320×240. Sample rendering results can be seen in Figure 3.15.

3.6.2 Markerless human motion transfer

Cheung *et al.* [43] approach differs from Carranza *et al.*'s in that Cheung *et al.* use a more detailed person-specific model (optimized over a sequence of images instead of just one) and the use of both silhouette and color information for motion tracking. A graphical overview of their technique for reconstructing the human shape, capturing the human motion, and transferring it to a different person, is shown in Figure 3.16.

The first step, kinematic modeling, consists of two parts. The first part involves acquiring the shape of the person from a video of the person on a rotating turntable using their "shape-from-silhouette across time" (SFSAT) algorithm [44]. The second part is the estimation of the joint skeleton. This is accomplished by recording the subject flexing a joint one at a time using an adapted version of their SFSAT algorithm. The human model used has 22 degrees of freedom, which includes global translation and rotation.

The second step is motion capture, and again the shape recovery is accomplished using the SFSAT algorithm. The alignment of the human model to the recovered shape is done hierarchically, initially by aligning the torso, followed by aligning each limb independently.

In the third step, the captured motion of one person (source) can be directly transferred to another person (target) by passing the joint motions. It is assumed that the reference views of the target have been captured for kinematic modeling; as a result, the color for a given point on the body surface can be extracted (ignoring occluded pixels). For a given time and known body posture, the appearance of each pixel in screen space is computed by ray-casting and calculating the weighted average of

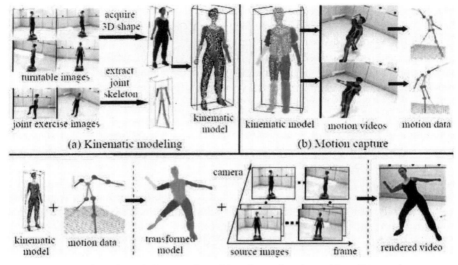

(a) Kinematic modeling

(b) Motion capture

(c) Image-based rendering (render subject #1 performing subject #2's motion)

Fig. 3.16. Overview of CMU modeling and rendering human movement, which includes movement transfer. Image (courtesy of Simon Baker and German Cheung) is from "Markerless human motion transfer," by G.K.M. Cheung, S. Baker, J. Hodgins, and T. Kanade, 2nd International Symposium on 3D Data Processing Visualization and Transmission, Sept. 2004, pp. 373-378. ©2004 IEEE.

See color plate section near center of book.

the color from the kinematic modeling step. The weights are determined based on proximity of the virtual view to the reference views.

3.6.3 Model-based multiple view reconstruction of people

Starck and Hilton [282] also recover human shape from multiple images using a human model. There are, however, differences with the approaches of Carranza *et al.*and Cheung *et al.*. For example, Starck and Hilton use not only silhouette information, but also stereo correspondences and feature cues. The feature cues are manually selected from the image and correspond to projected 3D locations of articulated joints and facial features such as eyes, ears, nose, and mouth. These are used to manually align the model to the images. This is a major drawback of the approach.

The aligned model information is subsequently used to constrain the search for stereo correspondence in a coarse-to-fine framework. The shape of the 3D mesh of the human is deformed to minimize an objective function that is the sum of the error in fitting the silhouette data, the error in stereo correspondence, the error in fitting the features, and a regularization term that encourages smooth 3D shape.

Starck and Hilton use nine cameras, eight of which form 4 stereo pairs that are positioned to provide 360-degree coverage of the subject and the ninth camera placed overhead. The cameras used are 3-CCD Sony DXC-9100P cameras, each with PAL resolution (that is, 720 × 576).

(a) Camera image   (b) Visual hull   (c) Voxel-coloring   (d) Merged stereo   (a) Model-based

Fig. 3.17. Results of Starck and Hilton's approach [282] compared to those of conventional approaches. Top row shows the shape reconstruction results and the bottom row shows rendering results. Images (courtesy of Adrian Hilton) are from "Model-based multiple view reconstruction of people," by J. Starck and A. Hilton, International Conference on Computer Vision, Oct. 2003, pp. 915-922. ©2003 IEEE.

Results of shape recovery and rendering are shown in Figure 3.17. Notice the improvements over methods that do not use any model prior. The voxel coloring technique used to produce results shown in Figure 3.17(c) is that of Seitz and Dyer [261]. The shapes in Figure 3.17(d) are obtained by fusing multiple stereo depth maps into a single surface model through volumetric fusion and iso-surface extraction [133]. A view-dependent rendering technique similar to that of Pulli *et al.* [237] is used. Here, view-dependent weighting is done on a per-vertex basis, which favors closer input camera views. Multi-pass texturing is used to blend the weighted textures.

3.7 Layer-based rendering

The work of Zitnick *et al.* [349] at Microsoft Research is inspired by Kanade *et al.*'s Virtualized Reality work [133]. Their system is much more modest in size, as only 8 cameras are used. However, the cameras used are of higher-resolution (1024 × 768 at 15 fps). In addition, as opposed to rendering only specific dynamic figures in past work, they render *entire dynamic scenes*. Photorealism is achieved using a two-layer representation that includes matting information.

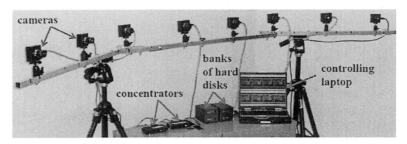

Fig. 3.18. Camera configuration for dynamic scene capture.

3.7.1 Hardware system

Figure 3.18 shows a configuration of Zitnick *et al.*'s video capturing system with eight cameras arranged along a horizontal arc. High resolution Firewire PtGrey color cameras are used to capture video at 15 fps. With the 8mm lenses used, a horizontal field of view of about 30° is obtained. Two "concentrator" units (built by PtGrey) are used to synchronize all the 8 cameras and pipe the uncompressed video streams directly into a bank of hard disks through fiber optic cables. The two concentrators are synchronized using a FireWire cable.

The cameras are precalibrated using a 36" × 36" calibration pattern mounted on a flat plate, which is moved around in front of all the cameras. The calibration technique of Zhang [346] is used to recover all the camera parameters necessary for Euclidean stereo recovery.

3.7.2 Image-based representation

Zitnick *et al.*choose a two-layer representation inspired by Layered Depth Images and sprites with depth [264]. This layer representation is also view-dependent, as local geometric proxies have been shown to be effective (e.g., [237, 60, 106]). To generate the representation, a stereo algorithm (described shortly) is used to compute the dense depth distribution for each image. Depth discontinuities are then detected in each depth map; boundary strips (around 8 pixels thick) are created around these depth discontinuities. The representation consists of two layers. The first layer is the boundary layer B, which has depth, color, and matting information associated with the foreground. The second layer is the main layer M, which consists of depth and color of pixels not part of the boundary layer pixels, *and* the background counterpart of B (depth and color, with matting information implicitly known from B). Thus, the pixel locations of B are a subset of M, with different color, depth, and matting information. M covers all the pixels in the image.

A variant of Bayesian matting [47] is used to automatically estimate the foreground and background colors, depths, and opacities (alpha values) within these strips. To reduce the data size, the multiple alpha-matted depth images are then compressed using a combination of temporal and spatial prediction.

Since the cameras are configured along a 1D arc, at rendering time, the two reference views nearest to the novel view are chosen, warped, and combined for view synthesis. The warped layers are combined based on their respective pixel depths, pixel opacity, and proximity of the reference view to the novel view.

3.7.3 Stereo algorithm

When developing a stereo vision algorithm for use in view interpolation, the requirements for accuracy are slightly different from those of standard stereo algorithms used for 3D reconstruction. In the work of Zitnick *et al.*, error in disparity is not as important as the error in intensity values for the interpolated image, i.e., the synthesized view has to be *plausible*.

Traditional stereo algorithms tend to produce erroneous results around depth discontinuities. Unfortunately, such errors produce some of the most noticeable artifacts in interpolated views, since they typically coincide with intensity edges. As a result, Zitnick *et al.*use a segmentation-based stereo in the same spirit as Tao *et al.* [293] (planar constraint for each segment) and Zhang and Kambhamettu [344] (segments used for local support).

The stereo algorithm used by Zitnick *et al.*consists of the following steps:

- *Segmentation.* For each time frame, each image is first smoothed using a variant of anisotropic diffusion [229], then segmented based on color using a variant of the mean shift [51].
- *Extraction of segment matching function.* All the camera images are used to extract depth; the global coordinate frame is chosen to coincide with the centrally located camera. Each segment is initially assumed frontal parallel (i.e., with constant disparity) with respect to this global coordinate frame. The segment matching error function is computed in a similar manner as the now-familiar *disparity space image* (DSI) [50]. This error function, called *disparity space distribution* (DSD), is subsequently refined by enforcing a smoothness constraint between adjacent segments and a consistency constraint between images.
- *Disparity smoothing.* In this step, the frontal parallel assumption is relaxed. This is done by iteratively computing local depth averaging, subject to projection consistency across images.

A result of the stereo algorithm is shown in Figure 3.19. Once the disparity map has been computed for each of the input images, boundary matting is performed as described earlier.

3.7.4 Rendering

In order to interactively manipulate the viewpoint, Zitnick *et al.*ported their rendering algorithm to the graphics processing unit (GPU). Because of recent advances in the

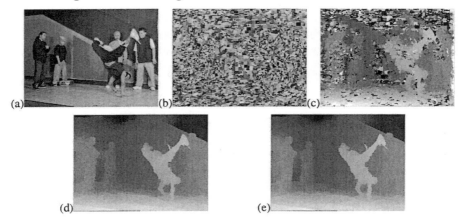

Fig. 3.19. Sample results from stereo reconstruction stage: (a) input color image; (b) color-based segmentation; (c) initial disparity estimates; (d) refined disparity estimates; (e) smoothed disparity estimates.

programmability of GPUs, they are able to render directly from the output of the file decompressor (more efficient version of JPEG) without using the CPU for any additional processing. The input to the renderer per view consists of 5 planes of data: main color, main depth, boundary alpha matte, boundary color, and boundary depth.

For a chosen virtual view, the nearest two cameras in the data set (say i and $i+1$) are picked. For each camera, the main data M_i and boundary data B_i are projected into the virtual view. The results are stored in separate buffers, each containing color, opacity, and depth. These are blended to generate the final frame. A block diagram of this process is shown in Figure 3.20.

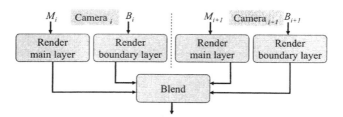

Fig. 3.20. Rendering system: the main and boundary images from each camera are warped (rendered) and then blended.

Each depth map is converted into a 3D mesh using a simple vertex shader program. The boundary data is fairly sparse, since only vertices with non-zero alpha values are rendered. Note that the boundary and main meshes share vertices at their boundaries in order to avoid cracks and aliasing artifacts.

Once all layers have been rendered into separate color and depth buffers, a custom pixel shader is used to blend these results. The blending shader is given a weight for each camera based on the camera's distance from the novel virtual view (similar to [61]). It also uses the alpha matte value for blending.

Figure 3.21 shows four intermediate images generated during the rendering process. It shows how the soft alpha-matted boundary elements and view-dependent blending are used to create high-quality results. Two more view interpolation results are shown in Figure 3.22. As can be seen, the quality of the interpolation is excellent.

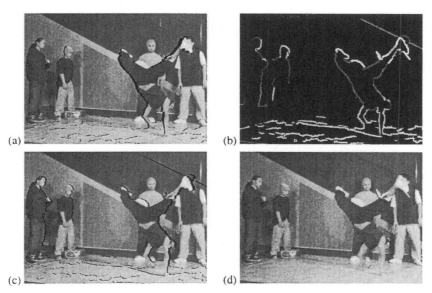

Fig. 3.21. Intermediate results from rendering stage: (a) rendered main layer from one view, with depth discontinuities erased, (b) rendered boundary layer, (c) rendered main layer from the other view, and (d) blended result.

3.8 Comparisons of systems

We see a wide variety of design decisions made for systems to render dynamic scenes in this chapter. There are differences in system setup and camera configuration, types of data extracted, scene representation, and rendering algorithm. The rendering algorithm used is intimately tied to the representation used, which also affects the rendering speed. The examples highlighted in this chapter provide us valuable lessons on what worked well and what worked less effectively, given the goal of *photorealistic* (that is, artifact-free) dynamic scene rendering.

Fig. 3.22. More video view interpolation results: (a,c) input images from vertical arc and (b) interpolated view; (d,f) input images from ballet studio and (e) interpolated view.

3.8.1 Camera setup

The camera setups range from dense configuration (Stanford Light Field Camera) to intermediate camera spacing (layer-based work at Microsoft Research) to wide camera distribution (Virtualized Reality^{TM}). Currently, only the Virtualized Reality^{TM} camera setup allows almost complete 360 degree range of virtual pan and tilt. However, the wider spacing between the cameras in this system provide more of a challenge in producing locally consistent geometries and hence photorealistic views. This is because occlusions become more of an issue and the non-rigid effects associated with non-Lambertian surface properties (specularities) are much more difficult to deal with.

A significantly denser camera configuration such as that of the Stanford Light Field Camera allows effects such as synthetic aperture and focusing [116]. Note that synthetic aperture imagery allows objects that are occluded with respect to any given camera to be seen. As demonstrated by the Light Field-related approaches for static scenes [91, 160], dense sampling permits photorealistic rendering with just either a simple planar geometric representation or a rough geometric approximation. However, the disadvantage is the large number of images required for rendering. This issue of the image-geometry trade-off was discussed in Chapter 5. The work of Zitnick *et al.*attempts to reduce the required number of input cameras and compensate for this by providing high-quality stereo data.

Resolution obviously plays an important role in achieving photorealism, but having a higher resolution will not help if rendering artifacts are not properly handled. These artifacts include boundary or cut-out effects, incorrect or blurred texturing, missing data, and flickering. Boundary or cut-out effects are caused by mixed foreground and background colors in object boundary pixels. Incorrect or blurred texturing can be caused by incorrect stereo extraction, occlusion, and non-rigid effects, while flickering sometimes occurs if temporal consistency is not accounted for. Unfortunately, humans are highly sensitive to high-frequency spatial and temporal artifacts. Although using a reduced resolution would conveniently help to mask or ameliorate such artifacts, it should *never* be viewed as a solution or an excuse.

3.8.2 Scene representation

The choice of scene representation is critical to the goal of photorealism. Since surfaces of a real scene tend to be non-Lambertian, using a single extracted 3D geometry to represent the scene is not recommended. An exception may be if the scene is highly structured as in the Façade work on modeling and rendering buildings [61]. In such a case, view-dependent texturing on a single geometry may be adequate. However, in general, we think the best choice would be to use *view-dependent geometries*. This has been demonstrated in a number of approaches, such as [237, 60, 106], with considerable success.

As Zitnick *et al.*have demonstrated, using view-dependent geometries as well as extracting the matting (alpha) information at the boundaries and using it for rendering have proven to be highly effective. Boundaries need to be handled correctly in order to avoid artifacts (blurring and/or cut-out effects), and using matting information has been shown to be effective. The difficulty is in avoiding the manual process of indicating the matting areas for subsequent matte extraction. The manual component is required in prior matte extraction techniques. Zitnick *et al.*use depth discontinuities to automatically indicate areas where foreground and background pixel colors exist, and apply an existing technique for matte extraction [47, 315]. A more systematic technique for simultaneously extracting matte information and refining depths at discontinuities uses 3D deformable contours as unifying structures [102].

The spatial-temporal view interpolation technique of Vedula *et al.* [309] is an appropriate approach to ensure temporal continuity and thus avoid flickering during rendering. Interestingly, Zitnick *et al.*showed that it is possible to produce flicker-free rendering without considering the time domain if the stereo data extracted is accurate enough (from the photoconsistency point of view). However, this feat will be difficult to replicate for general scenes with significant non-rigid effects such as specularities and translucencies.

3.8.3 Compression and rendering

While the multiple video streams used in acquiring the dynamic scene results in a substantial amount of data, there is a lot of redundancy. How would one capitalize on the redundancy? The topic of compressing IBR data is handled in Part III of this book, and more specifically in Chapter 13 in the context of dynamic scenes. The approaches described in this chapter do not explicitly handle the compression issue (except for Zitnick *et al.*).

In order for rendering to be real-time, the representation of the scene has to be amenable to hardware rendering. In most cases, this means converting the IBR data to either a texture-mapped global triangulated 3D model or multi-texturing using multiple triangulated depth images. (A depth image contains both texture and depth.)

3.9 Challenges

The challenges delineated for static scenes (Section 2.6) apply for dynamic scenes as well. Scene complexity, non-linear effects, ability to edit scenes, and feature correspondence all pose serious problems for creating *useful* representations of dynamic scenes. Rendering dynamic scenes has the additional requirement of temporal consistency. Solutions to handle temporal consistency exist (e.g., [307]), but they tend to rely on the brightness constancy assumption. This assumption is violated for non-Lambertian surfaces.

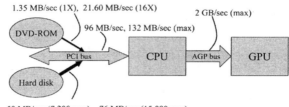

Fig. 3.23. Data transfer bottlenecks. The worst bottleneck is the DVD reader at 1X speed.

In order for the dynamic scene rendering technology to be feasible, the rendering system will have to read from a regular PC hard drive, or even better, from a DVD. Compression will thus be part of the critical path to a practical system. As can be seen in Figure 3.23, the DVD reader has the worst I/O bottleneck at about 1.35 MB/sec at 1X speed. The decoder and renderer must also be able to sustain the stream of data read from the medium.

The applications for a dynamic scene rendering system are numerous: games (football, hockey), performances (plays, circus acts), instructional videos (martial arts, golf), and DVD extras that accompany movie DVDs. The goal of online, real-time photorealistic viewing is very difficult to achieve today. Not only does the I/O bandwidth issue have to be addressed, the input video has to be processed in real-time as well. With rapid advancements in processor speeds, compression techniques, and other hardware technology, it is just a matter of time before a system capable of real-time video acquisition and scene rendering becomes a reality.

4

Rendering Techniques

The previous two chapters on IBR representations show that while dense representations that are image-rich tend to produce more visually-accurate view reconstruction, they tend to be fat. Trade-offs have been made to reduce this dependency on the large number of image samples with more geometric information; these trade-offs resulted in different representations and rendering techniques.

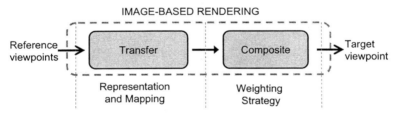

Fig. 4.1. Goals of rendering: establish mapping between representation and image screen, and blend.

For early image-based representations that are based just on image samples, their rendering techniques are simple—they tend to be either just image blending (as in panoramas) or interpolation (as in light field). As more sophisticated representations cropped up with different trade-offs between images and geometry, such as layered depth images and surface light fields, rendering techniques changed accordingly. Graphics hardware has also been exploited to accelerate the rendering process. Nonetheless, all rendering techniques have the same goal: to establish a mapping relationship between parts of the representation and the screen pixels, and composite these parts to produce the virtual view (Figure 4.1).

In this chapter, we focus on the rendering aspect of IBR. For didactic purposes, we revisit the IBR continuum and further categorize it based on the type of rendering (Section 4.1). Note that the rendering type is not mutually exclusive; the more detailed categorization, which we call the *geometry-rendering matrix*, is created to enable us to clearly associate rendering techniques to mapping types.

4.1 Geometry-rendering matrix

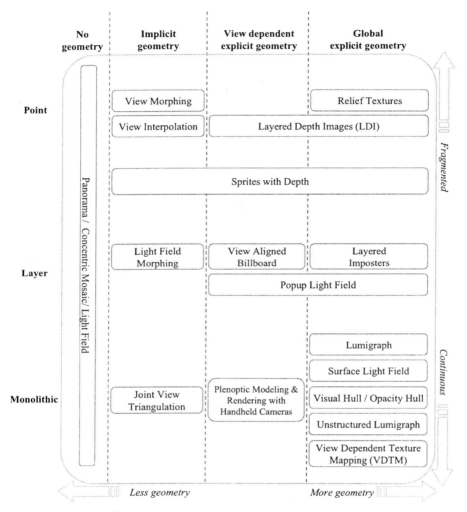

Fig. 4.2. Geometry-rendering matrix. This matrix shows the types of representations (along horizontal axis) and rendering (along vertical axis) in IBR.

We summarize the image-based representations and their rendering methods in the diagram shown in Figure 4.2. These methods are discussed in the context of the representation of the geometry: *no geometry, implicit/explicit, local (or view dependent),* and *global.* For the purpose of discussion, we consider the rendering to be either *point-based, layer-based,* and *monolithic.* There is the special case where no geometry is used, which we also discuss.

4.1.1 Types of rendering

Point-based rendering is applied to 3D point clouds (such as layered depth image [264] and relief texture mapping [216]), or point correspondences (used in techniques such as view interpolation [40] and view morphing [260]). Typically, each point is rendered independently. In point-based rendering, the target view is typically restricted to be near to the reference view. The point on the reference view is usually directly mapped to the target view with no compositing operation.

Monolithic rendering is done on single pieces of geometry, each of which usually being a contiguous triangular mesh. We refer to such geometry as monolithic geometry. The monolithic geometry is rendered as an entire object using texture mapping techniques. Therefore, the mapping between the geometry and target view can be easily determined through view projection. Since the global geometry is reconstructed from multiple images, the final rendering result is composited from multiple reference colors mapped on the same surface point. This category of techniques include view-dependent texture mapping [61], image-based visual hulls [190], opacity hulls [191], surface light field [323], Lumigraph [91], and unstructured Lumigraph [22]. Note that the joint view interpolation [162] also falls into this rendering category because it divides each reference view by triangles and rendered as an entire mesh.

Layer-based rendering is performed on a layer-by-layer basis, i.e., each layer is rendered independently and then composed to produce the final view. A planar geometry is usually assumed for each layer, which can be easily rendered as monolithic geometry. This category of techniques include layered imposters [255], sprites with depth [264], and pop-up light field [270]. Sprites with depth [264] use a view-aligned height map to represent the geometry details; they are rendered using techniques as discussed in the point-based category.

These three rendering types operate on representations that range from being continuous (triangle meshes) to fragmented (point clouds). The rendering of continuous representations has been well exploited using the conventional graphics pipeline. However, it is difficult reconstructing such continuous representations directly from image samples. On the other hand, more fragmented representations are significantly easier to reconstruct, using 3D scanners or vision techniques such as stereo and structure from motion. Unfortunately, fragmented representations tend to be view-dependent and limited by the input image resolution; occlusion and holes are particularly difficult problems to handle.

Note that there is one category of rendering techniques that do not require geometry information. This category includes QuickTime VR [41], light field [160], Concentric Mosaics (CMs) [267], and manifold hopping [274]. (Strictly speaking, both the 4D light field and CMs do have a geometric proxy, namely the focal plane and cylinder respectively, but they do not require any geometric reconstruction.)

4.1.2 Organization of chapter

We first discuss the IBR rendering methods using no geometry in Section 4.2, followed by point-based rendering in Section 4.3. In both cases, we focus on how pixels are mapped from the reference images to the target screen, i.e., the transfer stage. In Section 4.4, we discuss monolithic rendering and concentrate on the composite stage. Finally, in Section 4.5, we briefly review some layer-based rendering methods, which are based on either point-based or monolithic rendering techniques. Throughout the chapter, we point out the advantages and disadvantages of each category of techniques and possible hardware accelerations; we also consider issues such as discontinuity handling, hole-filling methods, and matte handling.

4.2 Rendering with no geometry

In this section, we discuss rendering techniques that do not require geometry information. These techniques organize the reference images in ray space. The mapping and composition step of rendering is simply implemented as ray-space interpolation. Specifically, each ray that corresponds to a target screen pixel is mapped to nearby sampled rays, which are then composited through ray-space interpolation. While the details in the rendering depend on the dimensionality and complexity of the ray space, these techniques all fundamentally target efficient organization and indexing of the reference rays for mapping onto screen space.

4.2.1 Ray space interpolation

Recall that the plenoptic function [2] represents the real world in a 7D function $P_7(V_x, V_y, V_z, \theta, \phi, \lambda, t)$, given 3D position (V_x, V_y, V_z), ray direction (θ, ϕ), wavelength λ, and time t. Rendering a virtual image is equivalent to fitting this function using a set of input images and interpolating the function at points corresponding to pixels in the virtual image. However, adequately fitting P_7 in real world requires a large amount of data, which makes data acquisition intractable and impractical. As surveyed in Chapters 2 and 3, various techniques have been used to try simplify the problem by limiting the degrees of freedom of the plenoptic function.

Plenoptic modeling [194] reduces P_7 into 5D by ignoring the wavelength and time. The light field [160] and Lumigraph [91] representations further simplify P_7 into 4D by assuming the colors along a ray direction do not change. As a result, the rays within an occlusion-free space can be indexed by two-plane parameterization as (u, v, s, t), consisting the pixels position (u, v) on the image plane and the reference viewpoint position (s, t) in the camera focal plane. As shown in Figure 4.3, any ray passing through two planes can be indexed by the two intersection points and subsequently rendered using quadratic linear interpolation of the neighboring 16 rays.

Concentric Mosaics (CMs) [267] further simplifies P_7 into 3D by restricting the viewpoint to within a 2D planar disc. It captures the plenoptic function with a set of multiperspective mosaics, which presents a 360° field view while providing a parallax experience when the virtual viewpoint is translated within a 2D plane. Since the plenoptic function is represented in a 3D volume by stacking multiperspective

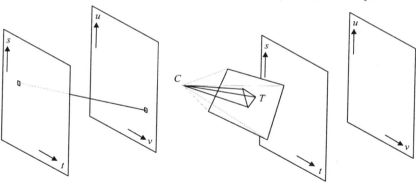

Fig. 4.3. Light field rendering.

panoramas, the search of the closest ray can be done efficiently by indexing columns, and the rendering can be simplified as trilinear interpolation in the volume.

Based on a similar representation, manifold hopping [274] (Chapter 14) can be regarded as a 2.5D representation since it restricts the viewpoint on a set of discrete concentric circles. It provides user experience of comparable quality to CMs, while using significantly less data. The rendering of manifold hopping is also simplified to 2D image warping.

QuickTime VR [40] is a 2D representation which does not allow viewpoint translation. It captures the plenoptic function from one optical center and allows the virtual camera rotating freely to any direction. This technology uses no more data than a large field of view image, and provides realistic user experience by simple and efficient image warping techniques.

The major advantage of using no geometry representation is the simplicity of its construction. The images are captured without knowing scene geometry. Rendering consists of simply indexing the sampled (simplified) plenoptic function data and interpolating the colors of the neighboring samples. However, this category of techniques (with the exception of panoramas and possibly Manifold hopping) usually require a large number of images for anti-aliased rendering, which makes such techniques less practical. Simple heuristics such as using constant depth manifold (e.g., CMs [267], manifold hopping [274]) or an appropriate focal plane (e.g., dynamically re-parameterized light field [116]) may improve the rendering quality significantly even when the image samples may not be adequate. It has been shown that there is a relationship governing the image sample density, geometric proxy errors, and screen resolution [33] (Chapter 5).

4.2.2 Other forms of interpolation

It is worth noting that interpolation in ray space does not necessarily involve sampling only the nearest rays. It is possible to avoid artifacts such as blurring, ghosting, and pixelation due to insufficient image samples or incorrectly placed geometric proxy, by using *image priors* [75]. In this technique, the output texture is constrained by local statistics of the input images—in other words, each local patch of the output must be similar to some patch in the input image. It is formulated in a Bayesian framework where the prior is a collection of 5×5 patch samples. Unfortunately, the (iterative) energy minimization process is computationally expensive, since it requires computing similarity with all sampled patches at every step. A faster version using a coarse-to-fine strategy was later proposed [325]. However, despite the two orders of magnitude rendering speed-up, it still takes seconds to render an 800×600 frame.

In the dynamically reparameterized light field [116], rendering can take place with a larger support around the target ray to emulate a large aperture optical lens. By using this technique of synthetic aperture, varying depths of field appearance can be simulated. Objects at the desired depth can be rendered sharply with objects at other depths appearing defocused. In addition, variable focus can also be simulated by varying the focal plane. For a constant number of cameras, the rendering complexity is related to aperture size and generally is $O(N^4)$, where N is the image width in pixels. Ng [212] proposed a method to decrease the rendering complexity into $O(N^2 \log N)$ by Fourier slicing in 4D frequency space.

4.2.3 Hardware rendering

Current commodity graphics hardware does not support 4D textures and quadratic interpolation. However, they usually support bilinear interpolation of 2D textures. Approaches such as [244] have been devised to exploit ways for rendering the light field using conventional graphics hardware. They typically decompose the 4D light field into a set of 2D textures, each of which represents a reference view. Multiple texture mapping is then used for interpolation. As shown in Figure 4.3, for a virtual viewpoint C, all reference view centers are projected onto the image plane of C and triangulated. The pixels inside a triangle T are rendered using the textures of three reference views corresponding to the three vertices of the triangle. Since the blending weight can be computed using barycentric distance, it is possible to use pixel shader to composite the three textures with proper weights in a single rendering pass [270]. The composition can be done using conventional graphics hardware through multipass rendering [277].

Barycentric coordinates

In computer graphics, barycentric coordinates are commonly used to characterize points within convex polygons. Consider a set of N points $S = \{P_1, ..., P_N\}$ and consider the set of all affine combinations taken from these points, i.e., $P = a_1 P_1 + ... + a_N P_N$. P is inside the convex hull of S if $a_1 + ... + a_N = 1$, with $a_i \geq 0$ for all $i = 1, ..., N$. The N-tuple $(a_1, ..., a_N)$ is the *barycentric coordinates* with respect to S. The barycentric weight or distance associated with P_i is a_i. More properties of the barycentric coordinates can be found in Section 13.7 (pages 216-221) in [55].

Fixed function vs. programmable

The graphics pipeline typically consists of two stages: the vertex processing stage and pixel (fragment) processing stage. In conventional fixed-function graphics hardware, the operations in both the vertex processing stage and the pixel processing stage are fixed. The user can only change the parameters of the rendering operations. More specifically, in the vertex processing stage, the triangle vertices are first transformed into camera space before lighting is applied to the vertices. These triangles are then projected to screen space and rasterized to pixels. In the pixel processing stage, for each rasterized pixel, the color is interpolated from the colors on the triangle vertices. Texture mapping is applied to each pixel; using depth comparison to remove occluded colors, the final color is obtained by compositing unoccluded colors in the color buffer.

Recent advances in graphics hardware have enabled it to be programmable, giving rise to vertex and pixel shaders. The programmable graphics hardware allows the user to customize the rendering process at different stages of the pipeline. Vertex shaders manipulate the vertex data values, such as 3D coordinates, normals, and colors. 3D mesh deformation can be done using vertex shaders, for example. On the other hand, pixel shaders (also known as fragment shader), affect the pixel processing stage of the graphics pipeline. They calculate effects on a per-pixel basis, e.g., texturing pixels and adding atmospheric effects. Pixel shaders often require data from vertex shaders (such as orientation at vertices or light vector) to work.

4.3 Point-based rendering

Point-based rendering is applied to representations that are created from 3D point clouds or 2D correspondences between reference images. Each point is usually mapped independently. Because of this flexibility, object details can be captured well. More importantly, point-based rendering is a more natural choice for data extracted using certain geometry acquisition methods such as a 3D scanner or active rangefinder, stereo reconstruction, and structure from motion techniques. Excellent surveys on point-based rendering can be found in [145, 251].

Points are mapped to the target screen through *forward mapping* or *backward mapping* (also referred to as *inverse mapping*). Referring to Figure 4.4, the mapping can be written as

$$X = C_r + \rho_r P_r x_r = C_t + \rho_t P_t x_t. \tag{4.1}$$

Here, x_t and x_r are homogeneous coordinates of the projection of 3D point X on target screen and reference images, respectively. C and P are camera center and projection matrix respectively. ρ is a scale factor. Point-based rendering is based on (4.1); the direction of mapping depends on which 2D coordinates are evaluated.

4.3.1 Forward mapping

Forward mapping techniques map each pixel on the reference view(s) to the target view using some form of geometry, e.g., depth map (explicit geometry) or correspon-

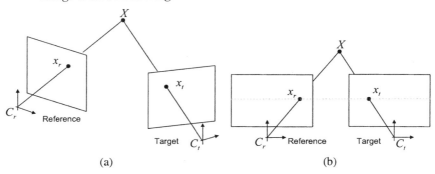

Fig. 4.4. Relationship between 2D points x_r and x_t in the reference and target images, respectively, and 3D point X. On the left shows an arbitrary virtual view while the right shows a lateral motion (with rectified geometry).

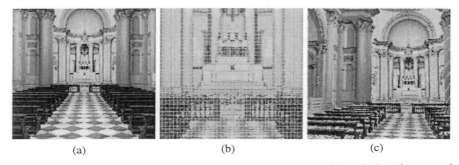

(a)   (b)   (c)

Fig. 4.5. Hole creation with forward mapping. (a) A viewpoint with no holes, (b) zoomed viewpoint with holes caused by significant changes in spatial footprint, (c) viewpoint with holes caused mostly by depth discontinuities (and missing data).

dences between views (implicit geometry). Using (4.1), we evaluate x_t:

$$\rho_t x_t = P_t^{-1}(C_r - C_t) + \rho_r P_t^{-1} P_r x_r \qquad (4.2)$$

Since P_t and C_t are known, ρ_t can be computed using the depth of X with respect to target camera C_t and focus length f_t: $\rho_t = (0, 0, 1/f_t)^{\mathrm{T}} \cdot P_t^{-1}(X - C_t)$. Therefore, given x_r and ρ_r, we can compute the exact position of x_t on the target screen and transfer the color from x_r to x_t. This process is called forward mapping.

x_t is almost always at a subpixel location. If we map pixels from reference images to the target screen using the nearest neighbor scheme, gaps may appear. Unfortunately, even if x_t is exactly at a pixel location, gaps may still appear. There are two other possible reasons for the gaps or holes in the target screen: magnification and disocclusion.

A straightforward reason for the occurrence of gaps is magnification due to the virtual camera moving closer to the scene. An example of holes created this way can be seen in Figure 4.5(b).

Splatting techniques [94, 161] have been proposed to handle the subpixel target location and alleviate the gap problem caused by the larger target footprint. A filter kernel is used to cover an area larger than a pixel to make up for the expected larger image footprint for the rendered scene. The shape and size of the kernel depend on the spatial relationship between the reference and target cameras and the distance of X to the target screen. The Gaussian filter kernel is the most commonly used, and the corresponding technique is called the Elliptical Weighted Average [94].

Splatting requires a post-processing stage to normalize contributing colors and opacities at each pixel. While this can be slow using pure software implementation, a recent effort has shown that hardware acceleration is possible [19, 245], speeding up rendering by at least an order of magnitude. Unfortunately, splatting tends to blur the target image. The work on surfel rendering [231] showed how to choose the kernel to achieve necessary hole filling yet avoid over-blurring of the target image.

Gaps can also occur in the target screen if there is disocclusion caused by depth discontinuity in the scene (see Figure 4.5(c)). Such gaps cannot be filled merely by splatting because the missing pixels on the target screen are not visible from the reference view. A typical solution is to rely on other reference views to fill in the missing information. The multi-view technique of Pulli *et al.* [237] shows how multiple textured range data sets are used to generate complete views of objects.

Apart from the gap or hole problem, we also have to contend with the issue of multiple pixels from the reference view landing on the same pixel in the target view. In this case, we need to decide which pixel or pixels to use in the final rendering. The most straightforward solution is to use the Z-buffer to make this decision. In [237], depth thresholds are used to pick the frontmost mapped pixels (whose colors are then linearly combined).

There is a more efficient rendering algorithm that obviates the need for the Z-buffer, namely the modified painter's algorithm [193]. The modified painter's algorithm uses the epipolar geometry to find the order in which pixels are scanned (Figure 4.6). This order, interestingly, is independent of the depth of the scene. To find the order, the epipole [72] **e** is first computed by projecting the camera center of the virtual view C_t onto the reference camera. If the virtual camera is behind the reference camera, we render the pixels away from **e**; otherwise, we render towards **e**.

In some cases, forward mapping can be simplified. For example, as shown in Figure 4.4(b), the target camera is a laterally translated version of the reference camera, so that scanlines are parallel to the camera offset $C_t - C_r$. In computer vision, the images are considered *rectified*. Here, $\rho_t = \rho_r = \rho = (0, 0, 1/f)^{\mathrm{T}} \cdot (X - C_r)$ and $P_t = P_r = P$. As a result, (4.1) can be simplified to

$$x_t = x_r + \frac{1}{\rho} P^{-1}(C_t - C_r) = x_r + u(x_r). \tag{4.3}$$

Here, $u(x_r)$ is the disparity associated with pixel x_r, which is proportional to the depth of 3D point X. We can then easily determine the position of x_t given x_r and its depth on reference image.

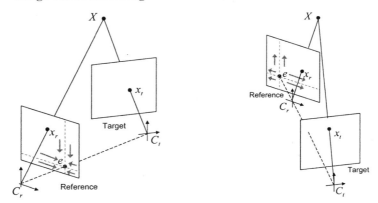

Fig. 4.6. The modified painter's algorithm can be used for forward mapping without requiring the Z-buffer. Left: Target camera is in front of reference (source) camera. Right: Target camera is behind the source camera.

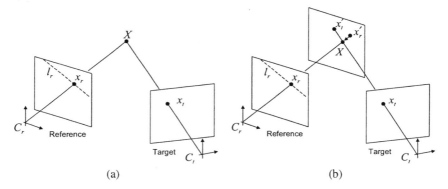

Fig. 4.7. Backward mapping from target screen to reference view.

Another interesting feature of this lateral-translate configuration is that only disparity $u(x_r)$ is needed for view transfer. This property is capitalized in techniques based on implicit geometry (i.e., point correspondences), such as Chen and Williams's view interpolation approach [40] and view morphing [260]. Rendering involves computing 2D pixel correspondences in the form of $x_t = x_r + u(x_r)$, without knowing any explicit 3D information. Since mapping occurs along the same scanline, rendering can also be simplified to 1D splatting. Szeliski and Cohen [290] suggested line drawing instead of splatting to fill the gaps. They also introduced a two-pass rendering method to reduce the gap filling operation; we describe this method later in this section.

4.3.2 Backward mapping

In backward mapping, also known as inverse mapping, the pixel mapping relationship is found by tracing the ray from the target view back to the reference view(s). Given a pixel on the target screen x_t, we can rewrite (4.1) as

$$\rho_r x_r = P_r^{-1}(C_t - C_r) + \rho_t P_r^{-1} P_t x_t, \tag{4.4}$$

which can be further simplified to

$$x_r \equiv \mathbf{H} x_t + d\mathbf{e}. \tag{4.5}$$

Here, $\mathbf{H} = P_r^{-1} P_t$ defines the 2D planar perspective transformation (also known as a homography) from target screen plane to reference camera plane. e is the epipole [72], and can be obtained by intersecting the line $C_t - C_r$ with the reference view image plane. d is a scale factor and $d\mathbf{e}$ defines a line called the epipolar line (shown as l_r in Figure 4.7). This line can be obtained by intersecting the reference camera plane with the plane defined by C_r, X and C_t (also called the epipolar plane).

Consequently, given x_t, x_r can be obtained by searching along the epipolar line for the pixel that fulfills (4.5) with minimum depth to target camera C_t. This process is called backward mapping or inverse mapping. Each pixel on target screen can be mapped to an unambiguous location x_r in the reference view, and can be rendered through resampling. This ensures that there are no gaps or holes in the target view. However, the search yields an invalid result if the pixel at x_t is occluded in the reference view.

In the special case where all 3D pixels are located on a 3D plane, the backward mapping process can be implemented as perspective texture mapping. Here, the mapping reduces to $x_r = \mathbf{H}' x_t$, where $\mathbf{H}'$ is defined by P_t, P_r and the location of the 3D plane. This is supported by current commodity graphics hardware and hence performed very efficiently.

In general, however, the 3D points do not lie on a plane. As a result, backward mapping involves a search for d, and is therefore typically slow. In the next section, we discuss a special case when the 3D points are reasonably close to a 3D plane, enabling a hybrid approach that uses forward mapping followed by backward mapping.

4.3.3 Hybrid methods

Backward mapping involves searching and is typically slower than forward mapping, unless the 3D object is a plane, in which case backward mapping degenerates into a perspective mapping (which is fast). On the other hand, forward mapping may be slowed down by the splatting process necessary for filling gaps and holes. When the geometry is represented as a depth field on the reference camera, forward mapping can be performed quickly with simple pixel offset and scanline-based splatting as described earlier.

The approaches in [264, 290] reformulated (4.5) as

$$x_t = \mathbf{H}'(x_r + d\mathbf{e}') = \mathbf{H}'x_t'. \tag{4.6}$$

As shown in Figure 4.7(b), rendering can be decomposed into two stages (referred to as the pre-warp stage and texture mapping stage [216]). In the pre-warp stage, x' is rendered using forward mapping from the reference view to an intermediate plane that is parallel to the reference camera plane. Since the geometry can be represented as a depth field, x_t' can be rendered quickly using 1D splatting and modified painter's algorithm as discussed above. Moreover, in order to make full use of scanline-based splatting, Oliviera *et al.* [216] proposed a two-pass pre-warp process, which forward maps from x_r to x_t' vertically and horizontally. After the pre-warp stage, the reference image is then warped to a 3D plane—which can then be very quickly mapped to the target screen using perspective mapping in the second (texture-mapping) stage. The overall performance of the hybrid two-pass technique is significantly better than traditional backward mapping.

There is a cost associated with the hybrid two-pass method: the reference images are resampled multiple times before finally rendered on the target screen. This causes the rendering result to look slightly blurrier; because of the multiple resampling, the filters used need to be more carefully designed.

4.3.4 Hardware acceleration

As described above, if the 3D geometry is just a 3D plane, backward mapping reduces to perspective texture mapping, which can capitalize on the conventional graphics pipeline. In general, however, hardware acceleration is not trivial to implement for point-based rendering on graphics hardware. The conventional graphics pipeline can be easily used for forward mapping, except for the hole filling process. In order to fill in the gaps caused by an increase in the object footprint, each pixel from the reference image is typically rendered using a micro facet larger than a pixel's area. Some techniques [189] build tiny triangular meshes on the reference image before rendering and allow the texture mapping engine to resample the texture and subsequently fill in the gaps. However, this usually involves a large number of vertices and is not practical unless top-of-the-line accelerators are used.

Approaches to hardware-accelerated surface splatting (for general non-planar points) are similar in that they involve three rendering passes. The first pass is *visibility splatting*; here, the object is rendered without lighting to fill the depth buffer only. This is followed by the *blending pass* where colors and weights (alphas) of pixels with small depth differentials are accummulated. The final *normalization pass* involves division of the weighted sum of colors by the sum of weights, which can be implemented on the GPU (e.g., [96]).

Coconu and Hege [48] implemented a version of hardware-accelerated splatting with restricted shape and size of filter kernels. Ren *et al.*[245] implemented a hardware-accelerated version of Elliptical Weighted Average (EWA) [94] surface splatting; they represent each splat by an alpha-textured quad in the splat rasterization stage. On the other hand, Botsch *et al.* [19] use per-pixel Phong shading and a simple approximation to the EWA.

If the geometry can be represented as a 3D plane plus a small amount of depth variation (e.g., sprites with depth [264] and relief texture [216]), the hybrid mapping methods discussed in Section 4.3.3 can be used to take advantage of conventional hardware acceleration of the projective texture mapping. As shown in Figure 4.8, the source image is forward mapped using the depth map to an intermediate texture, which can then be fed to a conventional graphics pipeline for final backward mapping (traditional texture mapping).

With programmable graphics hardware, backward mapping using view-aligned depth fields can also be accelerated (e.g., real-time relief mapping [233]). In this approach, the points are represented as a depth map and stored as texture. For each pixel on target screen, the search along the EPI is executed in the pixel shader. After finding the point that is projected to this pixel, its texture coordinate is then used to index the point color (which is stored in color texture) for final rendering.

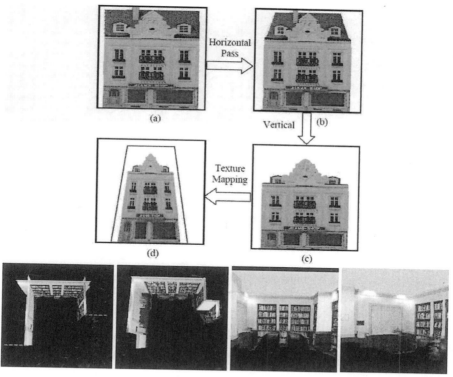

Fig. 4.8. Relief textures. (Images courtesy of Manuel Oliveira.) ©2000 ACM, Inc. Included here by permission.

4.4 Monolithic rendering

A monolithic geometry is usually represented as continuous polygon meshes with textures, which can be readily rendered using graphics hardware. This geometry can be obtained from 3D scanners, such as in those featured in the surface light field [323] work. Other sources include geometric proxies produced by interactive modeling systems (such as Façade [61]), convex hulls (e.g., visual hulls [190] and opacity hulls [191]), and reconstructed by stereo algorithms (such as joint view interpolation [162], structure from motion algorithms as used in the Lumigraph [91], unstructured Lumigraph [22], and plenoptic modeling with a hand-held camera [106]).

Rendering polygonal mesh model with textures has been well-explored. (For an excellent survey on texture mapping, see [104].) In IBR, view-dependent texture mapping (Figure 4.10) is usually necessary for photorealism. The major challenge for IBR with 3D polygonal models is in designing the compositing stage where the reference views and their blending weights have to be computed. We now describe the compositing stage for representations with implicit geometry, followed by the compositing stage for representations with explicit geometry.

4.4.1 Using implicit geometry model

As mentioned in Chapter 2, view synthesis techniques that are based on implicit geometry do not use 3D models. Instead, these technique rely on weakly calibrated parameters and feature correspondence to generate virtual views. Examples of such techniques include view interpolation [40], view morphing [260], and joint view triangulation (JVT) [162]. There is no Euclidean space representation of rays; reference views are usually chosen from the nearest neighboring views, and the weights used for blending the reference views are typically computed based on some measure of proximity between the virtual view and the contributing sampled view.

In the case of view morphing [260], the sampled viewpoints are assumed parallel with the relative motion being perpendicular to the viewing direction, as can be seen in Figure 2.17. The blending weight associated with the sample viewpoint is simple: it is inversely proportional to the distance of the virtual viewpoint to the sample viewpoint. This is easily seen from Section 2.2.2. The weights are used to blend the (warped) reference colors to produce the final virtual view. This strategy can be extended to three reference viewpoints as in JVT [162], using barycentric distance computed from the target camera position with respect to the three reference views in projective space.

Note that the linear blending strategy is only correct if the two input image planes are parallel to each other and parallel to the line defined by two camera centers. If the cameras are weakly calibrated in projective space, this condition is not guaranteed to hold. However, if the input cameras are close to each other, the linear distance assumption usually produces reasonable results. For two views, it is possible to first pre-warp (rectify) them to produce parallel image planes, as was done in view morphing [260] (see Figure 2.17).

4.4.2 Using explicit geometry model

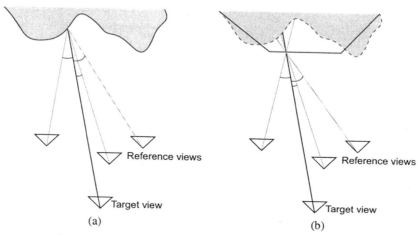

Fig. 4.9. Geometric proxy, and proximity of rays being based on angular distance at the surface of the geometric proxy.

IBR techniques which use explicit geometry (such as 3D surface mesh and depth maps) operate in Euclidean space. This makes spatial reasoning about rays easier: reasoning about ray and viewpoint proximity and ray-object interaction is more intuitive. As mentioned in Chapter 2, having explicit geometry reduces the number of input images required for high-quality view reconstruction. This explicit geometry is also known as a geometric proxy or impostor. The simplest case is when high precision geometry is available with only a limited number of input images; this is where view-dependent texture mapping is appropriate.

Debevec *et al.* [61] first implemented view-dependent texture mapping on architectural models in their Façade work. However, rendering required significant per-pixel computation on the CPU and did not scale well with the number of sampled images. Later, Debevec *et al.* [60] improved the efficiency of rendering substantially by precomputing visibility for each polygon and warping the closest textures through standard projective texture-mapping. Porquet *et al.* [234] achieved real-time rendering by precomputing textures of three nearest viewpoints and applying pixel shading on a decimated mesh.

If an object has a complex appearance (such as specular, glossy, or furry), having an accurate geometry but few images will typically be inadequate. To handle specular or glossy objects, Wood *et al.* [323] scanned highly accurate range data and acquired hundreds of images at a fixed lighting condition to create the surface light field. They used vector quantization and principal component analysis (PCA) to compress the data and represent its representation in a piecewise linear fashion. The result is a remarkably accurate visualization of the complicated object at interactive rates.

Matusik *et al.* [191] model objects that cannot be scanned accurately, such as objects with fur and feathers. They take thousands of images of the object at various poses and lighting directions against a plasma display, which acts as a green screen for matting. The estimated visual hull (with opacity) is used for rendering. They also compress the data by applying PCA on each set of common viewpoints (each set with varying illumination). Interpolation is performed over the four closest views.

In general, however, the geometry used is not always accurate. As Figure 4.9(b) shows, the geometric proxy may just be a rough approximation. Chapter 5 shows that there is an inverse relationship between how accurate the geometric proxy is and how densely sampled the input images should be for alias-free view reconstruction.

Choosing the rays from the input images to render at a virtual viewpoint is based on the notion of ray proximity—we ideally want to choose rays that are "close" to the virtual ray. The proximity of rays is not only related to the input viewpoints, but also determined by the geometric proxy itself. As shown in Figure 4.9(a), a natural strategy is to choose rays with smaller angle deviations with respect to the virtual ray on the geometry surface.

Other strategies for ray selection have been used. In the work of Debevec *et al.* [61], the same set of views are used to render a polygon. For each polygon, the weights are computed using the angles between the polygon normal and the viewing directions of the sampled views (the smaller the angle, the larger the weight, with a maximum of 1).

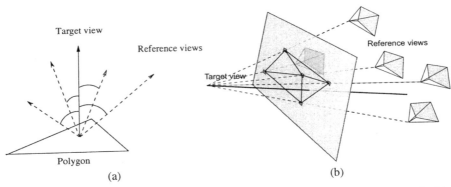

Fig. 4.10. Input view selection strategies. (a) In the work of [61], the weights associated with the input (reference) views are inversely proportional to the angle deviation. (b) In the work of [106], all the input camera centers are first projected to the target image, followed by 2D triangulation. Each triangle is associated with three input viewpoints corresponding to its vertices. These input viewpoints are the "nearest" three cameras for all the pixels that are contained within the triangle.

In [106], the images were obtained by moving the camera approximately within a 2D plane in a serpentine manner. Their technique for ray/view selection is to project all reference camera centers to the target camera and triangulate these points on the

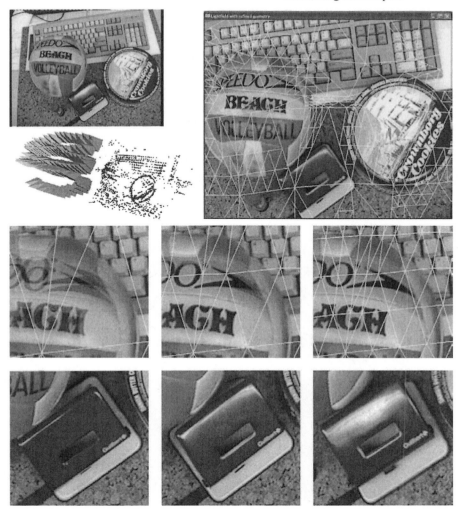

Fig. 4.11. Plenoptic modeling using a hand-held camera. Top left: Sample image of the scene and depiction of recovered camera poses and 3D points. Top right: View rendered using adaptive subdivision. A triangle that is considered too large is split into four triangles; the 3D locations of the mid-points of the original triangle are obtained using local depth maps. Middle row: Reduction of ghosting with refined geometry. Bottom row: Illustration of view-dependent effect. (Images courtesy of Marc Pollefeys.)

target screen (Figure 4.10(b)). For each pixel, the "nearest" three input views correspond to the vertices of the triangle that contains it. The blending weights assigned to these input views correspond the barycentric coordinates (see Section 4.2.3). View synthesis results can be seen in Figure 4.11.

The ray selection strategy of the unstructured Lumigraph [22] combines a number of properties: use of geometric proxies, epipole consistency (i.e., if a ray passes through the center of a source camera, the virtual ray can be trivially constructed), matching of resolution, virtual ray consistency (the same ray, regardless of the location of the virtual camera, should have the same set of "nearest" source cameras), and minimum angular deviation. To enable real-time rendering, Buehler *et al.* [22] compute the camera blending only at a discrete set of points in the image plane. These points are triangulated, and the dense blend weight distribution is subsequently computed through interpolation.

4.5 Layer-based rendering

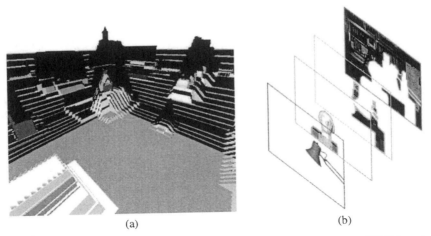

(a)   (b)

Fig. 4.12. Planar layers to approximate scene geometry. (a) Planar impostors [131] (courtesy of Stefan Jeschke), (b) pop-up light field [270] (also Chapter 16).

Layered techniques usually discretize the scene into a collection of planar layers, with each layer consisting of a 3D plane with texture and optionally a transparency map. Two versions of such a representation are shown in Figure 4.12.

Compared to point-based rendering, layer-based rendering is easier to implement using the GPU. The layers can be thought of as a discontinuous set of polygonal models, and as such, very amenable to conventional texture mapping (and to view-dependent texture mapping as well). In addition, compared to monolithic representations, a layer-based representation is usually easier to construct since no connectivity between layers is required. The lack of connectivity information is also a disadvantage: it can severely limit where the scene can be viewed.

Jeschke *et al.* [131] overcome the limitation in viewpoint range in two ways: use layers to represent only far-away portions of the scene (with the nearby parts being regular polygon meshes), and make the sets of layers location dependent (with the viewspace discretized into separate *view cells*). The layer representation described in [255] also adopts a view-dependent geometry solution; more specifically, when target view is too far from the current source views, it automatically generates a new set of source views with new layers. Other representations such as the pop-up light field [270] use texture synthesis to fill holes in the background layers.

Layer-based rendering usually consists of two phases. First, each layer is rendered using either point-based or monolithic rendering techniques as discussed in Sections 4.3 and 4.4, respectively. Subsequently, all rendered layers are composed in back-to-front order to produce the final view.

The painter's algorithm is often used to combine the layers, i.e., the layers are rendered from back to front relative to the target image plane. Occlusions are automatically handled this way. As with point splatting, the layers can also be rendered in an arbitrary order, with the help of the Z-buffer and A-buffer as described in [191]. This technique can be used when the order of layers relative to the target view is unknown. The layers are rendered to the color buffer with the Z-buffer dictating occlusion between layers; meanwhile, the A-buffer accumulates the alpha weights used for normalization in the final stage.

It is worth noting that when rendering a layer with semi-transparency using multiple textures from different source views [270], the colors should be pre-multiplied by the alpha value before blending for higher computational efficiency. Wallace [312] derived a recursive blending equation in which two semi-transparent images are combined to produce another semi-transparent image. Porter and Duff [235] later simplified the recursive blending equation by substituting the original colors with colors with pre-multiplied alpha. This recursive blending equation is significant for compositing three or more layers, because it preserves associativity. In other words,

$$\text{layer}_1 \oplus (\text{layer}_2 \oplus \text{layer}_3) = (\text{layer}_1 \oplus \text{layer}_2) \oplus \text{layer}_3,$$

with $\oplus$ being the compositing operator. Rather than applying a linear operation (e.g., scaling) to each of the layers prior to compositing, it is more efficient to composite the layers *first*, followed by applying the linear operation on the composited result.

4.6 Software and hardware issues

Geometry is often extracted as a means for reducing the number of source images required for high-quality rendering. In addition, explicit geometry models can be efficiently rendered by conventional graphics hardware. However, the process of recovering geometry is often performed on the CPU side, which is slow.

There is an emerging field on using the GPU to perform non-graphics related computation. This field is called "general-purpose computation using graphics hardware," otherwise referred to as GPGPU (see [223] and http://gpgpu.org/). As an example, Yang *et al.* developed a hardware-accelerated version (using pixel shaders) of the plane-sweep algorithm [50] to compute depth at interactive rates with multiple cameras [336]. Woetzel and Koch [320] later extended the multi-camera stereo system to incorporate shiftable windows and view selection for occlusion handling. The technique of [336] uses a winner-take-all strategy to select the depths, which is prone to noise. Another attempt on porting stereo algorithms to the GPU is [183] where binocular stereo based on a variational deformable model has been shown to run in interactive rates.

The conventional graphics pipeline supports IBR through texture mapping, especially with multi-texture extensions. However, this support is applicable to only a subset of IBR representations, and a fair amount of work is required to fully capitalize on the capabilities of the GPU. Why do not we merely rely on the CPU? While CPU speeds are getting faster, memory access speeds remain about the same. Unfortunately, IBR techniques tend to be memory intensive—as a result, it is critical for IBR systems to have fast memory access. In addition, for a hardware system to be more "IBR-compliant," it must be capable of forward mapping. Whitted [316] discussed various IBR-related software and hardware issues, and provided an outline of a generic forward mapping processor.

Part II

Sampling

So far, we have introduced various representations for image-based rendering. In the second part of the book, we study a fundamental problem in image-based rendering which we call *plenoptic sampling*, or the minimum number of images needed for anti-aliased image-based rendering. We show different approaches for minimum sampling including a spectral analysis based on the sampling theorem (Chapter 5), a geometric analysis (Chapter 6) and an optical analysis (Chapter 7). Also included in this part is an image-based rendering system called layered lumigraph (Chapter 7) that takes advantage of sampling analysis to optimize the rendering performance of image-based rendering.

From a spectral analysis of light field signals and using the sampling theorem, we mathematically derive the analytical functions to determine the minimum sampling rate for light field rendering in Chapter 5. The spectral support of a light field signal is bounded by the minimum and maximum depths only, no matter how complicated the spectral support might be because of depth variations in the scene. The minimum sampling rate for light field rendering is obtained by compacting the replicas of the spectral support of the sampled light field within the smallest interval. Given the minimum and maximum depths, a reconstruction filter with an optimal and constant depth can be designed to achieve anti-aliased light field rendering.

Chapter 6 presents a geometric approach to analyzing the minimum number of images needed for anti-aliased image-based rendering. In geometric terms, anti-aliased light field rendering is equivalent to eliminating the "double image" artifacts caused by view interpolation. We derive the same minimum sampling rate using geometric analysis as using the spectral analysis. Using geometric analysis, however, does not require the light fields to be uniformed sampled as in the case of spectral analysis. We thus present the minimum sampling analysis for Concentric Mosaics as well.

Chapter 7 shows that the minimum sampling rate for light field rendering can also be interpreted using an optical analysis. A light field is analogous to a virtual imaging system by defining its depth of field, focal plane, aperture and the circle of confusion. Therefore, the hyperfocal distance of the virtual optical system becomes a key parameter for light field rendering because it determines the relationship between the spacing of cameras and rendering resolution.

In Chapter 8, we describe an IBR representation called the layered Lumigraph. Given the output image resolution and the rendering platform (e.g., process speed, memory, etc.), the layered Lumigraph representation is configured for optimized rendering performance. Moreover, it is capable of level-of-detail (LOD) control using the image and geometry trade-off from the sampling analysis.

Additional Notes on Chapters

Chapter 5 first appeared as an article in ACM SIGGRAPH, July 2000, pages 307–318. The co-authors are Jin-Xiang Chai, Xin Tong, Shing-Chow Chan, and Heung-Yeung Shum.

Most of Chapter 6 first appeared in IEEE CVPR 2000 pages 588–579. The co-authors of this article are Zhouchen Lin and Heung-Yeung Shum. A more complete journal version of this paper was later published at the *International Journal of Computer Vision*, volume 58, number 2, July 2004, pages 121–138 by the same authors.

In Chapter 7, an optical analysis of light field rendering was introduced by Tao Feng and Heung-Yeung Shum at ACCV 2002, although part of its analysis first appeared briefly in the Plenoptic Sampling paper at ACM SIGGRAPH, July 2000, pages 307–318.

Chapter 8 describes the layered Lumigraph with LOD control, which was originally the subject of the article co-authored by Xin Tong, Jin-Xiang Chai, and Heung-Yeung Shum in *Journal of Visualization and Computer Animation*, volume 13, number 4, pages 249–261, 2002.

5

Plenoptic Sampling

This chapter studies the problem of *plenoptic sampling* in image-based rendering (IBR). From a spectral analysis of light field signals and using the sampling theorem, we mathematically derive the analytical functions to determine the *minimum sampling rate* for light field rendering. The spectral support of a light field signal is bounded by the minimum and maximum depths only, no matter how complicated the spectral support might be because of depth variations in the scene. The minimum sampling rate for light field rendering is obtained by compacting the replicas of the spectral support of the sampled light field within the smallest interval. Given the minimum and maximum depths, a reconstruction filter with an optimal and constant depth can be designed to achieve anti-aliased light field rendering.

Plenoptic sampling goes beyond the minimum number of images needed for anti-aliased light field rendering. More significantly, it utilizes the scene depth information to determine the *minimum sampling curve* in the joint image and geometry space. The minimum sampling curve quantitatively describes the relationship among three key elements in IBR systems: scene complexity (geometrical and textural information), the number of image samples, and the output resolution. Plenoptic sampling bridges the gap between image-based rendering and traditional geometry-based rendering.

5.1 Introduction

Previous work on IBR reveals a continuum of image-based representations (see Chapter 1 and [155, 134]) based on the tradeoff between how many input images are needed and how much is known about the scene geometry. At one end, traditional texture mapping relies on very accurate geometrical models but only a few images. In an image-based rendering system with depth maps, such as 3D warping [189], view interpolation [40], view morphing [260] and layered-depth images (LDI) [264], LDI tree [39], etc., the model consists of a set of images of a scene and their associated depth maps. When depth is available for every point in an image, the image can be rendered from any nearby point of view by projecting the pixels of the image to their proper 3D locations and re-projecting them onto a new picture.

At the other end, light field rendering uses many images but does not require any geometrical information. Light field rendering [160] generates a new image of a scene by appropriately filtering and interpolating a pre-acquired set of samples. The Lumigraph [91] is similar to light field rendering but it applies approximated geometry to compensate for non-uniform sampling in order to improve rendering performance. Unlike light field and Lumigraph where cameras are placed on a two-dimensional manifold, Concentric Mosaics [267] reduce the amount of data by only capturing a sequence of images along a circular path. Light field rendering, however, typically relies on oversampling to counter undesirable aliasing effects in output display. Oversampling means more intensive data acquisition, more storage, and more redundancy. Therefore, a fundamental problem in IBR is determining the lower bound or the minimum number of samples needed for light field rendering.

Sampling analysis in IBR is a difficult problem because it involves the complex relationship among three elements: the depth and texture information of the scene, the number of sample images, and the rendering resolution. The topic of prefiltering a light field has been explored in [160]. Similar filtering process has been previously discussed by Halle [99] in the context of Holographic stereograms. A parameterization for more uniform sampling [25] has also been proposed. From an initially undersampled Lumigraph, new views can be adaptively acquired if the rendering quality can be improved [256]. An opposite approach is to start with an oversampled light field, and to cull an input view if it can be predicted by its neighboring frames [107, 277]. Using a geometrical approach and without considering textural information of the scene, Lin and Shum [167] were the first to study the number of samples needed in light field rendering with constant depth assumption and bilinear interpolation.

In this chapter, we study *plenoptic sampling*, or how many samples are needed for plenoptic modeling [194, 2]. Plenoptic sampling can be stated as:

How many samples of the plenoptic function (e.g., from a 4D light field) and how much geometrical and textural information are needed to generate a continuous representation of the plenoptic function?

Specifically, this chapter tackles the following two problems under plenoptic sampling, with and without geometrical information:

- Minimum sampling rate for light field rendering;
- Minimum sampling curve in joint image and geometry space.

The plenoptic sampling analysis is formulated as a high dimensional signal processing problem. The assumptions made for this analysis are: Lambertian surfaces, and uniform sampling geometry or lattice for the light field. Rather than attempting to obtain a closed-form general solution to the 4D light field spectral analysis, we only analyze the bounds of the spectral support of the light field signals. A key analysis to be presented in this chapter is that the spectral support of a light field signal is bounded by only the minimum and maximum depths, irrespective of how complicated the spectral support might be because of depth variations in the scene.

Given the minimum and maximum depths, a reconstruction filter with an optimal and constant depth can be designed to achieve anti-aliased light field rendering.

The minimum sampling rate of light field rendering is obtained by compacting the replicas of the spectral support of the sampled light field within the smallest interval without any overlap. Using more depth information, plenoptic sampling in the joint image and geometry space allows a significant reduction in the number of images needed. In fact, the relationship between the number of images and the geometrical information under a given rendering resolution can be quantitatively described by a minimum sampling curve. This minimal sampling curve serves as the design principles for IBR systems. Furthermore, it bridges the gap between image-based rendering and traditional geometry-based rendering.

The sampling analysis was inspired by the work on motion compensation filter in the area of digital video processing, in which depth information has been incorporated into the design of the optimal motion compensation filter [295, 84]. In digital video processing, global constant depth and arbitrary motion are considered for both static and dynamic scenes. Meanwhile, the sampling analysis involves static scenes with arbitrary geometries and uniformly sampled camera setups.

The remainder of this chapter is organized as follows. In Section 5.2, a spectral analysis of 4D light field is introduced and the bounds of its spectral support are determined. From these bounds, the minimum sampling rate for light field rendering can be derived analytically. Plenoptic sampling in the joint image and geometry space is studied in Section 5.3. The minimum sampling curves are deduced with accurate and approximated depths. Experimental results and conclusions are presented in Sections 5.4 and 5.5, respectively.

5.2 Spectral analysis of light field

5.2.1 Light field representation

We begin by briefly reviewing the properties of light field representation. We will follow the notations in the Lumigraph paper [91]. In the standard two-plane ray database parameterization, there is a camera plane, with parameter (s, t), and a focal plane, with parameter (u, v). Each ray in the parameterization is uniquely determined by the quadruple (u, v, s, t). We refer the reader to Figure 2(a) of [91] for more details.

A 2D subspace given by fixed values of s and t resembles an image, whereas fixed values of u and v give a hypothetical radiance function. Fixing t and v gives rise to an epipolar image, or EPI [17]. An example of a 2D light field or EPI is shown in Figure 5.1. Note that in the sampling analysis, (u, v) is defined in the local coordinates of (s, t), unlike in conventional light field where (u, v, s, t) are defined in a global coordinate system.

Let the sample intervals along s and t directions be Δs and Δt, respectively. As a result, the horizontal and vertical disparities between two grid cameras in the (s, t) plane are given by $k_1 \Delta s f / z$ and $k_2 \Delta t f / z$, respectively. Here f is the focal length of the camera, z is the depth value and $(k_1\Delta s, k_2\Delta t)$ is the sample interval between two grid points (s, t).

Similarly, the sample intervals along u and v directions are assumed to be Δu and Δv, respectively. A pinhole camera model is adopted to capture the light field. What a camera sees is a blurred version of the plenoptic function because of finite camera resolution. A pixel value is a weighted integral of the illumination of the light arriving at the camera plane, or the convolution of the plenoptic function with a low-pass filter.

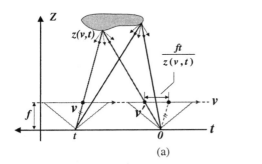 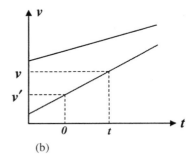

(a)   (b)

Fig. 5.1. An illustration of 2D light field or EPI: (a) a point is observed by two cameras 0 and t; (b) two lines are formed by stacking pixels captured along the camera path. Each line has a uniform color because of Lambertian assumption on object surfaces.

5.2.2 A framework for light field reconstruction

Let $l(u, v, s, t)$ represent the continuous light field, $p(u, v, s, t)$ the sampling pattern in light field, $r(u, v, s, t)$ the combined filtering and interpolating low-pass filter, and $i(u, v, s, t)$ the output image after reconstruction. Let L, P, R and I represent their corresponding spectra, respectively. In the spatial domain, the light field reconstruction can be computed as

$$i(u, v, s, t) = r(u, v, s, t) * [l(u, v, s, t)p(u, v, s, t)] \tag{5.1}$$

where $*$ represents the convolution operation.

In the frequency domain, we have

$$I(\Omega_u, \Omega_v, \Omega_s, \Omega_t) = R(\Omega_u, \Omega_v, \Omega_s, \Omega_t)(L(\Omega_u, \Omega_v, \Omega_s, \Omega_t)$$
$$*P(\Omega_u, \Omega_v, \Omega_s, \Omega_t)) \tag{5.2}$$

The problem of light field reconstruction is to find a reconstruction filter $r(u, v, s, t)$ for anti-aliased light field rendering, given the sampled light field signals.

5.2.3 Spectral support of light fields

In this section, we introduce the spectral supports of continuous light field $L(\Omega_u, \Omega_v, \Omega_s, \Omega_t)$ and sampled light field $L(\Omega_u, \Omega_v, \Omega_s, \Omega_t) * P(\Omega_u, \Omega_v, \Omega_s, \Omega_t)$.

5.2.3.1 Spectral support of continuous light field

The depth function of the scene is assumed to be equal to $z(u, v, s, t)$. As shown in Figure 5.1(a), the same 3D point is observed at v' and v in the local coordinate systems of cameras 0 and t, respectively. The disparity between the two image coordinates can be computed easily as $v - v' = ft/z$. Figure 5.1(b) shows an EPI image where each line represents the radiance observed from different cameras. For simplicity of analysis, the BRDF model of a real scene is assumed to be Lambertian. As a result, each line in Figure 5.1(b) has a uniform color.

The radiance received at the camera position (s, t) is given by

$$l(u, v, s, t) = l(u - \frac{fs}{z(u, v, s, t)}, v - \frac{ft}{z(u, v, s, t)}, 0, 0)$$

and its Fourier transform is

$$L(\Omega_u, \Omega_v, \Omega_s, \Omega_t) = \int_{-\infty}^{\infty} \int_{-\infty}^{\infty} \int_{-\infty}^{\infty} l(u, v, s, t) e^{-j\Omega^T x} dx$$
$$e^{-j(\Omega_s s + \Omega_t t)} ds dt \qquad (5.3)$$

where $x^T = [u, v]$ and $\Omega^T = [\Omega_u, \Omega_v]$.

Computing the Fourier transform (5.3) is very complicated and is beyond the scope of this book. Instead, we show the analysis of the bounds of the spectral support of light fields. Also for simplicity, it is assumed that samples of the light field are taken over the commonly used rectangular sampling lattice.

5.2.3.2 Spectral support of sampled light field

Using the rectangular sampling lattice, the sampled light field $l_s(u, v, s, t)$ is represented by

$$l_s(u, v, s, t) = l(u, v, s, t) \sum_{n_1, n_2, k_1, k_2 \in Z}$$
$$\delta(u - n_1 \Delta u)\delta(v - n_2 \Delta v)\delta(s - k_1 \Delta s)\delta(t - k_2 \Delta t) \qquad (5.4)$$

and its Fourier transform is

$$L_s(\Omega_u, \Omega_v, \Omega_s, \Omega_t) = \sum_{m_1, m_2, l_1, l_2 \in Z}$$
$$L(\Omega_u - \frac{2\pi m_1}{\Delta u}, \Omega_v - \frac{2\pi m_2}{\Delta v}, \Omega_s - \frac{2\pi l_1}{\Delta s}, \Omega_t - \frac{2\pi l_2}{\Delta t}) \qquad (5.5)$$

The above equation indicates that $L_s(\Omega_u, \Omega_v, \Omega_s, \Omega_t)$ consists of replicas of $L(\Omega_u, \Omega_v, \Omega_s, \Omega_t)$, shifted to the 4D grid points

$$(2\pi m_1/\Delta u, 2\pi m_2/\Delta v, 2\pi l_1/\Delta s, 2\pi l_2/\Delta t),$$

where $m_1, m_2, l_1, l_2 \in Z$, and Z is the set of integers.

These shifted spectra, or replicas, except the original one at $m_1 = m_2 = l_1 = l_2 = 0$, are called the alias components. If L is not bandlimited outside the Nyquist frequencies, some replicas will overlap with the others, creating aliasing artifacts.

In general, there are two ways to combat aliasing effects in output display when we render a novel image. First, we can increase the sampling rate. The higher the sampling rate, the less the aliasing effects. Indeed, uniform oversampling has been consistently employed in many IBR systems to avoid undesirable aliasing effects. However, oversampling means more effort in data acquisition and requires more storage. Though redundancy in the oversampled image database can be partially eliminated by compression, excessive samples are always wasteful.

Second, light field signals can also be made bandlimited by filtering with an appropriate filter kernel. Similar filtering has to be performed to remove the overlapping of alias components during reconstruction or rendering. The design of such a kernel is, however, related to the depth of the scene. Previous work on Lumigraph shows that approximate depth correction can significantly improve the interpolation results. The questions are: is there an optimal filter? Given the number of samples captured, how accurately should the depth be recovered? Similarly, given the depth information one can recover, how many samples can be removed from the original input?

5.2.4 Analysis of bounds in spectral support

5.2.4.1 A model of global constant depth

Let us first consider the simplest scene model in which every point is at a constant depth (z_0). The first frame is chosen as the reference frame, and $l(u, v, 0, 0)$ denotes the 2D intensity distribution within the reference frame. The 4D Fourier transform of the light field signal $l(u, v, s, t)$ with constant depth is

$$
L(\Omega_u, \Omega_v, \Omega_s, \Omega_t) = \int_{-\infty}^{\infty} \int_{-\infty}^{\infty} l(u, v, 0, 0) e^{-j(\Omega_u u + \Omega_v v)} du\, dv
$$

$$
\int_{-\infty}^{\infty} e^{-j(\frac{f}{z_0}\Omega_u + \Omega_s)s} ds \int_{-\infty}^{\infty} e^{-j(\frac{f}{z_0}\Omega_v + \Omega_t)t} dt
$$

$$
= 4\pi^2 L'(\Omega_u, \Omega_v) \delta(\frac{f}{z_0}\Omega_u + \Omega_s) \delta(\frac{f}{z_0}\Omega_v + \Omega_t),
$$

where $L'(\Omega_u, \Omega_v)$ is the 2D Fourier transform of continuous signal $l(u, v, 0, 0)$ and $\delta(\cdot)$ is the 1D Dirac delta function. For simplicity, the following discussion will focus on the projection of the support of $L(\Omega_u, \Omega_v, \Omega_s, \Omega_t)$ onto the (Ω_v, Ω_t) plane, which is denoted by $L(\Omega_v, \Omega_t)$.

Under the constant depth model, the spectral support of the continuous light field signal $L(\Omega_v, \Omega_t)$ is defined by a line $\Omega_v f / z_0 + \Omega_t = 0$, as shown in Figure 5.2(b). The spectral support of the corresponding sampled light field signals is shown in

Figure 5.2(c). Note that, due to sampling, replicas of $L(\Omega_v, \Omega_t)$ appear at intervals $2\pi m_2/\Delta v$ and $2\pi l_2/\Delta t$ in the Ω_v and Ω_t directions, respectively.

Figure 5.6(a) shows a constant depth scene (a1), its EPI image (a2), and the Fourier transform of the EPI (a3). As expected, the spectral support is a straight line.[1]

5.2.4.2 Spatially varying depth model

It is now straightforward to observe that any scene with a depth between the minimum z_{min} and the maximum z_{max} will have its continuous spectral support bounded in the frequency domain, by two lines $\Omega_v f/z_{min} + \Omega_t = 0$ and $\Omega_v f/z_{max} + \Omega_t = 0$. Figure 5.6(b3) shows the spectral support when two planes with constant depths are in the scene. Adding another tilted plane in between (Figure 5.6(c1)) results in no variations in the bounds of the spectral support, even though the resulting spectral support (Figure 5.6(c3)) differs significantly from that in Figure 5.6(c2). This is further illustrated when a curved surface is inserted in between two original planes, as shown in Figure 5.6(d1). Even though the spectral supports differ significantly, Figures 5.6(b3), (c3) and (d3) all have the same bounds.

Another important observation is that geometrical information can help to reduce the bounds of the spectral support in the frequency domain. As will be illustrated in the following section, the optimal reconstruction filter is determined precisely by the bounds of the spectral support. Furthermore, these bounds are functions of the minimum and maximum depths of the scene. If some information on the scene geometry is available, the scene geometry can be decomposed into a collection of constant depth models on a block-by-block basis. Each model will have a much tighter bound than the original model. How tight the bound is will depend on the accuracy of the geometry. Figure 5.3 illustrates the reduction in bounds, from $[z_{min}, z_{max}]$ to $\max([z_{min}, z_0], [z_0, z_{max}])$, with the introduction of another layer.

5.2.4.3 A model with truncating windows

Because of the linearity of the Fourier transform, the spectral support of the EPI image for a scene with two constant planes will be two straight lines. However, this statement is true only if these two planes do not occlude each other. For synthetic environments, we can construct such EPI images on different layers by simply ignoring the occlusion.

In practice, we can represent a complicated environment using a model with truncating windows. For example, we can approximate an environment using truncated and piece-wise constant depth segments. Specifically, suppose the depth can be partitioned as

$$z(v) = z_i, \text{ for } v_i \leq v < v_{i+1}, \ i = 1, \cdots, N_d$$

where v_1 and v_{N_d+1} are the smallest and largest v of interest respectively. Then

---

[1] The ringing effect in the vicinity of the horizontal and vertical axes is caused by convolving with $\sin(\Omega_v)/\Omega_v$ because of the rectangular image boundary.

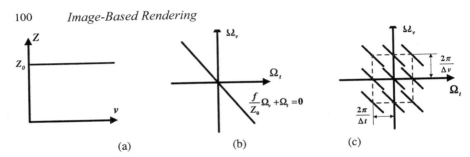

(a)   (b)   (c)

Fig. 5.2. Spectral support of light field signals with constant depth: (a) a model of constant depth; (b) the spectral support of continuous light field signals; (c) the spectral support of sampled light field signals.

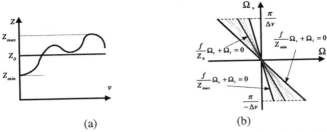

(a)   (b)

Fig. 5.3. Spectral support for light field signal with spatially varying depths: (a) a local constant depth model bounded by z_{min} and z_{max} is augmented with another depth value z_0; (b)spectral support is now bounded by two smaller regions, with the introduction of the new line of z_0.

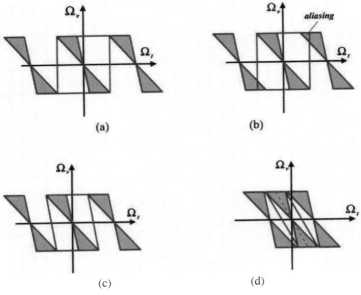

(a)   (b)

(c)   (d)

Fig. 5.4. Three reconstruction filters with different constant depths: (a) infinite depth; (b) infinite depth (aliasing occurs); (c) maximum depth; (d) optimal depth at z_c.

See color plate section near center of book.

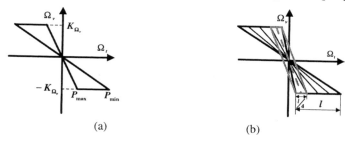

(a)   (b)

Fig. 5.5. (a) The smallest interval that replicas can be packed without any overlap is $P_{max}P_{min}$, determined by the highest frequency K_{Ω_v}. (b) A spectral support decomposed into multiple layers.

$$l(v,t) = l_i(v - ft/z_i, 0), \text{ if } v_i \leq v < v_{i+1}$$

and

$$L(\Omega_v, \Omega_t) = \sum_{i=1}^{N_d} \exp(-j\frac{v_i + v_{i+1}}{2}(\Omega_v + \Omega_t z_i/f))$$

$$\frac{2\sin(\frac{v_{i+1}-v_i}{2}(\Omega_v + \Omega_t z_i/f))}{f\Omega_v/z_i + \Omega_t} L_i(-\Omega_t z_i/f)$$

$$\equiv \sum_{i=1}^{N_d} Q_i(\Omega_v, \Omega_t) \tag{5.6}$$

where L_i is the $1D$ Fourier transform of l_i.

In (5.6), because the function $\frac{\sin x}{x}$ decays fast, and $L_i(-\Omega_t z_i/f)$ also decreases fast when $|\Omega_t|$ grows, the spectral support of $Q_i(\Omega_v, \Omega_t)$ will look like a narrow ellipse. Nevertheless, because of high frequency leak, cut-off frequency should be used in the sampling analysis.

An example of two constant planes in an environment is shown in Figures 5.6(b1) (original image), 5.6(b2) (EPI) and 5.6(b3) (spectral support). Note that the shape of each of the two spectral supports, i.e., two approximated lines, is not significantly affected by occlusion because the width of each spectral support is not too large.

5.2.5 A reconstruction filter using a constant depth

Given a constant depth, a reconstruction filter can be designed. Figure 5.4 illustrates three reconstruction filters with different constant depths. Aliasing occurs when replicas overlap with the reconstruction filters in the frequency domain (Ω_t and Ω_v), as shown in Figure 5.4(a)(b)(d). Anti-aliased light field rendering can be achieved by applying the optimal filter as shown in Figure 5.4(c), where the optimal constant depth is defined as the inverse of average disparity d_c, i.e.,

$$d_c = \frac{1}{z_c} = (\frac{1}{z_{min}} + \frac{1}{z_{max}})/2.$$

Fig. 5.6. Spectral support of a 2D light field: (a) a single plane; (b) two planes; (c) a third and tilted plane in between; (d) a curved surface in between.

See color plate section near center of book.

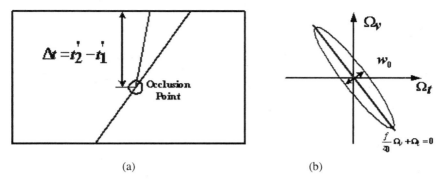

(a)   (b)

Fig. 5.7. Effects with truncation along the t direction: (a) truncation in an EPI image; (b) truncation effect in spectral support of a constant plane. Note that occlusion can be regarded as a special case of truncation as in (a).

Figure 5.8 shows the effect of applying reconstruction filters with different constant depths. As we sweep through the object with a constant depth plane, the aliasing effect is the worst at the minimum and maximum depths. The best rendering quality is obtained at the optimal depth (Figure 5.8(b)), not at the focal plane as has been commonly assumed in light field [160] or Lumigraph [91] rendering. In fact, the optimal depth can be used as a guidance for selecting the focal plane. For comparison, we also show the rendering result using average depth in Figure 5.8(c). Similar sweeping effects have also been discussed in the dynamically reparameterized light field [116].

5.2.6 Minimum sampling rate for light field rendering

With the above theoretical analysis, we are now ready to address the issue of the minimum sampling rate for light field rendering. Since we are dealing with rectangular sampling lattice, the Nyquist sampling theorem for 1D signal applies to both directions v and t. According to the Nyquist sampling theorem, in order for a signal to be reconstructed without aliasing, the sampling frequency needs to be greater than the Nyquist rate, or two times that of the Nyquist frequency. Without loss of generality, we only study the Nyquist frequency along the Ω_t direction in the frequency domain. However, the Nyquist frequency along the Ω_v direction can be analyzed in a similar way.

The minimum interval, by which the replicas of spectral support can be packed without any overlapping, can be computed as shown in Figure 5.5(a)

$$|P_{max}P_{min}| = K_{\Omega_v}fh_d = 2\pi K_{f_v}fh_d \tag{5.7}$$

where

$$h_d = \frac{1}{z_{min}} - \frac{1}{z_{max}},$$

(a)   (b)

(c)   (d)

Fig. 5.8. Sweeping a constant depth plane through an object: (a) at the minimum depth; (b) at the optimal plane; (c) at the average distance between minimum and maximum depths; (d) at the maximum depth. The best rendering quality is achieved in (b).

and

$$K_{f_v} f h_d = \min(B_v^s, 1/(2\Delta v), 1/(2\delta v))$$

is the highest frequency for the light field signal, which is determined by the scene texture distribution (represented by the highest frequency B_v^s), the resolution of the sampling camera (Δv), and the resolution of the rendering camera (δv). The frequency B_v^s can be computed from the spectral support of the light field. Rendering resolution is taken into account because rendering at a resolution higher than the output resolution is wasteful. For simplicity, we assume $\delta v = \Delta v$ from now on.

The minimum sampling rate is equivalent to the maximum camera spacing Δt_{max}, which can be computed as

$$\Delta t_{max} = \frac{1}{K_{f_v} f h_d}. \tag{5.8}$$

The minimum sampling rate can also be interpreted in terms of the maximum disparity defined as the projection error using the optimal reconstruction filter for rendering. From Equation 5.8, we have the maximum disparity

$$\Delta t_{max} f h_d / 2 = \frac{1}{2K_{f_v}} = \max(\Delta v, 1/(2B_v^s)). \tag{5.9}$$

Therefore, the disparity is less than 1 pixel (i.e., the camera resolution) or half cycle of the highest frequency ($1/B_v^s$ is defined as a cycle) presented in the EPI image because of the textural complexity of the observed scene.

If the textural complexity of the scene is not considered, the minimum sampling rate for light field rendering can also be derived in the spatial domain. For example, by considering the light field rendering as a synthetic aperture optical system, we present an optical analysis of light field rendering in Chapter 7.

The maximum camera spacing will be larger if the scene texture variation gets more uniform, or if the rendering camera resolution becomes lower. By setting the higher frequency part of the spectrum to zero so that $B_v^s < 1/(2\Delta v)$, we can reduce the minimum sampling rate. One way to reduce B_v^s is to apply a low-pass filter to the input v-t image. This approach is similar to prefiltering a light field (see Figure 7 in [160]).

In particular, the minimum sampling rate is also determined by the relative depth variation $f(z_{min}^{-1} - z_{max}^{-1})$. The closer the object gets to the camera, the smaller the z_{min} is, and the higher the minimum sampling rate will be. As f gets larger, the sampling camera will cover a more detailed scene, but the minimum sampling rate needs to be increased. Therefore, the plenoptic sampling problem should not be considered in the image space alone, but in the joint image and geometry space.

5.3 Minimum sampling in joint image-geometry space

In this section, we study the minimum sampling problem in the joint geometry and image space. Since the CPU speed, memory, storage space, graphics capability and network bandwidth used vary from user to user, it is very important for users to be able to seek the most economical balance between image samples and depth layers for a given rendering quality.

It is interesting to note that the minimum sampling rate for light field rendering represents essentially one point in the joint image and geometry space, in which little amount of depth information has been utilized. As more geometrical information becomes available, fewer images are necessary at any given rendering resolution. Figure 5.9 illustrates the minimum sampling rate in the image space, the minimum sampling curve in the joint image and geometry space, and minimum sampling curves at different rendering resolutions. Any sampling point above the minimum sampling curve (e.g., Figure 5.9b) is redundant.

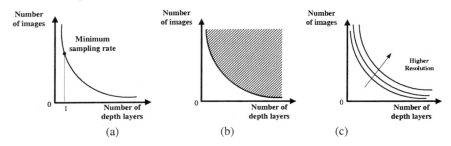

Fig. 5.9. Plenoptic sampling: (a) the minimum sampling rate in image space; (b) the minimum sampling curve in the joint image and geometry space (any sampling point above the curve is redundant); (c) minimum sampling curves at different rendering resolutions.

5.3.1 Minimum sampling with accurate depth

Given an initial set of accurate geometrical data, a scene can be decomposed into multiple layers of sub-regions. Accordingly, the whole spectral support can be decomposed into multiple layers (see Figure 5.5b) due to the correspondence between a constant depth and its spectral support. For each decomposed spectral support, an optimal constant depth filter can be designed. Specifically, for each depth layer $i = 1, \ldots, N_d$, the depth of optimal filter is described as

$$\frac{1}{z_i} = \lambda_i \frac{1}{z_{min}} + (1 - \lambda_i) \frac{1}{z_{max}}, \tag{5.10}$$

where

$$\lambda_i = \frac{i - 0.5}{N_d}.$$

Therefore, a depth value can be assigned to one of the depth layers $z = z_i$ if

$$\frac{-h_d}{2N_d} \leq \frac{1}{z} - \frac{1}{z_i} \leq \frac{h_d}{2N_d}. \tag{5.11}$$

The layers are quantized uniformly in the disparity space. This is because perspective images have been used in the light fields. If parallel projection images are used instead, the quantization should be uniform in the depth space [32].

Similar to Equation 5.8, the minimum sampling in the joint image and accurate depth space is obtained when

$$\frac{\Delta t}{N_d} = \frac{1}{K_{f_v} f h_d}, \quad N_d \geq 1, \tag{5.12}$$

where N_d and Δt are the number of depth layers and the sampling interval along the t direction, respectively. The interval between replicas is uniformly divided into N_d segments.

The number of depth layers needed for scene representation is a function of the sampling and rendering camera resolution, the scene's texture complexity, the spacing of the sampling cameras and the depth variation relative to the focal length.

5.3.1.1 Applications

Based on the above quantitative analysis in the joint image and depth space for sufficient rendering, a number of important applications can be explored:

- **Image-based geometry simplification.** Given the appropriate number of image samples an average user can afford, the minimum sampling curve in the joint space determines how much depth information is needed. Thus, it simplifies the original complex geometrical model to the minimum while still guaranteeing the same rendering quality.
- **Geometry-based image database reduction.** In contrast, given the number of depth layers available, the number of image samples needed can also be reduced to the minimum for a given rendering resolution. The reduction of image samples is particularly useful for light field rendering.
- **Level of details (LOD) in joint image and depth space.** The idea of LOD in geometry space can be adopted in the joint image and geometry space. When an object becomes farther away, its relative size on screen space diminishes so that the number of required image samples or the number of required depth layers can be reduced accordingly. Zooming-in onto and zooming-out of objects also demand a dynamic change in the number of image samples or depth layers.
- **Light field with layered depth.** A general data structure for the minimum sampling curve in the joint image and geometry space can be a light field with layered depth. With different numbers of images and depth layers used, the trade-off between rendering speed and data storage has to be studied.

5.3.2 Minimum sampling with depth uncertainty

Another aspect of minimum sampling in the joint image and geometry space is related to depth uncertainty. Specifically, minimum sampling with depth uncertainty describes the quantitative relationship between the number of image samples, noisy depth and depth uncertainty. It is important to study this relationship because in general the recovered geometry is noisy as modeling a real environment is difficult. Given an estimated depth z_e and its depth uncertainty $\Delta\eta$, the depth value should be located within the range $(z_e - \Delta\eta, z_e + \Delta\eta)$. The maximum camera spacing can be computed as

$$\Delta t_{max} = \min_{z_e} \frac{(z_e + \Delta\eta)(z_e - \Delta\eta)}{2fK_{f_v}\Delta\eta} = \frac{\min_{z_e} z_e^2 - \Delta\eta^2}{2fK_{f_v}\Delta\eta}. \tag{5.13}$$

In addition, geometrical uncertainty also exists when an accurate model is simplified. Given the correct depth z_0 and an estimated depth z_e, the maximum camera spacing can be computed as

$$\Delta t_{max} = \min_{z_e} \frac{z_e z_0}{2fK_{f_v}|z_e - z_0|}. \tag{5.14}$$

5.3.2.1 Applications

Knowledge about the minimum number of images under noisy depth has many practical applications, such as:

- **Minimum sampling rate.** For a specific light field rendering with no depth maps or with noisy depth maps, we can determine the minimum number of images for antialiased light field rendering. Redundant image samples can then be left out from the sampled database for light field rendering.
- **Rendering-driven vision reconstruction.** This is a very interesting application, considering that general vision algorithms would not recover accurate scene depth. Given the number of image samples, how accurately should the depth be recovered to guarantee the rendering quality? Rendering-driven vision reconstruction is different from classical geometry-driven vision reconstruction in that the former is guided by the depth accuracy that the rendering process can have.

5.4 Experiments

Table 5.1 summarizes the parameters of each light field data set used in the experiments. Here, it is assumed that the resolutions of the input image and output display are the same. It is also assumed that the highest frequency in images is bounded by the resolution of the capturing camera.

Different settings of focal length were used for the "Head," "Statue," and "Table" datasets. The focal plane was placed slightly in front of the "Head" object. A smaller focal length would have reduced the minimum sampling rate. For the "Statue" scene, the focal plane was set approximately at its forehead. In fact, the focal length (3000) was set very close to the optimal (3323). Because the "Table" scene has significant depth variation, a small camera focal length was used so that each image covered a large part of the scene.

First, various rendering qualities along the minimal sampling curve in the joint image and geometry space are compared. They are then compared with the best rendering quality obtainable with all images and accurate depth. According to the sampling theory (Equation (5.12)), the number of images is inversely proportional to the number of depth layers in use. The rendering results corresponding to five different image and depth combinations along the minimum sampling curve are shown in Figures 5.12(A)-(E). For example, C(7,8) represents the rendering result using 7 layers of depth and 8×8 images. In contrast, Figure 5.12(F) shows the best rendering output one can achieve from this set of data: accurate depth and all 32×32 images. The quality of the rendered images along the minimal sampling curve is almost indistinguishable[2] from that of using all images and accurate depth.

Figure 5.13(a) compares the rendering quality using different layers of depth and a given number of image samples. With 2×2 image samples of the "Head," images

---

[2] There exists little discrepancy because of the fact that we can not apply the optimal reconstruction filter in rendering.

(a)   (b)

Fig. 5.10. Comparison between conventional light field with 48×48 images and rendering with 16×16 images and 3 bits of depth: (a) artifacts are visible on the left with conventional rendering, (b) but not present with additional geometrical information because minimum sampling requirement is satisfied.

(A)-(E) in Figure 5.13(a) show the rendered images with different layers of depth at 4, 8, 10, 12, and 24. According to Eq (5.12), the minimum sampling point with 2×2 images of the "Head" is at approximately 12 layers of depth. Noticeable visual artifacts can be observed when the number of depth is below the minimal sampling point, as shown in images (A)-(C) of Figure 5.13(a). On the other hand, oversampling layers of depth does not improve the rendering quality, as shown in the images (D) and (E).

With the minimal sampling curve, we can now deduce the minimum number of image samples at any given number of depth layers available. For the Table scene, we find that 3 bits (or 8 layers) of depth information is sufficient for light field rendering when combined with 16×16 image samples (shown in image (D) of Figure 5.13(b)). When the number of depth layers is below the minimal sampling point, light field rendering produces noticeable artifacts, as shown in images (A)-(C) of Figure 5.13(b).

Given a single depth layer, the analysis (Equation 5.12) shows that the number of images for anti-aliased rendering of the table scene requires 124×124 images. Note that conventional light field may require even a larger number of images without using the optimal depth. This very large set of light field data is due to the significant depth variations in the Table scene. This perhaps explains why inside-looking-out light field rendering has not been used often in practice. The analysis also shows that using 3 bits (8 layers) of depth helps to reduce the number of images needed by a factor of 60, to 16×16 images. For comparison, Figure 5.10(a) shows conventional light field rendering with 48×48 images and Figure 5.10(b) shows the rendering result with 16×16 images plus 3 bits of depth. Visual artifacts such as double images at the edge of the wall are clearly visible in Figure 5.10(a). They are not present in Figure 5.10(b).

5.5 Conclusion and Discussion

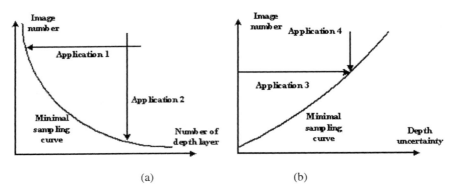

(a)   (b)

Fig. 5.11. Applications of plenoptic sampling: (a) minimum sampling with accurate depth; (b) minimum sampling curve in the joint image and geometry space. Application 1: image-based geometry simplification. Application 2: geometry-assisted image dataset reduction using accurate geometry. Application 3: rendering-driven vision reconstruction. Application 4: depth-assisted light field compression with noisy depth.

In this chapter, we studied the problem of plenoptic sampling. Specifically, by analyzing the bounds of spectral support of light field signals, we can analytically compute the minimum sampling rate of light field rendering. This analysis is based on the fact that the spectral support of a light field signal is bounded by only the minimum and maximum depths, irrespective of how complicated the spectral support might be because of depth variations in the scene. Given the minimum and maximum depths, a reconstruction filter with an optimal constant depth can be designed for anti-aliased light field rendering. The minimum sampling rate for light field rendering is obtained by compacting the replicas of the spectral support of the sampled light field within the smallest interval. The sampling analysis provides a solution to overcoming the oversampling problem in light field capturing and rendering.

We also showed how the minimum sampling curve can be derived through plenoptic sampling analysis in the joint image and geometry space. The minimum sampling curve quantitatively describes the relationship between the number of images and the information on scene geometry, given a specific rendering resolution. Indeed, minimum sampling curves with accurate depth and with noisy depth serve as the design principles for a number of applications. Such interesting applications include image-based geometry simplification, geometry-assisted image dataset reduction, rendering-driven vision reconstruction, in addition to depth-assisted light field compression, or the minimum sampling rate for light field rendering.

The analysis in this chapter has used the uniform and evenly spaced camera setup as in the original light fields. The minimum sampling rate, however, can be further reduced if the cameras can be packed more tightly, as shown by Zhang and Chen in a hexagon setup [338]. In Chapter 6, we describe a geometrical analysis of light fields that is equivalent to the spectral analysis. This geometrical analysis can be applied to IBR systems with more complicated camera configurations such as that for Concentric Mosaics.

With plenoptic sampling, there are a number of exciting areas for future work. For example, depth is used in this chapter to encode the geometry information. Depth is also used in image-assisted geometry simplification. However, surface normal is not considered. One can experiment with different techniques to generate image-assisted geometry simplification using geometrical representations other than depth. The efficiency of geometry simplification can be further enhanced by considering the standard techniques in geometrical simplification, e.g., visibility culling.

Current analysis of plenoptic sampling is based on the assumption that the surface is diffuse and little view-dependent variance can occur. It is conceivable that view dependent surface property will increase the minimum sampling rate for light field.

Another line of interesting work is on how to design a new rendering algorithm for the joint image and geometry representation. The complexity of the rendering algorithm should be proportional to the number of depth in use. In addition, error-bounded depth reconstruction should be considered as an alternative to traditional vision reconstruction, if the reconstruction result is to be used for rendering. Given the error bounds that are tolerable by the rendering algorithms, the difficulty of vision reconstruction may be alleviated. Such a system, called Layered Lumigraph, is described in Chapter 8.

	Focal length	Maximum depth	Minimum depth	(u,v) interval	(s,t) interval	Pixels per image	Image per slab	Spacing Δt_{max}
Head	160.0	308.79	183.40	0.78125	1.5625	256×256	64×64	4.41
Statue	3000.0	5817.86	2326.39	15.625	31.25	256×256	64×64	40.38
Table	350.0	3235.47	362.67	2.4306	7.29	288×288	96×96	5.67

Table 5.1. A summary of parameters used in three data sets in the experiments.

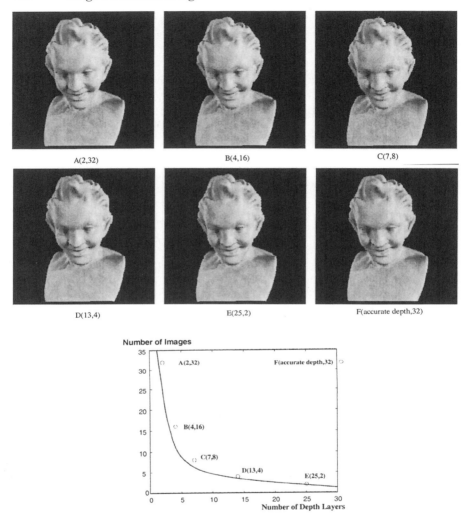

Fig. 5.12. Minimum sampling curve for the object "Statue" in the joint image and geometry space with accurate geometry. Sampling points in the figure have been chosen to be slightly above the minimum sampling curve due to quantization.

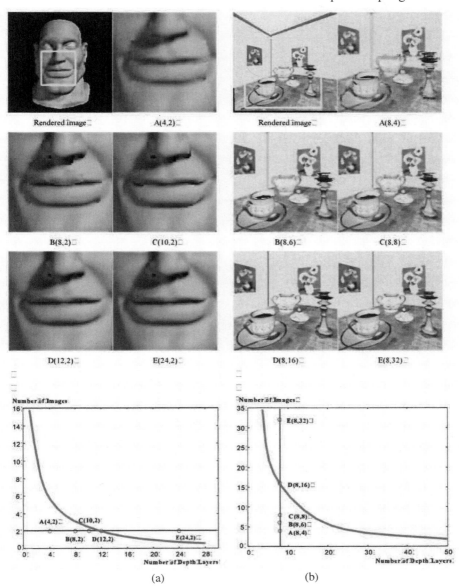

(a)   (b)

Fig. 5.13. Minimum sampling points in the joint image and geometry space: (a) for the "Head" object, when the number of images is 2×2; (b) for the "Table" scene, when the number of depth layers is 8.

See color plate section near center of book.

6

Geometric Analysis of Light Field Rendering

In the previous chapter, we studied the minimum sampling rate of light field rendering and the trade-off between the number of images and the amount of geometry information for image-based rendering. In this chapter, we continue our study on plenoptic sampling; we now adopt a geometric approach to analyzing the light field rendering and its sampling issues. Using a geometric analysis, the first-ever study on minimum sampling rates for light fields (and Concentric Mosaics) was presented in [167], and later in [168]. It was shown in [167] that the sampling rate is dependent on scene disparity variation and camera resolution.

We study the artifact of "double image" (a geometric counterpart of spectral aliasing), optimal constant depth, and maximum camera spacing from the geometric perspective. The geometric analysis is an alternative to the spectral analysis in the previous chapter. However, it is also applicable to irregular capturing and rendering configurations. For example, the results on Concentric Mosaics [167], which are difficult to obtain using a spectral analysis, can be easily obtained using a geometric analysis.

The remainder of this chapter is organized as follows. Section 6.1 describes the criterion for acceptable rendering quality. The minimum sampling rate for Concentric Mosaics rendering is studied in Section 6.2. The minimum sampling rate of the light field is also studied in Section 6.3. The issue of sampling with occlusion is discussed in Section 6.4.

6.1 Problem formulation

Before introducing the geometric approach, we describe the assumptions made about the camera, scene, and interpolation methods used in light field rendering.

6.1.1 Assumptions

The assumptions made in the geometric analysis are as follows:

- **Camera**: pin-hole with a finite resolution.
- **Scene**: occlusion-free and Lambertian.
- **Interpolation method**: bilinear.

6.1.1.1 Camera

The geometric analysis assumes a pin-hole camera model with a finite resolution. Thus, the camera records a blurred version of the plenoptic function or the light field. A pixel value is a weighted integral of the illumination of the light arriving at the camera plane. Alternatively, a pixel value is the convolution of the plenoptic function at the optical center with a low-pass filter. The shape of the filter is compactly supported, with the width of support being the angular resolution of camera. Equivalently, the camera simply samples the convoluted plenoptic function at the camera center. The value of a pixel is exactly the value of the blurred plenoptic function at the direction linking the pixel and the optical center.

Throughout this chapter, we use uniform angular resolution, in both vertical and horizontal directions. Both capturing and rendering cameras have the same resolution.

6.1.1.2 Scene

To simplify the geometric analysis, we study the characteristics of the scene element and bound its depth. The angular extent of a point is sufficiently small compared to the camera resolution, but not zero. Since a scene is composed of points, if every point can be correctly rendered, so can the scene. The scene points are first dealt with independently (i.e., ignoring occlusion) and the discussion on the sampling problem with occlusion is postponed until Section 6.4. The microscopic analysis methodology was inspired by common practices in physics where theories are often built on the analysis of independent particles and the interaction between particles. Moreover, by assuming that the scene is Lambertian, the analysis focuses on the scene geometry and ignores the illumination effects.

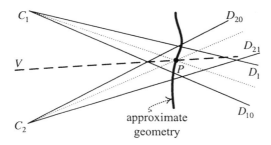

Fig. 6.1. Rendering with bilinear interpolation: the scene point along a ray VP is estimated at P and the ray is interpolated using the four nearest rays from cameras at C_1 and C_2.

6.1.1.3 Interpolation

The rendering process involves choosing "nearby" rays for each viewing ray and interpolating using them. Therefore, an estimate on the depth along the viewing ray is required for ray query. In fact, any approach to find "nearby" rays inherently involves an assumption of some scene depth. It is around where the rays intersect. For example, infinite depth is assumed in rendering with Concentric Mosaics, because interpolation using parallel rays is equivalent to infinite depth. In rendering with the light field, the depth is always implicitly assumed to be at the focal plane.

Usually, the nearest samples are the most important to reconstruct a signal. Bilinear interpolation has been commonly used in existing light field rendering systems (e.g., [40, 160, 267]) because it is simple and can produce good rendering quality. In the presence of more accurate depth information, better rendering results can be obtained by interpolating more light rays.

Figure 6.1 illustrates the rendering process. Suppose that we want to render a view at V, and the viewing ray intersects the approximate geometry at P. C_1 and C_2 are two nearby positions of the camera that are closest to VP, and $C_i D_{ij}$ ($i = 1, 2; j = 0, 1$) are nearby rays in camera C_i that are closest to the ray $C_i P$. Then the pixel value of VP can be bilinearly interpolated from rays $C_i D_{ij}$, e.g. [326], by assigning weights w_{ij} to rays $C_i D_{ij}$ in the following manner:

$$w_{10} = \frac{\angle VPC_2 \cdot \angle PC_1 D_{11}}{\angle VPC_1 + \angle VPC_2}, \quad w_{11} = \frac{\angle VPC_2 \cdot \angle PC_1 D_{10}}{\angle VPC_1 + \angle VPC_2},$$

$$w_{20} = \frac{\angle VPC_1 \cdot \angle PC_2 D_{21}}{\angle VPC_1 + \angle VPC_2}, \quad w_{21} = \frac{\angle VPC_1 \cdot \angle PC_2 D_{20}}{\angle VPC_1 + \angle VPC_2}.$$

6.1.2 Anti-aliasing condition

In this section, we investigate the visual artifacts caused by rendering with interpolation and inaccurate depth. We show that the anti-aliased light field rendering is equivalent to eliminating "double images" for each scene point.

Widening of intensity contribution after interpolation

We first consider within-view (intra-view) interpolation. As shown in Figure 6.2(a), camera C_1 is taking a snapshot of a point L. $C_1 D_{10}$ is the nearest sampling ray to $C_1 L$ while $C_1 D_{11}$ and $C_1 D_{12}$ are two nearby rays. Figure 6.2(b) maps the intensity contribution of L as a function of camera angular position, where

- the vertical line at 0 represents the ray $C_1 L$,
- δ is the angular resolution of the camera,
- the parabola-like curve is the intensity contribution of L in the continuous case (or the shape of low-pass filter for the blurred plenoptic function),
- and $C_1 D_{10}$ is displaced by angle ε ($-\frac{\delta}{2} \leq \varepsilon < \frac{\delta}{2}$) from $C_1 L$.

As a result, the contribution of L to the pixel that corresponds to the ray C_1D_{10} is just the value of the continuous contribution at ε. Since the angles of C_1D_{11} and C_1D_{12}, corresponding to $\varepsilon-\delta$ and $\varepsilon+\delta$ in Figure 6.2(b), respectively, are outside the interval of $[-\frac{\delta}{2}, \frac{\delta}{2})$, neither intensities of the two corresponding pixels are affected by the point L. Subsequent to linear interpolation, the intensity contribution of the point L becomes a wedge of width 2δ. However, it is the *width*, not the shape of the intensity contribution, that matters.

Suppose C_1 is one of the nearby cameras to the novel view V (Figure 6.3(a)), C_1L intersects the approximate geometry at P_1, and VP_1 intersects the focal plane at L_1. Then the part of the contribution of L to the novel view, transferred by camera C_1, centers around L_1 (Figure 6.3(b)).

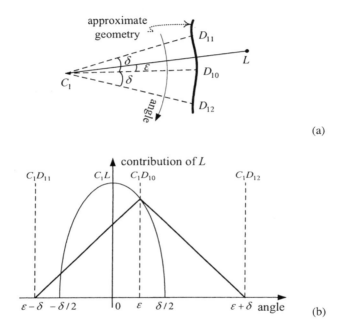

(a)

(b)

Fig. 6.2. The change of intensity contribution. (a) The camera C_1 is imaging a scene point L. (b) The intensity contribution of L changes from parabola-like to wedge-shape due to finite camera resolution and linear interpolation.

Rendering quality vs. geometry information

Now we consider the between-view (inter-view) interpolation. As shown in Figure 6.4(a), C_1 and C_2 are two nearby cameras, V is the novel view, and L is the scene point of interest. Suppose C_iL intersects the approximate geometry at P_i and

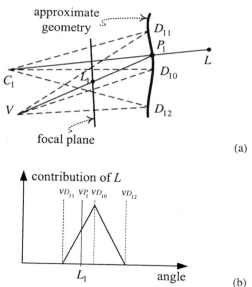

Fig. 6.3. Using the rays in camera C_1 generates a wedge of intensity contribution around L_1 in the novel view.

VP_i intersects the focal plane at L_i ($i = 1, 2$). Then around L_i lie two intensity contributions of the point L (Figure 6.4(b)) on the plane.

The contribution of L to the viewing rays is interpolated between these two intensity contributions. Obviously, the rendering quality depends on the distance between them. In Figure 6.5, the relative positions between the two intensity contributions are shown, where the horizontal axis represents the angular position of the rays in view V and the vertical axis represents the amount of contribution. The thick vertical dash-lines represent the viewing rays. When they nearly overlap (Figure 6.5(a)), there is only one pixel strongly influenced by L. So L will appear sharp on the rendered image. When they partially overlap (Figure 6.5(b)), then L contributes to two consecutive pixels and L will become blurred. When they no longer touch (Figure 6.5(d)), some viewing rays can fall in the gap between them. If the contrast between L and its neighborhood is large, the intensities of in-between viewing rays are different from the intensity of L. As a result, there will be two separate images of L on the rendered image. This phenomenon is the "double image" artifact.

The distance between the two intensity contributions is dependent on the sample spacing and the geometry, so is the rendering quality. When the sample spacing becomes larger and larger, or when the geometry information becomes less and less accurate, the rendered point gradually changes from being sharp to being blurry and further to becoming a double image artifact. The double image artifact is a result of an inadequate sampling rate.

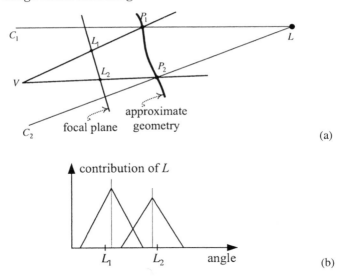

(a)

(b)

Fig. 6.4. Bilinear interpolation generates two wedges of intensity contribution around L_1 and L_2. Double images of L may appear on the rendered image if the two wedges do not overlap.

Strictly speaking, the change from being sharp to having double images is continuous. Nevertheless, the condition of two intensity contributions touching each other (Figure 6.5(c)) is a critical one.

When is rendering quality acceptable?

We now consider the circumstances under which a rendered image is considered acceptable. A rendered scene point may either be sharp, be blurry, or have double images. Sharpness is certainly what we desire. Thus it is important that we make a distinction between blurring and double images.

The phenomenon of double images has also been observed by Levoy and Hanrahan [160] and Halle [99]. Double images are the most salient and visually disturbing artifact, particularly when the sampling rate is low. Human perception is more tolerant to blurring than to double images. People are accustomed to seeing blurred objects in photos, which correspond to off-focus locations. It stands to reason that an object should only appear blurred with no double images should the geometry be inaccurate.

The above human perception has been summarized by the causality principle in scale-space theory [70], i.e., no "spurious detail" should be generated when smoothing an image. The light field rendering should obey this principle because both the intra- and inter-view interpolations are smoothing processes on a single image. The intra-view case is obvious. For the inter-view case, under the Lambertian premise and ignoring the visibility problem, the value of every pixel on the interpolated image can be computed by weighting several pixels on only one of the images chosen for in-

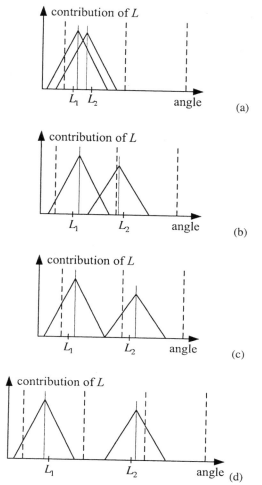

Fig. 6.5. The rendering quality depends on the distance between the two intensity contributions. The viewing rays are indicated by the thick vertical dash-lines. (a) When the intensity contributions are very close, the rendered point appears sharp. (b) When the intensity contributions partially overlap, it looks blurred. (c) That the two intensity contributions meet their ends is the critical condition that double images may occur. (d) Double images may occur when the two intensity contributions do not overlap.

terpolation. The weighting template is pixel-dependent, but it is still a smoothing operation because all the weights are non-negative. As a result, blurring is harmless but double images should be eliminated.

From the viewpoint of signal processing, light field rendering is a reconstruction process of a 4D signal. When the sampling rate is inadequate, aliasing shows up as double images on the rendered image. Indeed, we can prove that eliminating double images in the geometric viewpoint (or the overlap between two successive intensity contributions on the focal plane) is equivalent to the anti-aliasing condition in the viewpoint of signal processing.

First, we show the equivalence between the constant overlap of the wedge-shape intensity contributions in discrete images and the constant overlap of those in continuous ones (the parabola-like curve in Figure 6.2(b)). On one hand, it is apparent that the overlap of continuous intensity contributions guarantees the overlap of discrete intensity contributions (Figure 6.6(a)). On the other hand, as shown in Figure 6.2(b), if the offset ε of one of the intensity contributions is close to $-\delta/2$ and the other offset is close to $\delta/2$, then the overlap of these two wedge-shape intensity contributions requires that the two parabola-like ones at least touch each other (Figure 6.6(b)). In fact, the transfer from (6.15) to (6.16) in Section 6.3.1 has demonstrated such equivalence.

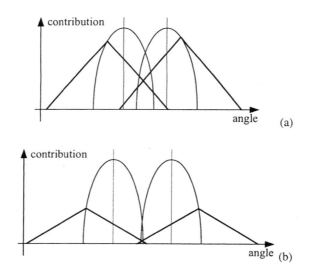

Fig. 6.6. Equivalence between types of overlap. (a) The overlap of parabola-like intensity contributions ensures the overlap of wedge-shape ones. (b) The overlap of wedge-shape intensity contributions at an extreme case implies the overlap of parabola-like ones.

Next, we prove that the constant overlap of parabola-like intensity contributions is equivalent to the anti-aliasing condition in the frequency domain. For a point at depth z_0, its perceived velocity in the video is f/z_0 [332], where f is the focal length. Using a constant depth Z is equivalent to estimating its velocity at f/Z. The parabola-like intensity contributions from other frames backprojected from the constant-depth plane to the current frame are the predicted positions of the point, using the estimated velocity f/Z. So the actual intensity contribution and the predicted one must overlap. From Figure 6.7(a), in spatial-temporal domain, the "time" duration d must satisfy

$$d|f/z_0 - f/Z| \leq \Delta, \qquad (6.1)$$

in order to ensure the overlap, where $\Delta = \delta f$ is the sample spacing on the focal plane.

Now let us consider the anti-aliasing condition in the frequency domain. The 2D light field is parameterized by t and v, where t is the "virtual time". Since the actual velocity is constant, the spectrum of the point is simply a slant line segment, with the highest frequency in v being $1/(2\Delta)$ [295]. Using the estimated velocity is equivalent to using a motion-compensated reconstruction filter to reconstruct the video. The filter is a parallelogram shown in Figure 6.7(b). Then the anti-aliasing condition is that the spectrum of the video must completely lay inside the parallelogram as we have shown in the previous chapter, or

$$1/(2d) \geq 1/(2\Delta)|f/z_0 - f/Z|.$$

We see that the above inequality is identical to (6.1).

In conclusion, rendering quality is acceptable when all objects in the rendered image appear either sharp or blurred. No double images are allowed.

6.2 Minimum sampling rate of Concentric Mosaics

6.2.1 Review of Concentric Mosaics

Concentric Mosaics form a 3D plenoptic function by collecting all rays from a rotating camera on a plane [267]. At each rotation angle, an image with multiple verticle lines is captured. The nth Concentric Mosaic is created by putting together the nth verticle lines in all the images captured. Concentric Mosaics index all input image rays naturally in 3 parameters: radius, rotation angle and vertical elevation. It has been shown that any novel view inside the visible region can be rendered without any 3D reconstruction. As shown in Figure 6.8, a camera swings on the circle S_1. And a constant-depth circle (or cylindrical surface) S_2 is assumed for rendering. To render a novel ray (e.g., OP_2), first we find its intersection point (P_2) with the constant-depth surface. Then two nearest rays ($C_1 P_2$ and $C_2 P_2$) from nearby cameras (C_1 and C_2) are interpolated to generate the rendering result.

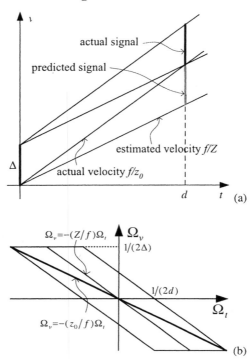

Fig. 6.7. The anti-aliasing conditions in the spatial-temporal domain and spectral domain. (a) In the spatial-temporal domain, the condition is the overlap of parabola-like intensity contributions. (b) In the frequency domain, the condition is that the motion-compensated reconstruction filter contain the entire spectrum of the video.

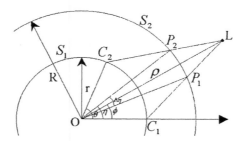

Fig. 6.8. Geometry for Concentric Mosaics.

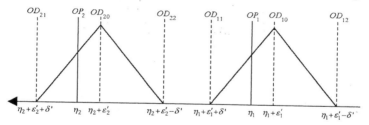

Fig. 6.9. Double images happen when two wedges do not touch: the angle between successive positions of the camera is too wide or the point is too far from the constant-depth surface.

6.2.2 Minimum sampling condition

We now move from the cylindrical coordinate in Figure 6.8 to Figure 6.9, where the horizontal axis represents the angle of the ray starting from the cylinder center O, the two wedge-like intensity distributions of the point L might not overlap if C_1 and C_2 are not sufficiently close and L is not near S_2. As shown in Figure 6.9,

1. η_i is the position of OP_i,
2. $\eta_i + \varepsilon'_i$ is the position of OD_{i0}, where $C_i D_{i0}$ is the nearest ray (see $C_1 D_{10}$ in Figure 6.2(a)) to $C_i L$ viewed from C_i (to save notation, we assume that D_{ij} ($i = 1, 2$; $j = 0, 1, 2$) are on S_2), and
3. $\eta_i + \varepsilon'_i \pm \delta'$ are positions of OD_{i1} and OD_{i2} (see $C_1 D_{11}$ and $C_1 D_{12}$ in Figure 6.2(a)) respectively,

where ε'_i is the angle between OP_i and OD_{i0}, and δ' is the angle between OD_{i0} and OD_{ij} ($j = 1, 2$)[1].

In this case, if $\eta_2 + \varepsilon'_2 - \delta' > \eta_1 + \varepsilon'_1 + \delta'$, when the viewer is at O and the viewing-rays are indicated by the thick dashed-lines in Figure 6.9, the rendered image of the point L will appear double.

Therefore, to avoid double images, that $\eta_2 + \varepsilon'_2 - \delta' \leq \eta_1 + \varepsilon'_1 + \delta'$ must be fulfilled, or equivalently

$$\eta_2 - \eta_1 \leq \varepsilon'_1 - \varepsilon'_2 + 2\delta'. \tag{6.2}$$

This is the condition when L is outside S_2. If L is inside S_2, it becomes

$$\eta_1 - \eta_2 \leq \varepsilon'_2 - \varepsilon'_1 + 2\delta'. \tag{6.3}$$

Because L is random, both (6.2) and (6.3) must be fulfilled. Furthermore, $\varepsilon'_1 - \varepsilon'_2$ can take an arbitrary value in $[-\delta', \delta')$. Therefore, the final condition to avoid double images is

$$|\eta_2 - \eta_1| \leq \delta'. \tag{6.4}$$

---

[1] Note that since δ is extremely small and the FOV of the camera is fairly small, these four angles are nearly equal.

6.2.3 Lower bound analysis

Referring to Figure 6.8, suppose in polar coordinate $L = (\rho, \phi)$, $C_1 = (r, 0)$, $C_2 = (r, \theta)$, $P_2 = (R, \eta)$, then η satisfies:

$$\frac{R \sin \eta - r \sin \theta}{\rho \sin \phi - r \sin \theta} = \frac{R \cos \eta - r \cos \theta}{\rho \cos \phi - r \cos \theta},$$

since P_2 is on the line $C_2 L$. The above equation can be written as:

$$\frac{1}{r} \sin(\eta - \phi) - \frac{1}{\rho} \sin(\eta - \theta) = \frac{1}{R} \sin(\theta - \phi). \tag{6.5}$$

Because the angle of the point is between two successive positions of the camera, the three angles $\eta - \phi$, $\eta - \theta$ and $\theta - \phi$ are all small, (6.5) can be linearized to:

$$\frac{1}{r}(\eta - \phi) - \frac{1}{\rho}(\eta - \theta) \approx \frac{1}{R}(\theta - \phi).$$

Hence

$$\eta \approx \frac{\rho(R - r)\phi - r(R - \rho)\theta}{R(\rho - r)}. \tag{6.6}$$

Therefore, for successive positions of the camera (with same R, r, ϕ and ρ),

$$|\Delta \eta| = |\eta_1 - \eta_2| \approx \left| \frac{\frac{\rho}{R} - 1}{\frac{\rho}{r} - 1} \Delta \theta \right|. \tag{6.7}$$

Next, we set out to find the relationship between δ' and δ. Referring to Figure 6.10, for the triangle $\triangle OCQ$, using the law of sines, we have

$$\frac{|CQ|}{\sin \varphi'} = \frac{R}{\sin \varphi} = \frac{r}{\sin(\varphi - \varphi')}, \tag{6.8}$$

Again, because φ is relatively small (typically $|\varphi| \leq \frac{\pi}{10}$, and $|\sin \frac{\pi}{10} - \frac{\pi}{10}| < 0.0052$) and $\varphi' < \varphi$, the above relation can be linearized as

$$\frac{|CQ|}{\varphi'} \approx \frac{R}{\varphi} \approx \frac{r}{\varphi - \varphi'},$$

hence $|CQ| \approx R - r$. Therefore, the relationship between δ' and δ is

$$\delta' = \frac{\widehat{PQ}}{R} \approx \frac{|CQ|\delta}{R} \approx \frac{R - r}{R}\delta = (1 - \frac{r}{R})\delta. \tag{6.9}$$

Combining (6.4), (6.7) and (6.9), we obtain:

$$|\Delta \theta| \leq \left| \frac{(\frac{\rho}{r} - 1)(\frac{r}{R} - 1)}{\frac{\rho}{R} - 1} \right| \delta.$$

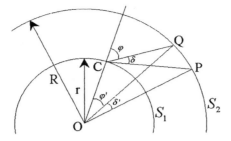

Fig. 6.10. The relationship between δ and δ'.

If the depth of the point L is constrained between A and B ($r < A \leq \rho \leq B$ and $A \leq R \leq B$), then the maximum value of $\left| \frac{\frac{\rho}{R} - 1}{(\frac{\rho}{r} - 1)(\frac{r}{R} - 1)} \right|$ is

$$m = \frac{1}{1 - \frac{r}{R}} \cdot \max \left\{ -\frac{\frac{A}{R} - 1}{\frac{A}{r} - 1}, \frac{\frac{B}{R} - 1}{\frac{B}{r} - 1} \right\} \tag{6.10}$$

Therefore a bound for the number of pictures is:

$$N_1 = \left\lceil \frac{2\pi m}{\delta} \right\rceil,$$

where $\lceil x \rceil$ denotes the smallest integer not less than x.

On the other hand, since the FOV of the camera is limited, the patches that every picture projected onto the cylinder must cover the cylinder, otherwise the bilinear interpolation could not be properly carried out. Accordingly, the number of pictures should also be larger than:

$$N_2 = \left\lceil \frac{2\pi}{\Phi} \right\rceil,$$

where $\Phi = \frac{1}{2}FOV - \arcsin\left(\frac{r}{R} \sin\left(\frac{1}{2}FOV\right)\right)$, using the second equality in (6.8).

Finally, the lower bound we attain is

$$N = \max\{N_1, N_2\}. \tag{6.11}$$

Note that we are not saying that if the number of pictures is greater than N then the visual quality will be acceptable. Rather, if the number of pictures is less than N then even the simplest scene with only a point could not be properly rendered. The theoretical lower bound should be higher than the one we have deduced above. However, when the scene becomes more complex, one might not notice too much artifact due to the characteristics of human vision. Consequently, the actual lower bound will not deviate too much from the one we estimated. The above analysis only considers the horizontal resolution; increasing the vertical resolution does not improve the minimum sampling rate.

6.2.4 Optimal constant-depth R

Instead of simply choosing constant-depth $R = \frac{A+B}{2}$, (6.10) can guide us to find a better constant-depth R, such that m is minimized, and the number of pictures required becomes smaller. Such an R satisfies

$$-\frac{\frac{A}{R}-1}{\frac{A}{r}-1} = \frac{\frac{B}{R}-1}{\frac{B}{r}-1}$$

$$R = \frac{2AB - (A+B)r}{A+B-2r}. \qquad (6.12)$$

One can easily check that $A \leq R \leq \frac{A+B}{2}$, and the equalities hold only for $A = B$. This choice of R is reasonable because closer objects will be distorted more and thus need more accurate depth information.

One should be cautious to provide relatively accurate minimum and maximum depths so that the computed optimal constant depth can really take effect. Fortunately, this is not a difficult task to measure them. Moreover, since $B > A$, R is much less sensitive to B. Therefore, only the minimum depth needs to be accurate.

It is interesting to rewrite (6.12) as:

$$\frac{2}{R-r} = \frac{1}{A-r} + \frac{1}{B-r},$$

which means that the optimal constant depth is exactly the harmonic mean of the minimum and maximum depths. It is also worth noting that such choice of R can make the objects at the minimum depth and the maximum depth be rendered equally sharply.

6.2.5 Validity of bound

1. If the scene is truly at a constant-depth, e.g., a painted cylinder, then $A = B = R$. In this case $N = N_2$, which is true.
2. If the scene is infinitely far away, let R chosen as (6.12), then

$$m = \frac{r(\frac{1}{A} - \frac{1}{B})}{2(1 - \frac{r}{A})(1 - \frac{r}{B})}. \qquad (6.13)$$

When $A \to \infty$ and $B \to \infty$, $m \to 0$. Again $N = N_2$, which is also true.
3. If $r \to 0$, then the Concentric Mosaics will reduce to a panorama, and the number of pictures needed is N_2. In this case, $m \to 0$ and $N = N_2$. The lower bound is correct again.
4. The above examples are all extreme cases. The geometric analysis has also been verified with real data. Figure 6.11 illustrates the top view of a real scene. The scene is enclosed by an ellipse and the center of the camera rig is placed near one of the focal points of the ellipse. The scales are labeled in the figure. The radius

of the rig is 1.7m. The horizontal FOV of the camera is $43°$ and each picture taken by the camera is 360 pixels by 288 pixels. Then $A = 3.4$, $B = 16.6$, $r = 1.7$, and $\delta = \frac{43}{360} \cdot \frac{\pi}{180} \approx \frac{\pi}{1500}$. To achieve the best quality, R is chosen as (6.12), thus $R = 4.75$. Then $N = 1329$. Figure 6.12 is a panoramic view of part of the scene. Figure 6.13 compares the details between the rendered scenes with two different sampling rates. We can see that when the number of pictures is 1479, the scene is rather satisfactory, while clear double images appear in the scene reconstructed from 986 pictures. In this experiment, the lower bound we computed is fairly accurate.

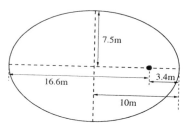

Fig. 6.11. Top view of a real scene that is used to capture Concentric Mosaics. The center of camera rig is placed at the dark dot.

6.3 Minimum sampling rate of light field

We now study the minimum sampling rate for light field rendering without any geometric information. Specifically, we analyze the maximum allowable distance between successive camera locations by assuming a globally constant depth for rendering. Note that the interpolation method used in the original light field rendering [160] implicitly assumes a globally constant depth at the focal plane.

6.3.1 Maximum camera spacing

Without loss of generality, we set up a coordinate system as in Figure 6.15, where

1. $L = (x_0, z_0)$ is a point in the scene,
2. C_1 and C_2 are two adjacent cameras, and
3. C_1D_{10} and C_2D_{20} are two rays in the light slab that are nearest to C_1L_1 and C_2L_2, respectively.

The global constant-depth plane, at a depth Z, is parallel to but might not be identical to the focal plane. This resembles a dynamically reparameterized light field [116]. We omit the projection from the constant-depth plane to the focal plane of the novel views as they are parallel.

We now examine the intensity contribution of L on the constant-depth plane in Figure 6.15. The details are shown in Figure 6.16, where

Fig. 6.12. Part of a panoramic view of a real scene.

Fig. 6.13. A close-up view of the scene reconstructed: from 986 pictures (top) and 1479 pictures (bottom).

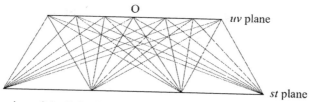

Fig. 6.14. Top view of the light slab in the lion light field. The constant-depth is implicitly assumed to be on the st plane.

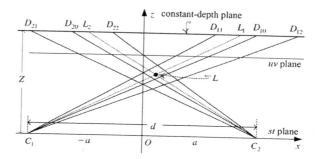

Fig. 6.15. Rendering a point L with a light field.

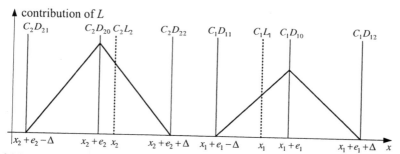

Fig. 6.16. The details of the intensity contributions of L on the constant-depth plane. The horizontal axis indicates the ray position and the vertical axis represents the intensity value of the rays.

1. x_i is the coordinate of L_i $(i = 1, 2)$ on the constant-depth plane,
2. $x_i + e_i$ is the coordinate of D_{i0}, where $C_i D_{i0}$ is the nearest sampling ray to $C_i L$ viewed from C_i, and
3. $x_i + e_i \pm \Delta$ are the coordinates of D_{i1} and D_{i2}, respectively,

in which e_i is the offset between L_i and D_{i0}, and $\Delta = \delta Z$ is the sample spacing on the constant-depth plane.

To avoid double images, the two wedge-shape intensity contributions of the point L must overlap. Hence, D_{11} must be on the left of D_{22}, or

$$x_1 + e_1 - \Delta \leq x_2 + e_2 + \Delta, \tag{6.14}$$

It is easy to compute that the coordinates of L_1 and L_2 are:

$$x_1 = -a + Z(x_0 + a)/z_0, \quad \text{and} \quad x_2 = a + Z(x_0 - a)/z_0,$$

respectively. Therefore, (6.14) becomes

$$\frac{2a(Z - z_0)}{z_0} \leq 2\Delta + (e_2 - e_1). \tag{6.15}$$

Since the position of L is arbitrary, $e_2 - e_1$ can vary between $-\Delta$ and Δ. Therefore the following condition must be satisfied:

$$\frac{2a(Z - z_0)}{z_0} \leq \Delta. \tag{6.16}$$

The above condition is deduced when $z_0 \leq Z$. If $z_0 > Z$, the corresponding condition is

$$\frac{2a(z_0 - Z)}{z_0} \leq \Delta.$$

Summing up, the sample spacing must satisfy:

$$d = 2a \leq \Delta \cdot \frac{z_0}{|z_0 - Z|} = \delta \cdot \frac{z_0 Z}{|z_0 - Z|}. \tag{6.17}$$

If z_0 is bounded between $z_{\min}(\leq Z)$ and $z_{\max}(\geq Z)$, then the minimum value of the right hand side of (6.17) is the maximum allowable distance between two locations of the camera, namely

$$d_{\max} = \delta \min_{z_{\min} \leq z_0 \leq z_{\max}} \left\{ \frac{z_0 Z}{|z_0 - Z|} \right\} = \delta Z \cdot \min \left\{ \frac{z_{\min}}{Z - z_{\min}}, \frac{z_{\max}}{z_{\max} - Z} \right\} \tag{6.18}$$

We may rewrite (6.17) as

$$\left| \frac{1}{z_0} - \frac{1}{Z} \right| d \leq \delta.$$

This means that when Z is the estimate of exact depth, the disparity error viewed from nearby cameras must not exceed *one* pixel. Since (6.18) is the minimization of (6.17) over all scene points, it means that if the globally constant depth Z is chosen for the scene, then the sample spacing should ensure that the disparity errors of all scene points between successive views must not exceed one pixel.

6.3.2 Optimal constant depth

(6.18) indicates that d_{max} is maximized when Z_{opt} satisfies:

$$\frac{z_{min}}{Z_{opt} - z_{min}} = \frac{z_{max}}{z_{max} - Z_{opt}},$$

or

$$Z_{opt} = \frac{2z_{min}z_{max}}{z_{min} + z_{max}}. \tag{6.19}$$

One can easily check that $z_{min} \leq Z_{opt} \leq \frac{1}{2}(z_{min} + z_{max})$, and the equalities hold only for $z_{min} = z_{max}$. This choice of Z_{opt} is reasonable because closer objects require more accurate depth information.

One should be cautious to provide relatively accurate minimum and maximum depths so that the computed optimal constant depth can really take effect. Fortunately, it is possible to estimate or measure them in practice. Moreover, Z_{opt} is less sensitive to z_{max} than to z_{min}.

6.3.3 Interpretation of optimal constant depth

It is interesting to note that (6.19) can be rewritten as:

$$\frac{1}{Z_{opt}} = \frac{1}{2}\left(\frac{1}{z_{min}} + \frac{1}{z_{max}}\right).$$

In this formulation, the optimal constant depth is exactly the harmonic mean of the minimum and maximum depths. (6.19) can also be written as:

$$\frac{1}{z_{min}} - \frac{1}{Z_{opt}} = \frac{1}{Z_{opt}} - \frac{1}{z_{max}}.$$

This implies that the nearest and farthest objects can be rendered with equal sharpness.

The optimal constant depth can be determined graphically. Referring to Figure 6.17, where

1. C_i ($i = 1, 2$) are on the camera plane,
2. F is the mid-point of C_1C_2,
3. L and L' are the farthest and the nearest points in the scene, and
4. C_2L' intersects C_1L at L_1 and C_1L' intersects C_2L at L_2.

Then it is guaranteed that L_1L_2 is parallel to C_1C_2. From projective geometry [8], one can prove that $\{L, E, L', F\}$ is a harmonic set of points, where E is the intersection point of LL' and L_1L_2. Therefore

$$\frac{|EL| \cdot |FL'|}{|FL| \cdot |EL'|} = 1,$$

which gives

$$\frac{(z_{max} - Z)z_{min}}{z_{max}(Z - z_{min})} = 1.$$

Thus this is a convenient way of finding the optimal constant depth Z_{opt}.

With the optimal constant depth, the sample spacing in (6.18) becomes:

$$d_{max} = \frac{2\delta}{z_{min}^{-1} - z_{max}^{-1}}. \qquad (6.20)$$

It is totally determined by the disparity variation of the scene and the camera resolution. The above equality can also be written as:

$$\left(\frac{1}{z_{min}} - \frac{1}{z_{max}}\right) d_{max} = 2\delta.$$

This means the maximum allowed sample spacing should make the disparity variation of the scene between successive cameras be 2 pixels.

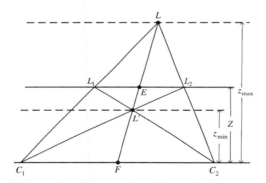

Fig. 6.17. Graphical determination of the optimal constant depth.

6.3.4 Prefiltering the light field

Let the sample spacing be d. If this sampling rate is inadequate, i.e., $d > d_{max}$, where d_{max} is given by (6.18), then it is necessary to prefilter the light field in order to eliminate the artifact of double images. As mentioned by Levoy and Hanrahan [160], the prefiltering can be done on the camera or focal plane, or both. Filtering on the focal plane reduces the image resolution, whereas filtering on the camera plane reduces the depth of focus [116]. It is easy to see that the size for focal-plane prefiltering should be d/d_{max} pixels, and the focal-plane prefiltering can be done more conveniently by postfiltering on the rendered images. However, in theory the camera-plane prefiltering cannot be effective because the samples are not taken after low-pass filtering the camera plane.

6.3.5 Disparity-based analysis

In fact, the minimum sampling rate can be found by simply observing the maximum and minimum disparities in pixels in two captured images, without any measurement over the scene or any camera calibration. So do rendering and prefiltering.

Suppose the maximum and minimum disparities found in two images are $N_{\max}^D$ and $N_{\min}^D$ pixels, respectively. From Section 6.3.3, we know that the sampling rate must make the maximum disparity difference in the scene be 2 pixels. Because the disparity variation is proportional to the sample spacing, the sampling interval should be shrunk by $(N_{\max}^D - N_{\min}^D)/2$ times so that the disparity variation is 2 pixels. Therefore $N = \text{ceil}((N_{\max}^D - N_{\min}^D)/2) + 1$ sample images are required for the interval between the two cameras.

The rendering can also be done conveniently with epipolar images [33], where a global motion compensation vector is needed for picking out appropriate pixels in different sample images and blending among them. The optimal constant depth now corresponds to the optimal motion compensation vector, which is $(N_{\min}^D + N_{\max}^D)/(2(M-1))$ pixels between successive sample images, where M is the number of sample images uniformly taken in the interval.

Finally, the size of prefiltering is $(N_{\max}^D - N_{\min}^D)/(2(M-1))$ pixels.

6.3.6 Experiments

Table 6.1. The data for the light field "Toy."

st sampling rate	$N_{\min}^D$	$N_{\max}^D$	minimum sampling rate	optimal const. depth
610×455	27	59	17×17	$1.3721 z_{\min}$

We now verify the geometric analysis described in the previous sections. In the "Toy" light field, two images were captured with a large distance between them. For these two images, the minimum and maximum disparities are 27 and 59 pixels, respectively. We then know that $(59-27)/2+1=17$ images are required for the interval. The relevant data are listed in Table 6.1. Figure 6.18(a) is one of the sample images. In order to detect the existence of double images, sharp features, which appear as thin vertical lines, are added to the scene. The upper and lower boxes are at the maximum and minimum depths, respectively.

First, the effectiveness of the optimal constant depth was tested. With the optimal constant depth chosen at $1.3721 z_{\min}^2$, the light field can be correctly rendered with equal sharpness at $z_{\min}$ and $z_{\max}$ (Figure 6.19(b)). If the optimal constant depth is not chosen, e.g., if the mean depth at $1.5926 z_{\min}$ is used instead, the scene points at $z_{\min}$ appear to be very blurry (Figures 6.19(c)), though those at $z_{\max}$ look sharper.

---

[2] The optimal motion compensation vector is $(59+27)/(2\times(17-1))=2.6875$ pixels between successive sample images. The real value of $z_{\min}$ is insignificant.

Second, the sufficiency of the minimum sampling rate was tested. When the sampling rate doubles, the visual improvement is nearly unnoticeable (Figures 6.19(b) and (d)). However, when the sampling rate is reduced by half, double images are clearly visible at both $z_{\min}$ and $z_{\max}$ (Figures 6.19(e)).

Third, the effectiveness of the prefiltering size was tested. The double image effect in Figures 6.18(e) disappeared when a prefiltering size of 2 pixels on the focal plane was chosen (Figures 6.19(f)). However, the rendered image became blurred.

6.4 Dealing with occlusion

In previous sections, the scene is always assumed to be occlusion-free. In a real scene, occlusion always exists and sampling with occlusion must be analyzed.

Occlusion destroys the desirable scene property that all scene points are visible from all views. It is easy to see that in general the spectrum of a light field with occlusion is *not* bandlimited, thus demanding the reconstruction of a completely artifact-free (in the sense of signal reconstruction) light field from discrete samples is impossible for a complex scene. Therefore, the minimum sampling rate of a scene with occlusion can only be found according to how much aliasing is tolerable. In practice, there is no visually unacceptable artifacts when scenes with occlusion are rendered at the minimum sampling rate for occlusion-free scenes. Ignoring occlusion effects, all scenes with the same disparity variation theoretically share the same minimum sampling rate.

Unfortunately, the minimum sampling curve no longer exists for scenes with occlusion. As shown in Figure 6.20, under the Lambertian premise, if there is no occlusion in the scene, interpolating $C_1 P_1$ with $C_2 P_2$ is equivalent to interpolating $C_1 P_1$ and $C_3 P_2$, or $C_4 P_2$, etc., because the pixel values of rays $C_2 P_2, C_3 P_2$, etc., are identical. This is the basis for different combinations of sampling rate and geometry information producing the same rendering result. On the other hand, if occlusion exists, the pixel values of rays $C_2 P_2, C_3 P_2$, etc., may not be identical. Therefore interpolating $C_1 P_1$ with $C_2 P_2$ is not equivalent to interpolating $C_1 P_1$ and $C_3 P_2$, or $C_4 P_2$, etc.. As a result, the sampling rate cannot be traded for geometry information.

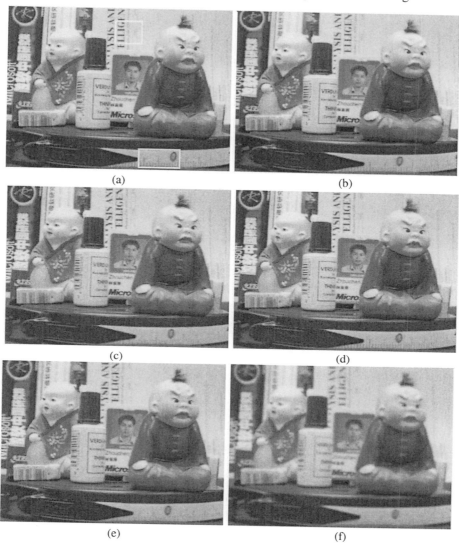

(a)   (b)

(c)   (d)

(e)   (f)

Fig. 6.18. The sample and rendered images of the light field "Toy." (a) A sample image of the light field "Toy." The vertical lines in the white boxes are for the detection of double images. (b) The rendered image from 17×17 images. (c) The same as (b) with the mean constant depth chosen. (d) Rendered from 30×30 images. (e) Rendered from 9×9 images. (f) Image (e) prefiltered with size of 2 pixels. (b), (d)~(f) all use the optimal constant depth.

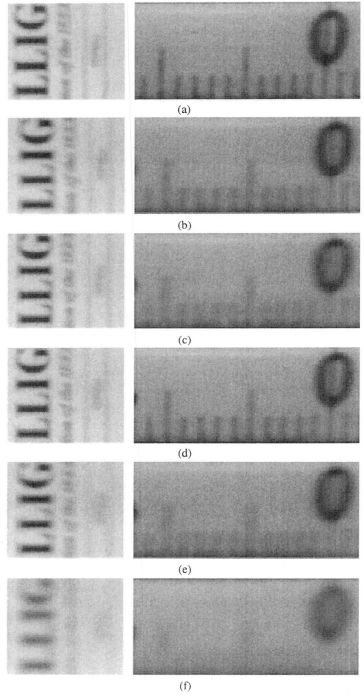

(a)

(b)

(c)

(d)

(e)

(f)

Fig. 6.19. Blow-up of the sampling and rendered images of the light field "Toy."

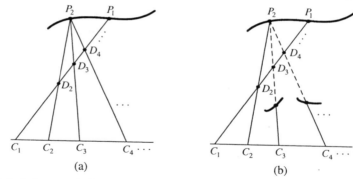

Fig. 6.20. Only in an occlusion-free scene can the sampling rate be traded for geometry information. (a) When the scene is occlusion-free and Lambertian, the interpolation between $C_1 P_1$ and $C_2 P_2$ is equivalent to those between $C_1 P_1$ and $C_3 P_2$, and between $C_1 P_1$ and $C_4 P_2$, etc.. Therefore, different combinations of sampling rate and geometry information can give the same rendering result. (b) When occlusion exists, interpolating $C_1 P_1$ with $C_2 P_2$ is not equivalent to interpolating $C_1 P_1$ and $C_3 P_2$, or $C_4 P_2$, etc. Therefore different combinations of sampling rate and geometry information cannot produce the same rendering result.

7

Optical Analysis of Light Field Rendering

We studied plenoptic sampling using a spectral analysis and a geometric analysis in the previous two chapters. In this chapter, we present an optical analysis to determine the minimum sampling rate for light field rendering. A light field can be considered as a virtual optical imaging system. Analogous to a real optical system, we can define the optical parameters of the light field rendering system, such as the depth of field, the aperture, the circle of confusion and the hyperfocal distance[1] of the virtual imaging system.

7.1 Introduction

Treating light field rendering system as a synthetic optical system is not entirely new. The original light field system regards the sampling interval as the aperture [160]. Aperture filtering has also been proposed. Isaksen *et al.* [116] presented a dynamically reparameterized light field by changing the synthetic aperture (i.e., the number of cameras) and the focal planes for environment. Kunita *et al.* [149] used "equivalent depth of field" to characterize the maximum acceptable depth variation in a standard constant depth light field rendering system by measuring the fidelity of the synthesized images.

The optical analysis quantitatively describes the relationship among three key elements in light field rendering: the scene complexity, the number of images, and the output resolution. Not only does the optical analysis allow us to compute the required number of images for anti-aliased light field rendering given output resolution, it also allows us to deduce how much geometrical information is necessary if the input image number is insufficient. Therefore, the optical analysis can be applied to guide the design of image-based rendering systems.

Based on the optical analysis of the virtual imaging system, we describe the relationship among the depth variation of the scene (depth of field), the constant depth

---

[1] In photography, the hyperfocal distance is the distance setting that produces the greatest depth of field. It has also been defined as the point of focus where everything from half that distance to infinity falls within the depth of field. These definitions are synonymous.

(perfectly focused plane), the spacing of cameras (aperture) and the rendering resolution (circle of confusion). Specifically, the hyperfocal distance of the virtual optical system, as a key parameter for light field rendering, determines the relationship between the spacing of cameras and rendering resolution. Given the minimum and maximum depths of the scene, the optimal constant depth and the hyperfocal distance are derived to achieve the best rendering quality or minimum number of images. The minimum number of images required for anti-aliasing rendering (i.e., the rendering error is smaller than the circle of confusion) can be further reduced by segmenting the depth into multiple depth layers. A quantitative relationship between hyperfocal distance, number of layers, and depth variation of the scene is described.

While similar results on the minimum sampling rate/curve of light field rendering have been obtained in the previous two chapters through geometric analysis and by a spectral analysis of 2D plenoptic function, the optical analysis from this chapter is more intuitive to understand.

The remainder of this chapter is organized as follows. In Section 7.2, we discuss an ideal thin lens imaging system and introduce a conventional image formation model commonly used in computer graphics and computer vision. In Section 7.3, we formulate the light field rendering system as a virtual optical system, and define its optical parameters. Furthermore we describe the imaging law of the constant depth rendering of the light field system. Then, we present the optical analysis and study the relationship among the elements of light field rendering system. Optimal constant depth and the minimal sampling rate are deduced in Section 7.4. Moreover, the optimal depth segmentation to extend the depth of field and the trade-off between the amount of geometrical information and the number of images needed is studied. Concluding remarks are given in Section 7.5.

7.2 Conventional thin lens optical system

7.2.1 Ideal thin lens model

Figure 7.1 shows the basic image formation geometry of a conventional thin lens optical system. With the perfectly converging thin lens and the aperture diameter A, all light rays radiated from an object point P (on the object plane) that pass through the aperture are refracted by the lens to converge at the point Q on the image plane. One can view the lens imaging system as transforming each point in the scene to a single focused point behind the lens. For the thin lens, the relationship between the object distance O, focal length of the lens f, and the image distance I is described by the Gaussian lens law:

$$\frac{1}{f} = \frac{1}{O} + \frac{1}{I} \tag{7.1}$$

Each unoccluded point on the object plane is projected onto a single point on the image plane, causing a focused image to be formed. The film (or sensor) plane must coincide with the image plane to record a sharp image. Points in front of object plane and behind object plane are not focused perfectly and therefore are distributed over a patch (blurred image) on the film plane.

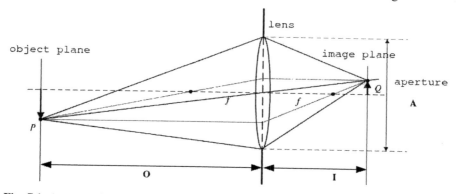

Fig. 7.1. A conventional thin lens imaging system with focal length f and aperture A. An object P at distance O away is filmed at point Q on the image plane.

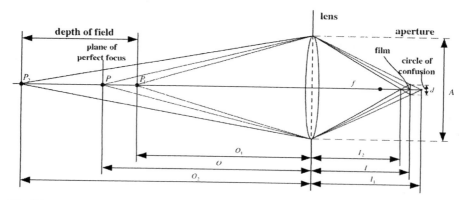

Fig. 7.2. The depth of field of a conventional optical system is defined by the distance between points P_1 and P_2, which are filmed on the image plane with the diameter of circle of confusion d. The point P is perfectly focused.

7.2.2 Depth of field and hyperfocal distance

As shown in Figure 7.2, the point displaces from the object plane image as a circle of confusion (CoC) on the film. The diameter of the circle of confusion C is determined by the congruent triangles formed by the rays passing through the aperture, i.e.,

$$\frac{A}{I'} = \frac{C}{\|I - I'\|} \tag{7.2}$$

$$\frac{1}{f} = \frac{1}{I} + \frac{1}{O} \tag{7.3}$$

$$\frac{1}{f} = \frac{1}{I'} + \frac{1}{O'}, \tag{7.4}$$

$$(7.5)$$

where O' and I' are the object distance and image distance of the point. (In Figure 7.2, either O_1 or O_2 can be O'; similarly for I'.)

The C in (7.5) can be computed as

$$C = \frac{\|O' - O\|}{O'} \frac{f}{O - f} A. \tag{7.6}$$

In practice, films have a finite resolution. Films cannot resolve details smaller than the minimum grain separation of the film emulsion (or pixel size of CCD). Points with the circle of confusion smaller than the resolution of film are "in focus." The depth of field (DOF) is defined as the total range of in focus zone, namely

$$C = \frac{Af}{O - f} \frac{\|O' - O\|}{O'} \le d, \tag{7.7}$$

where d is the maximum acceptable circle of confusion.

The nearest and farthest points that are acceptable are found at the distances,

$$O_f = \frac{d_H O}{d_H - (O - f)} \tag{7.8}$$

$$O_n = \frac{d_H O}{d_H + (O - f)} \tag{7.9}$$

respectively, where

$$d_H = \frac{A}{d/f} \tag{7.10}$$

is the hyperfocal distance [11], or the ratio of diameter of aperture to maximum acceptable angular blur. The depth of field is then computed as

$$DOF = O_f - O_n = \frac{2d_H O(O - f)}{d_H{}^2 - (O - f)^2}. \tag{7.11}$$

7.3 Light field rendering: An optical analysis

7.3.1 Overview of light field rendering

The sampling of the 4D plenoptic function in light field [160] and Lumigraph [91] can be represented by a two-plane parameterization. A ray passing through a light slab, as specified by a line connecting a point on the (s, t) plane and another point on the (u, v) plane, can be uniquely determined by a quadruple (u, v, s, t). A pinhole camera is adopted to capture the light field with the center of projection located on the (s, t) plane (camera plane).

To render a given ray, the line parameters (u, v, s, t) is computed and then the light slab is resampled and interpolated by a certain bandpass filter to reconstruct the

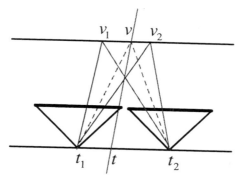

Fig. 7.3. The ray interpolation of the light field rendering: The desired ray (u, s) is interpolated by the nearest neighboring rays $(u_1, s_1), (u_1, s_2), (u_2, s_1),$ and (u_2, s_2).

desired ray. The 2D case of interpolation is illustrated in Figure 7.3. We refer the reader to the original light field and Lumigraph papers for more details.

For the sake of simplicity, we now discuss light field rendering in 2D space. As shown in Figure 7.4, cameras are aligned so that their centers of projection are located on the t axis with the same intervals D. To render novel view images, a virtual rendering camera C with infinite resolution is placed on the desired viewpoint behind the t axis at a distance I. Assume that there is only one ideal *object point* Q [11, 167] in the scene, which is placed Z units in front of the t axis.

7.3.2 Imaging law of light field rendering with constant depth

A constant depth plane that is closer to the object is selected to improve the rendering image quality. With the constant-depth assumption, all the rays captured by cameras are hypothesized emitted from points located on the constant depth plane, which is parallel to the camera plane. Figure 7.4 illustrates the 2D case of light field rendering process. The constant depth line is defined to be parallel to the t axis. Let us denote the distance between the constant depth line and t axis by Z_c. The virtual rendering camera is located behind the t axis with distance I. The rays captured by the cameras C_i through C_j are employed in rendering. Recall that, for constant depth assumption, all rays captured by the camera is hypothesized as being emitted from a point lying on the constant depth plane. To camera C_i, the ray emitted from Q is hypothesized as being emitted from a virtual point Q_i, which is the intersection of constant depth plane and the ray (C_i, Q). Thus, on the novel view image, the ray (C_i, Q) is rendered as image point q_i, which is the image of virtual point Q_i. Similarly, for the camera C_j, Q is rendered as q_j.

Accordingly, the light field rendering system with constant depth correction can be considered as a virtual optical system, whose imaging law is described above. The points q_i and q_j are called the *image points* of the object point Q. Only the points exactly on the constant plane are perfectly in focus.

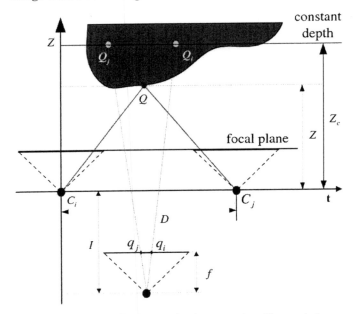

Fig. 7.4. Light field rendering with constant depth assumption. The rendering error $|q_i - q_j|$ incurred by the constant depth assumption (Z_c) is projected by $|Q_i - Q_j|$. C_i and C_j are neighboring capturing cameras, while C at the bottom is the rendering camera.

In Figure 7.4, from the congruent triangles $\triangle QQ_iQ_j$ and $\triangle QC_iC_j$, we have the following relationship:

$$\frac{|Q_i - Q_j|}{|C_i - C_j|} = \frac{|Z_c - Z|}{Z} \qquad (7.12)$$

where Z_c is the constant depth, Z is the depth of the object point Q.

From the congruent triangles $\triangle CQ_iQ_j \; \triangle Cq_iq_j$, we have

$$\frac{|Q_i - Q_j|}{|q_i - q_j|} = \frac{Z_c + I}{f}, \qquad (7.13)$$

where q_i and q_j are the virtual points on the constant depth line corresponding to cameras C_i and C_j, respectively. Also, f is the focal length of the virtual rendering camera C.

Substituting (7.13) for (7.12), we have

$$|q_i - q_j| = \frac{|Z_c - Z|}{Z} \frac{f}{Z_c + I} |C_i - C_k| \qquad (7.14)$$

(7.14) is the basic equation of the light field rendering that describes the quantitative relationship among the key elements of light field system and rendering image quality. From the above analysis, we can consider a light field rendering system as a virtual optical imaging system, whose imaging process depends on the constant depth assumption and rendering algorithm. Levoy and Hanrahan [160] suggested that the light field rendering system be considered as a *discrete aperture* imaging system, with the diameter of the aperture being equal to the spacing between camera locations. Similarly, a discrete aperture camera has also been discussed in [149]. In this section, however, we will derive important qualitative properties of the virtual optical system of light field rendering, much like the conventional thin lens optical system.

7.3.3 Depth of field

As shown in previous section, in the light field rendering system, images of an ideal object point could be more than one point on the desired image plane. To alleviate the aliasing, pre-filtering and post-filtering are applied by Levoy and Hanrahan. In the optical analysis, it is equivalent to blurring the image points to avoid the double image [167]. In this subsection, we will determine the depth variation range of scene, where the rendered image is spread less than a given length for all object points. From (7.14), we define the acceptable rendering quality as the largest range of $|q_i - q_j|$. That is,

$$|q_i - q_j| = \frac{|Z_c - Z|}{Z_c + I} f |C_i - C_j| \le d, \qquad (7.15)$$

where d is a predefined acceptable rendering quality (rendering resolution).

We define the aperture of the virtual optical system as $A' = |C_i - C_j|$, or the distance between two consecutive cameras. The nearest and farthest depths whose diameters of blurred circle are within d are specified as two solutions of (7.15):

$$Z_{min} = \frac{D_H Z_c}{D_H + (I + Z_c)} \qquad (7.16)$$

$$Z_{max} = \frac{D_H Z_c}{D_H - (I + Z_c)}, \qquad (7.17)$$

where

$$D_H = \frac{A'}{d/f}. \qquad (7.18)$$

We can easily verify that, for all Z ($Z_{min} \le Z \le Z_{max}$), $|q_i - q_j| \le d$ is satisfied. Therefore, the *depth of field* (DOF) of the virtual optical system is defined as the acceptable range of focus. That is,

$$Z_{DOF} = Z_{max} - Z_{min} = \frac{2D_H Z_c (I + Z_c)}{D_H^2 - (I + Z_c)^2}. \qquad (7.19)$$

The geometrical interpretation of (7.19) is illustrated in Figure 7.5. Z_c is the distance from the constant depth plane to U axis, Z_{min} and Z_{max} are the minimum and the maximum depth ranges for acceptable rendering quality. Note that the object points lie on Z_{min} and Z_{max} planes cause the same circle of confusion on the desired image. Figure 7.6 shows the depth of field variation due to changes in D_H and Z_c.

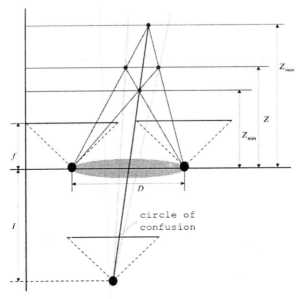

Fig. 7.5. The acceptable range (depth of field) of light field rendering, given the tolerable rendering error (circle of confusion).

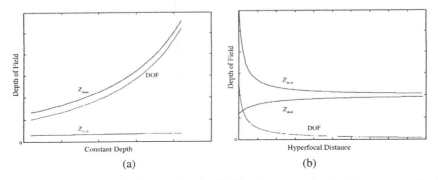

(a)   (b)

Fig. 7.6. (a) The relationship between depth of field and constant depth with given hyperfocal distance; (b) The relationship between the hyperfocal distance and the depth of field with a fixed constant depth.

7.3.4 Rendering camera on ST plane

A special configuration of light field rendering is of particular interest, when the rendering camera is located on the same plane with the capturing cameras $((s,t)$ plane), as shown in Figure 7.7. Substituting $I = 0$ for (7.14), we have

$$|q_i - q_j| = \frac{|Z_c - Z|}{Z} \frac{f}{Z_c} |C_i - C_k| \tag{7.20}$$

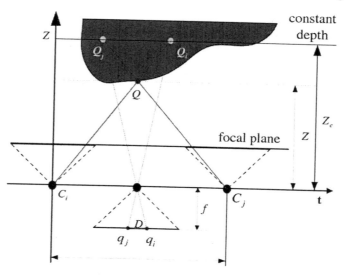

Fig. 7.7. When the rendering camera is located on the capturing camera plane, the light fielding rendering system becomes an ideal thin lens optical system. Objects lying on the constant plane are perfectly focused on the image plan. The focal length of the virtual lens is $f' = fz_c/(f + z_c)$.

As illustrated in Figure 7.7, we assign the image distance the focus length f of the rendering camera C, the object distance is the constant depth Z_c. Then an virtual focus length f' of the optical system is obtained by the Gaussian law (see (7.1)),

$$f' = \frac{Z_c f}{Z_c + f}. \tag{7.21}$$

Substituting (7.21) for (7.20), we have

$$|q_i - q_j| = \frac{f'}{Z_c - f'} \frac{|Z_c - Z|}{Z} |C_i - C_j|. \tag{7.22}$$

Notice that (7.7) has exactly the same form as that of (7.22). Therefore, *when the rendering camera is located on the (s,t) plane, a light field rendering system with constant depth correction can be regarded as an ideal thin lens imaging system.*
We can also verify that the depth of field has the same form as that of (7.11):

$$Z'_{DOF} = \frac{2D_H Z_c^2}{D_H^2 - Z_c^2} = \frac{2z_c^2 A' f/d}{(A' f/d)^2 - z_c^2} = \frac{2d'_H z_c(z_c - f')}{d'^2_H - (z_c - f')^2}, \tag{7.23}$$

where $d'_H = A' f'/d$.
In summary, analogous to the conventional optical system, the key elements of a light field can be defined in term of a virtual optical system:

- Aperture. $A' = |C_i - C_j|$
- Focal length. $f' = \frac{fz_c}{f + z_c}$
- Hyperfocal distance. $d'_H = A'f'/d$
- Depth of field. $z'_{DOF} = \frac{2d'_H z_c (z_c - f')}{d'_H{}^2 - (z_c - f')^2}$

In the virtual optical system, the aperture is defined as the intervals of cameras. Object distance is the distance between object point and (s, t) plane. Image distance is the distance between rendering image plane and (s, t) plane. Focal length is defined as a virtual value decided by the object distance and the image distance.

The hyperfocal distance plays an important role in minimum sampling of light field rendering because it describes the relationship between cameras spacing (the number of cameras) needed for capturing and the rendering quality (rendering resolution). Given rendering resolution, the higher D_H means fewer raw images needed for anti-aliasing rendering. The hyperfocal distance is proportional to the rendering resolution, and inversely proportional to the number of images used. Given a fixed hyperfocal distance, the relationship between the rendering resolution and the number of images is linear.

7.4 Minimum sampling of light field

7.4.1 Optimal constant depth

Given the minimum depth Z_{min} and maximum depth Z_{max} of a scene, we again determine the optimal constant depth Z_{opt} for minimum sampling rate or best rendering quality. By the definition of hyperfocal distance, it is equivalent to maximizing the hyperfocal distance d'_H or D_H. The optimal constant depth Z_{opt} satisfies

$$Z_{opt} = arg \max_z \{D_H\}$$

$$s.t. \ Z_{min} \geq \frac{D_H Z}{D_H + Z} \tag{7.24}$$

$$Z_{max} \leq \frac{D_H Z}{D_H - Z}.$$

The maximum D_H arrives when

$$\frac{1}{Z_{opt}} = \frac{1}{2}[\frac{1}{Z_{min}} + \frac{1}{Z_{max}}], \tag{7.25}$$

with

$$\frac{1}{D_H} = \frac{1}{2}[\frac{1}{Z_{min}} - \frac{1}{Z_{max}}]. \tag{7.26}$$

The geometrical interpretation of Z_{opt} is that the optimal constant depth is the harmonic mean of the minimum and maximum depths. The maximum D_H is decided by (7.26), which means the relationship between the spacing of the cameras and output quality is determined by the scene depth variation. In other words, (7.26) completely determines the minimum sampling of light field rendering.

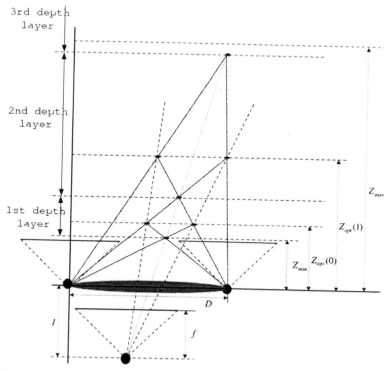

Fig. 7.8. Multiple layers of constant depth are recursively constructed for minimum sampling of light field rendering. Given the minimum and maximum depths of a scene, and the rendering resolution, the number of depth layers needed for anti-aliased light field rendering is computed recursively.

7.4.2 Multiple depth layers segmentation

From the above analysis, we know that the hyperfocal distance is specified by the maximum and minimum depths of the scene. We understand that, the larger the depth of field, the more input images needed for anti-aliased light field rendering. More images mean more storage and less efficient rendering. Therefore, segmenting the depth of the scene into multiple depth layers effectively decreases the depth of field and decreases the number of cameras needed for rendering. The number of layers is determined by the trade-off between rendering efficiency and the storage of image data.

Figure 7.8 illustrates how the scene is recursively segmented into multiple layers. From (7.17), we get the following relationship between the layers:

$$\begin{cases} \dfrac{1}{Z_{max}(n)} = \dfrac{D_H - I}{D_H} \dfrac{1}{Z_{opt}(n)} - \dfrac{1}{D_H} \\[2ex] \dfrac{1}{Z_{min}(n)} = \dfrac{D_H + I}{D_H} \dfrac{1}{Z_{opt}(n)} + \dfrac{1}{D_H} \quad , \\[2ex] Z_{max}(n) = Z_{min}(n+1) \end{cases} \qquad (7.27)$$

where $Z_{opt}(n)$, $Z_{min}(n)$, and $Z_{max}(n)$ are the optimal constant depth, the minimum depth and the maximum depth of the nth layer, respectively.

From (7.27), we have

$$\frac{1}{Z_{opt}(n)} = \left(\frac{D_H - I}{D_H + I}\right)^n \left(\frac{1}{Z_{opt}(0)} + \frac{1}{I}\right). \qquad (7.28)$$

A special case of (7.28) is that $I = 0$. When the rendering camera is located on the U axis, we get the following equations:

$$\frac{1}{Z_{opt}(n)} = \frac{n}{2}\left[\frac{1}{Z_{min}} + \frac{1}{Z_{max}}\right] \qquad (7.29)$$

$$\frac{1}{D_H} = \frac{1}{N+1}\left[\frac{1}{Z_{min}} - \frac{1}{Z_{max}}\right]. \qquad (7.30)$$

The above equations determine the minimum sampling in the joint image and geometry space. Specifically, the minimum sampling problem in the joint image and geometry space is described by the relationship among the number of images, the output resolution and the number of constant depth layers. Depth layer segmentation effectively increases the hyperfocal distance.

7.5 Summary

We have described an optical analysis of the light field rendering system. Here, light field rendering is regarded as a synthetic aperture optical system with a constant depth assumption for the scene. From the optical analysis of light field sampling, we obtain the relationship among the depth variation of the scene (depth of field), the number of images (aperture) and the rendering resolution (circle of confusion). Specifically, this relationship is completely described by the hyperfocal distance of the virtual optical system. The optical analysis is applied to estimate the optimal constant depth for a given scene for the best rendering quality. The minimum sampling rate is then derived. To extend the optical analysis to cover significantly larger depth variation, we presented the optimal depth segmentation using fewer number of images, without loss of rendering quality.

Analogous to the Gaussian optical system, we defined the following optical parameters:

- Focal length f;

- Smallest resolvable feature (on the image plane) d;
- Aperture D. Distance between two adjacent cameras;
- Circle of confusion $c = d/f$;
- Hyperfocal distance $D_H = D/c$.

Let the plane of perfect focus be at the distance z_{opt}, the minimum and maximum distances at which the rendering is acceptable be z_{min} and z_{max}, respectively. The following relations exist ([11], vol. 1, p.1.92):

$$z_{min} = \frac{D_H z_{opt}}{D_H + z_{opt}}, \quad \text{and} \quad z_{max} = \frac{D_H z_{opt}}{D_H - z_{opt}},$$

which lead to,

$$\frac{1}{z_{opt}} = (\frac{1}{z_{min}} + \frac{1}{z_{max}})/2$$

$$\frac{1}{D_H} = (\frac{1}{z_{min}} - \frac{1}{z_{max}})/2.$$

As a result, to have the best rendering quality, no matter which optical system is used, the focus should be always at z_{opt}. Moreover, to guarantee the rendering quality, D_H has to be satisfied, i.e.,

$$\frac{D}{d/f} = (\frac{1}{z_{min}} - \frac{1}{z_{max}})/2. \tag{7.31}$$

In other words, given the minimum and maximum distances, the maximum camera spacing can be determined in order to meet the specified rendering quality. The hyperfocal distance describes the relationship among the rendering resolution (circle of confusion), the scene geometry (depth of field) and the number of images needed (synthetic aperture). Intuitively, the minimum sampling rate is equivalent to having the maximum disparity less than the smallest resolvable feature on the image plane, e.g., camera resolution or one pixel, i.e., $d = \delta_v = 1$. The same result was also obtained by Lin and Shum [167] using a geometrical approach.

Equation (7.31), not surprisingly, is the same as the one from geometric analysis in the previous chapter, and is almost exactly the same as Equation (5.8) because $D_H = 2/h_d$. However, the approach using spectral analysis of light field signals incorporates the textural information in the sampling analysis.

While the experimental results are encouraging, the sampling rate of plenoptic sampling could be further reduced if the characteristics of human vision are considered. For example, manifold hopping (Chapter 14 and [273]) can greatly reduce the size of the input database.

8

Optimizing Rendering Performance using Sampling Analysis

From the sampling analyses in previous three chapters, it is evident that the rendering performance of an IBR system is determined by the number of images and the amount of geometrical information used. In this chapter, we describe the layered Lumigraph representation proposed by Tong *et al.* [297]. What is interesting about this representation is that, given the output image resolution and the rendering platform (e.g., process speed and memory), it is configured for optimized rendering performance based on the sampling analysis. The layered Lumigraph is produced by classifying all pixels into a number of depth layers. Based on the plenoptic sampling analysis, the layered Lumigraph is constructed to achieve the same rendering quality along the minimum sampling curve by balancing the number of images and depth layers. For a given rendering platform, the best rendering performance can be obtained by choosing the optimal number of images and depth layers. Moreover, the layered Lumigraph is capable of level-of-detail (LOD) control using the same image geometry trade-off. Therefore, the layered Lumigraph fully exploits the inherent constraints between the number of images, depth complexity, and output resolution. Finally, a backward warping technique is designed to efficiently render the layered Lumigraph by taking advantage of texture mapping hardware.

8.1 Introduction

As different types of rendering environment call for different IBR representations, IBR systems should be optimized in the joint image-geometry space. Plenoptic sampling analysis reveals that many IBR systems with different combinations of images and geometry can be used to achieve anti-aliased rendering. However, the existing IBR systems are typically not flexible enough to represent different combinations in the joint image-geometry space.

Moreover, to reduce rendering time, sampling resolution in the joint image-geometry space should also match the output resolution. As a result, the scene needs to be represented at different levels of details (LOD). Although the LOD control in either image space [318] or geometry space [108] has been explored in previous work, efficient LOD control in the joint image-geometry space is still nontrivial.

In this chapter, we describe an IBR representation called the layered Lumigraph with LOD control [297]. This representation permits efficient LOD control in both image and geometry space for the best rendering performance. The layered Lumigraph extends the conventional Lumigraph concept by classifying all pixels into a number of depth layers. It offers the following features:

- Optimal rendering performance across different rendering platforms. Because the layered Lumigraph describes the minimum sampling curve in the joint image-geometry space, representations with different trade-offs between the number of images and the number of depth layers can be easily constructed.
- Efficient LOD control in the joint image-geometry space. Both the image resolution and the number of depth layers are dynamically adapted to the output resolution.
- Rendering using an efficient backward warping algorithm implemented on texture mapping hardware.

The remainder of this chapter is organized as follows. We first give an overview of the related work in the next section. Plenoptic sampling analysis in the joint image-geometry space is then briefly reviewed. Following that, we describe the layered Lumigraph in the joint image-geometry space. The construction, optimization and hierarchical representation of the layered Lumigraph are also discussed. We then discuss how the layered Lumigraph can be efficiently rendered, with LOD control, using a backward warping technique. Experimental results are then presented, followed by concluding remarks.

8.2 Related work

8.2.1 Image-based representation

The light field technique [160] densely and uniformly samples the radiance of a scene or object as a 4D plenoptic function of position and direction without incorporating any geometry or depth information. The Lumigraph [91] is similar to light field rendering but it applies approximate geometry to improve rendering quality. The dynamically reparameterized light field [116] extends the light field technique to include a variable focus plane or surface and therefore reduces the sampling density required. Schirmacher et al. [257] introduced an interactive Lumigraph rendering algorithm by augmenting the light field with its corresponding depth map. When accurate depth values are used, the image can be rendered by directly warping the nearby input image according to the pixels' depth. To overcome the inefficiency of 3D warping, Schaufler [255, 64] introduced the concept of layered imposters to approximate the object geometry. For a given viewpoint, the scene is divided into a series of frontal-parallel layers from back to front. Each layer contains an image of the scene that belongs to this layer. For a new viewpoint, the scene can be rendered from these layers on texture mapping hardware. To deal with occlusion problems, Layered Depth Images (LDI) [264] stores multiple pixels and their depths along each line of sight that emits from a single viewpoint. For a real scene, the LDI can be generated by warping images from multiple reference views to a common view.

In this chapter, multiple reference cameras are set in a light field configuration while each ray is augmented with a quantized depth. The layered Lumigraph is similar to [257]. There are, however, several differences between these two representations. Firstly, the motivation of the layered Lumigraph is to provide flexibility in the joint image-geometry space so that it can be optimized for different types of computing environment and output resolutions. Secondly, in the layered Lumigraph, we not only can use a higher resolution depth map to reduce the number of image samples required, we can also use a large number of image samples to reduce depth complexity. Thirdly, while previous methods do not deal with hierarchical image-based representations, developing an automatic LOD control in the joint image-geometry space is an important goal of this chapter.

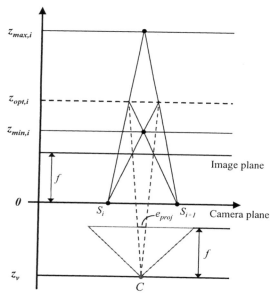

Fig. 8.1. The LOD control in the joint image-geometry space. The projection error becomes smaller when the view moves farther away from the camera plane. By assuming the maximum projection error allowed in the rendering result, the projection error can be used as the error metric for LOD control.

8.2.2 Minimum sampling curves for different output resolutions

In the previous chapters, we have learnt about the optimal depth and minimum sampling rate, and the minimum sampling curve in image-geometry space. To match the rendering complexity with output resolution, we need to construct levels of details in the joint image-geometry space. Plenoptic sampling also encodes the interaction between the output resolution and minimum sampling curve in the joint image-geometry space, but only for new views on the camera plane.

Figure 8.1 shows how a projection error occurs at a displaced view from the camera plane. Here we assume the minimum, maximum and optimal depthes of the current layer i to be $z_{min,i}$, $z_{max,i}$, $z_{opt,i}$ respectively. Also, let z_v be the depth distance between the current viewpoint and camera plane. The projection error is then

$$e_{proj} = \Delta s f(z_{opt}/z_{min} - 1)/(z_v + z_{opt}). \qquad (8.1)$$

When the current viewpoint lies on the camera plane and its focal length equals to those of the reference cameras, the minimum sampling rate can be derived as in the case of spectral analysis of plenoptic sampling. In this case, the maximum projection error is assumed to be Δt. Texture distribution and input resolution are however ignored here. In practice, a MipMap [318] can be used to make the input image resolution comparable to the output resolution. Moreover, if we have additional information about texture distribution, we can relax the above constraints further.

8.2.3 Hierarchical image-based representations

To deal with the sampling issue of LDI, Chang *et al.* [39] introduced a LDI tree which combines a hierarchical space partition scheme with the concept of LDI. The image caching technique [265] used the same hierarchical structure as the LDI tree but each space partition has an imposter instead of an LDI. Oliveira *et al.* [216] also incorporated a texture pyramid into relief texture mapping in order to keep the warping cost proportional to the size of the output image.

The layered Lumigraph representation is similar to [39, 216]. There are, however, two important differences. Firstly, its hierarchical representation is based on individual reference views rather than a combination of these views into a single center of projection. Therefore, it does not incur any loss of variations in scene appearance. Resampling errors are minimized because the original input images are not combined into a single reference view. The direct use of multiple reference views also reduces the sampling resolution required in the depth space. Secondly, its LOD control is embedded not in the image space but in the joint image-geometry space. If the screen resolution decreases or if the viewer moves farther away from the scene, the image resolution and geometrical details (the number of depth layers) can both be reduced without loss of visual quality.

8.2.4 Image warping

The image warping described in [189] is a forward mapping process. The pixels of the reference images are traversed and warped to the output image in the order they appear in the reference images. The disadvantage of the forward mapping algorithm is its tendency to produce hole artifacts due to self-occlusion or insufficient sampling. Backward warping is used as a rendering mechanism to avoid hole artifacts. However, it is computationally expensive. Drawing each pixel in the output image requires searching the entire epipolar line in the reference image. In [216], the 3D warping procedure is factored into a 1D forward mapping step and a traditional texture mapping step with the support of standard graphics hardware. In view-dependent texture mapping (VDTM) [60], the scene is represented as polygonal models with multiple texture maps, which are blended at render time.

The layered Lumigraph can be considered as a hierarchical extension to the method of VDTM [60]. Frontal-parallel sprites, rather than polygonal models, are used to represent the scene geometry. In addition, the number of layers used for rendering is automatically optimized for the available computing resources and the current output resolution by incorporating LOD control in the joint image-geometry space.

8.3 Layered Lumigraph

The relationship between the number of image samples and the number of depth layers described in the previous plenoptic sampling analysis leads to the layered Lumigraph. In this section, we first provide an overview of the layered Lumigraph rendering system. This is followed by a description of the layered Lumigraph preprocessing procedure.

8.3.1 System overview

Table 8.1. Data structure of the layered Lumigraph.

ColorPixel:	unsigned char ColorRGB [3];
ColorImage:	colorPixel Pixels[ResU][ResV];
DisparityIndexImage:	unsigned char DepthPixels[ResU][ResV];
Layered Lumigraph	
{	
int	NumberOfLayer;
int	ResS, ResT;
int	ResU, ResV;
float	LayerDisparityTable[NumberLayer];
ColorImage	ColorImages[ResS][ResT];
DisparityIndexImage	DisparityImages[ResS][ResT];
}	

The layered Lumigraph extends the conventional Lumigraph/light field representation by assigning a depth layer for each pixel. Like a Lumigraph/light field, the

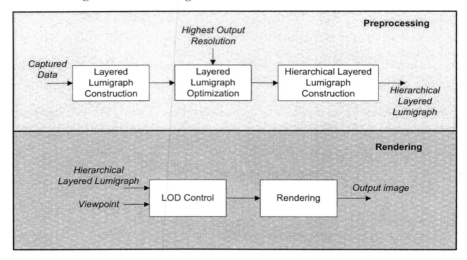

Fig. 8.2. The layered Lumigraph system.

layered Lumigraph is also represented by a two-plane parameterization. Because all sample cameras lie on the same plane, the depth layers used in a layered Lumigraph are planes parallel to the camera plane. The pseudo-code in Table 8.1 describes the data structure of the layered Lumigraph. In practice, Tong *et al.* [297] found that one byte is generally sufficient to encode the index of depth layers. If more depth layers (more than 256) are involved, more bits (16 or 32) should be used to index the depth layers.

As shown in Figure 8.2, the layered Lumigraph rendering system consists of two stages. In the preprocessing stage, the layered Lumigraph is first constructed from evenly sampled images. For the highest output resolution, this original layered Lumigraph is optimized to maximize the rendering performance on a given platform. The hierarchical layered Lumigraph is then generated from the optimal layered Lumigraph for efficient LOD control.

In the rendering stage, the LOD of the layered Lumigraph is first tailored to the current output resolution and view position. Then, the tailored layered Lumigraph is rendered via the texture mapping hardware.

8.3.2 Layered Lumigraph generation and optimization

To construct the layered Lumigraph, the raw Lumigraph/light field images of the scene are first captured at each sample grid on the camera plane. The depth maps from the same sample cameras are either captured simultaneously or computed later using a stereo algorithm. According to plenoptic sampling analysis, the disparity of the scene is uniformly divided into the maximum number of layers allowable by the system. Finally, the disparity value for each pixel is classified according to the

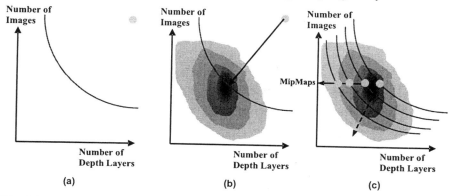

Fig. 8.3. Layered Lumigraph optimization: (a) original layered Lumigraph, shown as a point above the minimum sampling curve, (b) constructing the optimal layered Lumigraph for the given output resolution, (c) constructing LOD for the layered Lumigraph. The contour plot in (c) illustrates the rendering performance for different combinations of images and depth layers. The darker the region, the better the rendering performance.

layer it belongs to. As shown in Figure 8.3, instead of storing all the combinations of images and depth layers along the minimum sampling curve for the highest output resolution, only the images in the original Lumigraph and the maximum number of depth layers are required for storage.

As shown in Figure 8.3, given the highest output resolution specified by a user, the minimum sampling curve can be determined first in the joint image-geometry space. Along the minimum sampling curve, the best rendering performance for any given rendering platform is achieved at a specific point, which is defined as the optimal layered Lumigraph.

After the minimum sampling curve has been determined by the highest output resolution, a *constrained* optimal layered Lumigraph is constructed from the original layered Lumigraph. The constraints are based on the fact that the images are presampled. Since the rendering performance is determined by many issues (graphics hardware, CPU, memory, etc.), it is difficult to get a closed-form solution for the layered Lumigraph optimization.

Tong *et al.* [297] did the optimization empirically. They first derived a set of layered Lumigraphs with different combinations of images and depth layers from the original layered Lumigraph, each of which corresponds to a point on the minimum sampling curve of the highest output resolution. After the rendering speed of these layered Lumigraphs on the given platform is tested, the one with the best rendering performance is the constrained optimal layered Lumigraph for the given platform. For example, for the BLOCK data shown in the experimental results section, they tested the rendering performance of different combinations of images and depth layers on two platforms. As shown in Table 8.5, on System A, the optimal layered Lumigraph for the BLOCK scene contains 4×4 images. The optimal layered Lumigraph of the same scene for System B contains 8×8 images.

8.3.3 LOD construction for layered Lumigraphs

As shown in Figure 8.3c, for each output resolution, an optimized layered Lumigraph can be extracted from the original layered Lumigraph. All these optimal layered Lumigraphs comprise the optimal LOD for layered Lumigraphs in the joint image-geometry space (shown as the dashed line in Figure 8.3). In practice, it is time-consuming to construct the optimal layered Lumigraph for each output resolution. It is also difficult to control the LOD by switching among these optimal layered Lumigraphs at rendering time.

Therefore, based on the optimal layer Lumigraph for the highest output resolution, the LOD is constructed for different output resolutions with the fixed number of images (solid line in Figure 8.3c). In the image space, a series of image pyramids are defined for each image in the optimal layered Lumigraph. More specifically, every reference image is organized into a Gaussian pyramid, with each level containing successively lower-passed spatial frequency color and depth data in the input image. The original image forms the lowest level of the Gaussian pyramid, i.e., $G_0 = I$. Each successive level of the pyramid is produced by convolution with a Gaussian kernel followed by down-sampling by a factor of two: $G_{n+1} = 2 \downarrow (Gn \otimes K_{Gaussian})$, where $2 \downarrow (\cdot)$ is the two-times down-sampling operation and G_n is the nth level of the pyramid, which is $1/2^n$ the size of the original in each dimension.

For efficient rendering, an efficient transformation from the finest depth resolution to varying depth resolutions is also needed. Instead of generating the LOD for depth in the preprocessing stage, the number of depth layers is dynamically adjusted at rendering time.

8.4 Layered Lumigraph rendering

Given a specific rendering platform, the pyramids of the color images and depth maps in the optimal layered Lumigraph are loaded into the texture memory and defined as the texture MipMaps. The depth map is defined as the palette texture or alpha texture with a texture lookup table so that the depth maps can be updated quickly during rendering. For systems without palette texture or texture lookup table support, the depth texture must be reloaded into the texture memory for any layer adjustment.

As illustrated in Table 8.2, rendering proceeds in two steps. In the first step, the LOD of the layered Lumigraph is tailored to the current output resolution and viewpoint position. In the second step, by making use of texture mapping hardware, the layers are rendered in a back-to-front order. For each layer, the output image is divided into several regions, each of which can be rendered by texture mapping four neighboring reference images. In the rendering algorithm, the rendering time mainly depends on the number of layers and the output resolution. Details of the rendering algorithm are discussed in the following subsections.

Table 8.2. Rendering algorithm for hierarchical layered Lumigraph.

```
Preprocessing:defining color and palette texture maps;
Rendering:
for (novel view point)
      compute new depth layers for new view point;
      if (depth layers are different)
            load new depth layer map;
      for (each rectangle Rs on camera plane)
            project Rs to output image plane to get Ro;
      endfor
      for (each depth layer j back to front)
            project output region Ro to current layer to get region Rc;
            setting the alpha test;
            for (each of four sample cameras for Rs)
                  load texture map corresponds to current sample camera;
                  render the Rc at the optimized distance;
            endfor
      endfor
endfor
```

8.4.1 LOD control in joint image-geometry space

Given the viewpoint and its desired output resolution, the LOD of the layered Lumigraph is adapted to provide adequate image resolution and depth layers. In layered Lumigraph rendering, since the rendering hardware automatically selects the proper MipMap levels for texture mapping, the image resolution used for rendering is always consistent with the output resolution. In the algorithm, the number of depth layers is adjusted by merging a set of neighboring layers from the optimal layered Lumigraph into new layers.

As shown in Table 8.3, the original layers in the optimal layered Lumigraph are swept from back to front. A sweeping layer is merged with the new layer only if after merging, the projection error of the new layer (see (8.1)) is not larger than the output resolution.

After merging, the optimal disparity value is computed for each new layer. The mapping between the original layer and the new layer is constructed and defined by the texture lookup table. By mapping the index of the original layers to that of the new layers, the number of layers is also changed according to the output resolution.

Alternatively, the new layers could be computed from the layer configuration of the last frame so that the coherence between the frames is utilized. Here, Tong *et al.* [297] directly derived the new configuration from the original layers because this operation is very fast.

Table 8.3. LOD control for depth layers

```
merge(original layer, new layer)
{
        sweep_layer = the farthest original layer;
        new_layer = sweep_layer;
        for( sweep_layer != the nearest original layer)
                if(valid_merge (new_layer, sweep_layer))
                        new_layer = merge(new_layer, sweep_layer);
                else
                        new_layer = sweep_layer;
                sweep_layer = next original layer;
        endfor;
}

valid_merge(new_layer, sweep_layer, viewpoint, output resolution)
{
        test_layer=merge(new_layer, sweep_layer);
        if (error_proj(test_layer, viewpoint) < output resolution)
                return true;
        else
                return false;
}
```

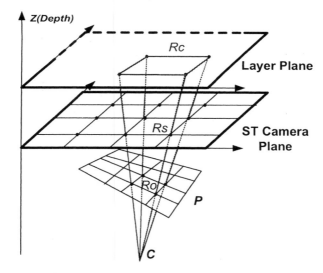

Fig. 8.4. Geometric explanation for rendering. For each region R_o on the output image P from view C, only images from the four sampling cameras at the corners of region R_s are used for rendering. In each sample image, only region R_c on each depth layer contributes to the output region R_o.

8.4.2 Rendering output images

Given a new viewpoint C and desired image plane P, a straightforward rendering algorithm traces each ray from C that passes through the pixel on P to a point in the scene geometry. Then the rays from the four nearest cameras to this point are used to interpolate the output ray.

Instead of rendering the output image pixel by pixel, we can directly project the layers to the output image because the scene is represented as a number of plane layers. The output image plane is divided into several output regions by projecting the sampling camera grids onto the output image plane P. As shown in Figure 8.4, for each region R_o of the output view C, images from four sampling cameras on the corresponding grid points are used to render the layers from back to front. For each sampling camera, a set of polygons parallel to the image plane is placed at the optimized distance of each layer. These polygons are bounded by the view frustum of the corresponding sample cameras. Moreover, for each layer j, we only need to render the region R_c, which is the projection of the output region R_o onto the layer j. Any pixel out of the region R_c makes no contribution to the region R_o.

For each sampling camera, the polygons in the current layer are first clipped by the region R_c's boundary. These polygons are rendered with the corresponding color and depth texture maps. The depth texture map is converted to the new layers first and then is applied to the alpha channel of the polygon. The MipMap level of the images (texture maps) is controlled by the graphics API automatically. After texture mapping, the alpha test is applied to guarantee that only pixels within the current depth layer are rendered into the image buffer. The pseudo code of the rendering is listed in Table 8.2.

To reduce the overhead of texture mapping, each sample image is uniformly subdivided into smaller rectangular texture regions of size $n \times m$. More specifically, the depth range of the original layers contained in each small texture region is recorded in the preprocessing stage. In the rendering stage, instead of defining a large polygon for the whole texture image, a set of small polygons is used for rendering, each of which corresponds to one texture region. For each rendering layer, if the original layers contained in the current new layer fall into the range of the layer in a texture region, the small polygon corresponding to the texture region is used for rendering.

8.4.3 Performance of layered Lumigraph rendering

The performance of layered Lumigraph rendering is affected by several factors. Firstly, if more layers are used in the layered Lumigraph, more polygons are defined and rendered so that more texture mapping operations are executed during rendering. As described above, by subdividing the images into smaller regions, we can achieve a trade-off between texture mapping overhead and the transformations and clipping operations required for the polygons. Secondly, although using the layered Lumigraph with more images and fewer depth layers reduces the number of depth layers, it would take more time to load texture maps into the graphics hardware. Increasing the number of images would also result in an increase in the number of polygons and geometry clipping operations. This adversely affects the performance of geometry processing.

The benefit of LOD control in the joint image-geometry space is that the number of polygons used for rendering is proportional to the number of depth layers. Thus, when rendering a layered Lumigraph with lower LOD, the fewer number of layers alleviates the load of the whole graphics pipeline. On the contrary, if LOD control is applied only to the image space, only the texture mapping stage benefits from LOD control. As we will show in the next section, the LOD control for the layered Lumigraph is more efficient than a simple LOD control in image space.

8.5 Experimental Results

Tong *et al.* [297] implemented the layered Lumigraph rendering in C++ and OpenGL on two PCs with 256MB main memories. The first PC (System A) was configured with a PIII 733MHZ CPU and a Nvidia Geforce256 graphics card. The second PC (System B) was configured with the PIII 866MHZ CPU and a Matrox G400 graphics card.

Table 8.4. Configuration of three layered Lumigraph data sets.

Scene	Size of UV plane	Size of ST plane	F	Z_{min}	Z_{max}	Max Num of layers	Max Num of images
NETFERT	200.0	100.0	160.0	183.4	308.8	256	65×65
STATUE	400.0	200.0	300.0	232.6	581.8	256	65×65
BLOCK	80.0	40.0	60.0	35.4	134.0	256	65×65

Fig. 8.5. Rendering results of the scenes used in the experiments. Left: BLOCK; Middle: STATUE; Right: NETFERT

Three synthetic scenes (STATUE, NETFERT, and BLOCK) were used for experimentation. The image resolution in the data sets is 256×256. The reference color images and accurate depth images were generated by ray tracing using 3D Studio Max. After quantizing the accurate depth into 256 layers, the reference images were converted into the original layered Lumigraph data sets. For simplicity, only a single light slab of the scene is captured and used. The configurations of the original layered Lumigraph for the three scenes are summarized in Table 8.4. Figure 8.5 shows the rendering results of the data sets.

Table 8.5. A comparison on the average rendering time (ms) of the data sets.

Scene	Rendering environment	Number of images		
		2×2	4×4	8×8
NETFERT	System A	19.2	13.5	14.5
	System B	56.8	35.2	29.1
STATUE	System A	12.4	10.6	12.5
	System B	35.7	24.0	24.4
BLOCK	System A	22.1	16.7	19.6
	System B	64.1	43.1	38.8

For a different platform, the configuration of the optimal layered Lumigraph can be different. To obtain the optimal layered Lumigraph for each data set, the rendering speed was measured for different combinations of images and depth layers; each combination corresponds to a point on the minimum sampling curve for the highest output resolution 400×400. As shown in Table 8.5, for the NETFERT scene and System A, the optimal layered Lumigraph contains 4×4 images. Whereas for the same scene, the optimal layered Lumigraph on System B contains 8×8 images. Comparing with the other representations [160, 264] that correspond to some specific points in the joint image-geometry space, the layered Lumigraph is flexible enough to provide different configurations along the minimum sampling curve. Thus the best rendering performance can be achieved on any platform with the optimal layered Lumigraph.

To test the efficiency of the LOD control used in the rendering algorithm, the viewpoint was placed at several distances to the camera plane and computed the average rendering time for 100 frames at each distance. Two algorithms are executed in the experiments to render the optimal layered Lumigraph on System A. The rendering algorithm with LOD control, labelled as LOD Algorithm, dynamically adjusts the number of depth layers and the image resolution when the viewpoint changes. On the other hand, the conventional MipMap algorithm only adjusts the image resolution to adapt to viewpoint movement.

As illustrated in Figure 8.6, when the viewpoint lies on the camera plane, both the LOD algorithm and MipMap algorithm use the finest resolution for rendering. So the rendering time of the two algorithms for all data sets are the same. When the viewpoint moves away from the camera plane, both the image resolution and the number of depth layers are reduced in the LOD algorithm. Since the overhead of the whole graphics pipeline is reduced, the rendering time is decreased accordingly. For

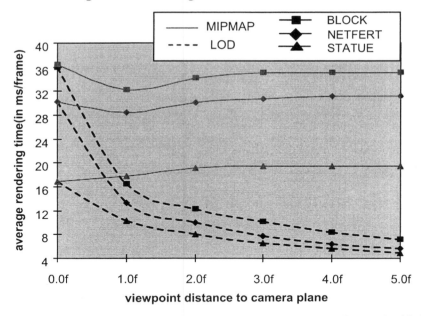

Fig. 8.6. The rendering time (in ms/frame) of the three data sets on System A with different LOD control algorithms, where f is the focal length of the sampling camera. As the viewpoint moves away from the camera plane, the rendering time of the algorithm decreases accordingly. In contrast, the rendering time of the MipMap algorithm remains relatively constant for all viewpoints.

comparison, the rendering time of the MipMap algorithm is nearly constant for all viewpoints. For the viewpoint whose position is $4.0f$ (where f is the focal length of the sampling camera) away from the camera plane, the rendering time of the LOD algorithm is only one third of the rendering time of the MipMap algorithm. Obviously, the LOD control in the joint image and geometry space is more efficient than conventional LOD control in the image space alone.

Figure 8.7 compares the rendering results of the NETFERT scene at different LODs, which are rendered by the LOD algorithm and MipMap algorithm respectively. Although fewer depth layers are used in the LOD algorithm for rendering, the rendering quality of the two algorithms are visually indistinguishable.

8.6 Summary

The layered Lumigraph with LOD control is an efficient image-based representation that is well-suited for different tradeoffs in the joint image-geometry space. Based on plenoptic sampling, the layered Lumigraph can be efficiently optimized to achieve the best rendering performance for different rendering platforms. An efficient backward warping algorithm is introduced to render the output view by taking advantage of common texture mapping hardware. With automatic LOD control in the joint image-geometry space, the rendering performance of the layered Lumigraph is further optimized according to the output resolution. The key idea presented here is not limited to the configuration of uniform light field sampling and frontal-parallel depth layers.

Fig. 8.7. Rendering results of the NETFERT scene at several levels of details rendered with different LOD control algorithms. Above images are rendered by the LOD algorithm. Below images are rendered by MipMap algorithm. These images are visually indistinguishable.

Part III

Thus far, we have described the characteristics of various types of image-based representations as well as rendering issues. It is clear that image-intensive representations such as light fields, Lumigraphs, and CMs are capable of photo-realistic rendering, but this is achieved at the expense of large storage and transmission bandwidth. To overcome these problems, a significant amount of work has been done on effective compression and transmission of image-based representations. Although image and video compression have been studied extensively and many advanced algorithms and international standards are now available [30, 118, 123, 119, 120, 125, 126], there are specific important requirements in IBR that need to be addressed.

There are two particularly related important attributes for efficient IBR compression: random access and selective decoding. In conventional videos, random access at the group-of-picture level is usually provided (e.g., in MPEG-2 videos) to support fast forward, backward, and jumping to selected locations or chapters. However, IBR representations such as the light field and Lumigraph require random access at the pixel level while CMs require line level random access. As most existing compression algorithms use entropy coding (such as Huffman or arithmetic coding) for better compression ratio, the compression sizes of data chunks are variable. Without random access (and hence selective decoding), extracting data for rendering will be dependent on the location of the data and typically expensive. This is because sequential access require the retrieval and decoding of all pixels or lines located in front of those requested.

There is a substantial body of work on data compression. It is beyond the scope of this book to cover all the topics and issues in conventional data compression. Instead, we provide some fundamental background on compression and focus on compression techniques specifically geared for image-based rendering. Information on basic compression techniques such as lossless compression and subband coding are readily available in resources such as [15, 23, 54, 76, 82, 129, 185, 198, 208, 242, 247, 284, 294, 304, 311, 314].

A brief introduction to the basic concept and fundamental techniques in image and video compression are first given in Chapters 9 to 11. Then, in Chapters 12 and 13, the problem of compressing various static and dynamic image-based representations will be discussed. Chapters 9 to 11 mainly serve as a quick review of image and video compression in order for the readers to understand the materials to be discussed in Chapters 12 and 13. Readers with prior exposition to these topics may proceed directly to Chapters 12 and 13.

9

Introduction to Compression

An important problem of digital representation of signals and IBR in particular is its large amount of digital storage and bandwidth for transmission. As an example, consider the transmission of a video with a resolution of (352×288) and frame rate of 30 frames/second. The bits per second (bps) or bit-rate required for transmitting this video is 37 Mega bits/sec (Mbps). Using advanced video compression algorithms such as the H.261 or H.263 video coding standards, it is possible to reduce the data rate to 128-384 kbps and lower. Table 9.1 summarizes the typical data sizes of several IBR representations and videos before and after compression. Because of the large data size of common IBR representations, data compression becomes an essential part of practical IBR systems.

9.1 Waveform coding

Data compression is also closely related to source coding in digital communications. Figure 9.1 shows the general structure of a digital transmission system where a signal $x(n)$ is transmitted to the receiver through a communication channel. The source encoder explores the redundancy in the input signal (or a data file) and reduces its bit-rate for transmission over the medium (or storage). The output of the source encoder is a binary representation of the original signal. The original signal can be reconstructed at the receiver by a source decoder. Depending on the nature of the source coding, the reconstructed signal $\hat{x}(n)$ might or might not be identical to the original signal. If they are identical, the coding is referred to as *lossless*, because there is no loss in information after the coding or compression. Lossless coding/compression is desirable for data files and medical images, where the integrity of the data has to be preserved. On the other hand, for speech, images, videos, and IBR, slight errors or distortion of the reconstructed signal are usually introduced by the source encoder in exchange for a more compact representation and hence a higher compression ratio. The coding process is referred to as *lossy*, because some information is lost after compression.

Table 9.1. Typical data sizes of IBR before compression and after compression. Frames per second : fps, Kilobits per second : kbps, Megabits per second : Mbps. Note : The YCrCb (4:2:0) color component has been used in the calculation, which is half of that for 24-bit RGB color system.

Application	Data Size/Rate YCrCb(4:2:0)	
	Uncompressed	Compressed
Concentric mosaic (3D) with 1463 normal view images (352 × 288) × 1463	1.78 Gb	29-36 Mb [269]
32 × 32 Light field (4D) (256 × 256) × 32 × 32	805 Mb	8-12 Mb [341]
Panoramic videos (3D) (2048 × 768), 25 fps	472 Mbps	9-18 Mbps [210]
Plenoptic videos (4D) with 8 cameras BT.601: (720 × 480), 30 fps × 8	995 Mbps	24-48 Mbps
Video conferencing CIF format: (352 × 288), 30 fps	37 Mbps	128-384 kbps (H.261,H.263)
High quality video distribution BT.601 : (720 × 480), 30 fps	124 Mbps	4-8 Mbps (MPEG-2 video)
HDTV SMPTE296M : (1280 × 720), 60 fps	664 Mbps	20-45 Mbps (MPEG-2 video)

Techniques for lossy compression of digital signals can be classified broadly into *waveform coding* techniques and *model* or *content-based* techniques. In waveform coding, we try to approximate the input signal $x(n)$ by its encoded value $\hat{x}(n)$, subject to certain bit-rate or distortion constraints. One commonly used criterion for measuring the distortion between $x(n)$ and $\hat{x}(n)$ is the mean squared error (MSE) (to be described later). Waveform coding is widely used in encoding speech, images, videos and IBR. In fact, a number of international standards for coding images and videos, such as JPEG [118], JPEG-2000 [123], H.261 [30], H.263 [126], H.264 [125], MPEG-1 [119], MPEG-2 [120] are based on waveform coding techniques. Most IBR compression techniques we describe in Chapters 12 and 13 also make use of waveform coding. In order to provide higher compression ratio and more flexibility in content manipulation, model- and content-based techniques are gaining more attention recently. In model-based coding, models of the objects to be encoded are employed to further improve the coding performance. Interested readers are referred to [314] for a comprehensive description of content-based approaches.

Figure 9.1(b) shows the general structure of a waveform coding system. It consists of three functional blocks: decorrelation, quantization and entropy coding. The main reason for performing decorrelation is that signals such as speech, images and videos are highly correlated. This means that adjacent samples in a speech signal, or adjacent image pixels in an image or videos usually resemble each other. If the image pixels are directly compressed losslessly by entropy coding techniques such as Huffman or Arithmetic codes, the compression ratio is usually very limited. To improve

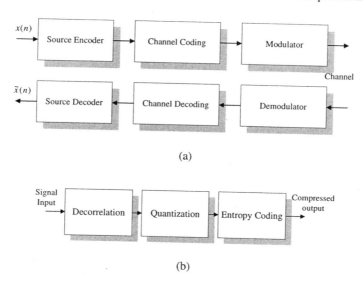

Fig. 9.1. (a) Structure of a data transmission system. (b) General structure of a waveform coding system.

the compression ratio, the input signal is usually decorrelated using decorrelation techniques such as orthogonal transformation, subband filters, linear prediction, or a combination of them. These techniques reduce the redundancy in the input signal and produce a more compact representation of the input signal. For example, in transform coding (Section 10.2), the input is first divided into non-overlapping vectors. Each vector will undergo a linear signal transformation so that most of its energy is concentrated into a few significant transform coefficients, while the amplitudes of the remaining coefficients are usually very small. Significant compression can be achieved by thresholding these small amplitude coefficients to zero and approximating the others with fewer number of bits. This lossy process is called quantization (Section 9.3) and it leads to a very high compression ratio. The quantized signals and other auxiliary information, called the symbols, are then losslessly encoded using entropy coding. The basic idea of entropy coding is to assign codewords with different length to symbols with different probabilities so that the average wordlength of the symbols to be transmitted can be minimized. Ideally, we would like the coding process to produce a codeword of length $\log_2(1/p)$, called the information content, for a symbol of probability p. Therefore, more probable symbols are given shorter codewords. In Huffman code [111], the length of each codeword is an integer and it increases with the information content of the corresponding symbol. For example, suppose we have five source symbols s_1 to s_5 with the following probabilities:

$$p(s_1) = 0.4, \quad p(s_2) = 0.2, \quad p(s_3) = 0.15, \quad p(s_4) = 0.15, \quad p(s_5) = 0.10.$$

Their Huffman codes will be given by $s_1 = 1$, $s_2 = 000$, $s_3 = 001$, $s_0 10 = 1$ and $s_5 = 011$. Interested readers are referred to standard texts [314] for the generation and decoding of Huffman code and its properties. Since the codewords are of variable lengths, it is also called a variable length code (VLC).

One might notice that in order to access a particular symbol in the compressed bit stream, all the previous symbols have to be decoded. This is an important problem in IBR applications, where fast random access to image pixels or line of pixels is required. Usually a group of symbols will be coded independent of the other such that they can be accessed in groups using pointers or other structure. This will however degrade the coding performance to some extent. More detail will be given in Chapters 12 and 13.

In arithmetic coding, each possible symbol is assigned a subinterval between [0,1) with a length equal to its probability. During encoding, each additional symbol is used to specify a new subinterval within the previous one and the subdivision is done in a way identical to the partitioning in the symbol sub-division. When a series of symbols is processed, the length of the sub-interval associated with the encoded symbols gets smaller and smaller. After encoding the input message, a subinterval between [0,1] will be obtained. The code is a binary number having the minimum number of bits that lies inside that subinterval. For example, if we have five symbols with probabilities:

$$p(s_1) = 0.1, \quad p(s_2) = p(s_3) = 0.15, \quad p(s_4) = 0.2, \quad p(s_5) = 0.4.$$

For the message $\{s_5, s_5, s_4, s_3, s_2, s_1\}$, the number $V = 0.9124908447$ with the following 16-bit binary representation will be sent to the receiver.

$$2^{-1} + 2^{-2} + 2^{-3} + 2^{-5} + 2^{-8} + 2^{-9} + 2^{-12} + 2^{-13} + 2^{-16} = (1110100110011001).$$

To encode the same symbols, Huffman code requires 14 bits, while fixed length code requires 18 bits. It can be seen that arithmetic code produces a single codeword for a set of symbols. Further, according to the symbol sub-interval ranges, more probable data sets will correspond to larger sub-intervals and hence it requires less precision or wordlength to specify a codeword in that interval than less probable data sets. Arithmetic code usually performs better than Huffman code because multiple message symbols are coded together in a very simple manner.

In image coding standards like JPEG and JBIG (Joint binary image expert Group [122]), binary arithmetic coding with binary alphabet is adopted. It is because binary alphabet allows simple approximation to be made in the interval scaling to eliminate the need for multiplication. Another reason is that simple probability estimation technique can be developed for binary coders to adapt to the changing probabilities. Interested readers are referred to [228, 199] for more details. In addition to Huffman and arithmetic codes, pattern based coders such as Lempel-Ziv-Welsh coding (LZW) [351, 350] are also frequently used in lossless compression of data file.

In the following, we describe some basic terminology in data compression. A brief summary of quantization techniques will be given in Section 9.3.

9.2 Basic concept and terminology

We now describe a few basic concepts and terminologies in data compression.

9.2.1 Compression ratio

One important consideration in designing a data compression system is the performance of the compression algorithm. This is usually measured by the compression ratio and the quality of the signal or information after reconstruction. Compression ratio is simply the ratio of the amount of storage or transmission bandwidth (usually in terms of bits (or Bytes) for storage and bits per second for transmission bandwidth) before compression to that after compression. For instance, the 4D light field of the Buddha statue [160], which consists of an (32×32) array of (256×256) 24-bit per pixel images, has a total size of 192 MB. If the data size after compression is 1.92 MB, then the compression ratio is 100.

As mentioned earlier, loss of information is usually undesirable in compressing data files and other information such as medical images. Because of this limitation, the compression ratio of lossless compression algorithms is usually limited from 1 to 10, depending on the characteristics of input data. Greater compression can be achieved if loss or distortion of the information to be compressed is allowed. In these lossy compression algorithms, the reconstructed signal quality can be traded for a higher compression ratio and appropriate measures are needed to quantify the distortion due to data compression.

9.2.2 Distortion measures

Common measures of signal quality include signal-to-noise ratio (SNR), peak signal-to-noise-ratio (PSNR), and mean opinion score (MOS).

Signal-to-noise ratio

Suppose that a signal $x(n)$ is encoded to $\hat{x}(n)$. The difference between $x(n)$ and $\hat{x}(n)$ is called the reconstruction (quantization) error or noise $e(n)$, i.e. $e(n) = x(n) - \hat{x}(n)$. A standard objective measure of distortion is the SNR (usually expressed in decibels (dB)), which is defined as the ratio of signal energy $E[x^2(n)]$ to that of the quantization error $E[e^2(n)]$:

$$\text{SNR(dB)} = 10 \log_{10} \left(\frac{E[x^2(n)]}{E[e^2(n)]} \right),$$

where $E[\cdot]$ denotes the expectation or averaging operator. The quantity $E[e^2(n)]$ is also referred to as the mean square error (MSE). As an example, for a 2D image of size $(N_1 \times N_2)$, $\{x(n_1, n_2), n_1 = 0, 1, ..., N_1; n_2 = 0, 1, ..., N_2\}$,

$$\text{MSE} = \frac{1}{N_1 N_2} \sum_{n_1=1}^{N_1} \sum_{n_1=1}^{N_1} e^2(n_1, n_2).$$

Table 9.2. Five-point adjectival scales for quality and impairment, and the corresponding number scores.

Number scores	Quality scale	Impairment scale
5	Excellent	Imperceptible
4	Good	(Just) Perceptible but not Annoying
3	Fair	(Perceptible and) Slightly Annoying
2	Poor	Annoying (but not Objectionable)
1	Unsatisfactory (Bad)	Very Annoying (Objectionable)

The higher the value of SNR, the smaller $E[e^2(n)]$ will be and hence the closer will be the reconstructed value $\hat{x}(n)$ to its original $x(n)$.

Peak signal-to-noise ratio (PSNR)

Another objective measure, which is commonly used in image or video coding, is the PSNR. It is defined as

$$\text{PSNR(dB)} = 10 \log_{10} \frac{(255)^2}{\text{MSE}}.$$

Instead of computing the signal energy (energy of an image or video), the maximum value of an 8-bit precision pixel, i.e. 255, is used to simplify the computation.

Mean opinion score (MOS)

MOS is a subjective measurement of coder performance. It involves an ensemble of subjects (person). Each of the subjects classifies a stimulus (coder outputs are compared) on an N-point quality factor. An example of five-point adjectival scale for signal quality or signal impairment is shown in Table 9.2.

MOS is frequently employed in subjective evaluation of speech, audio, image and video coding. In particular, it was found that MOS is more relevant to the quality of the speech codecs than to SNR. A codec with sufficiently high MOS but low SNR can still produce speech with good intelligibility for communication purposes (communications quality). Interested readers are referred to [129] on other subjective measurement of coder performance.

9.2.3 Signal delay and implementation complexity

From Figure 9.1, we can see that the input signal $x(n)$ has to go through a number of system components before it can be reconstructed as $\hat{x}(n)$. A certain time delay is thus unavoidably experienced in the receiving end. A long time delay can adversely affect the users' experience in interactive applications such as voice- or video- phones. The total time delay consists of two major components: coding and transmission delays.

Coding (or algorithmic) delay is the time delay associated with the processing of the signal at the encoder and decoder. It arises from the inherent delay of the coding algorithm and the processing of the signal by physical hardware. For example, in MPEG-2 video coding, if successive reference frames are 3 pictures apart, then 5 pictures need to be stored in order to predict the 3 pictures in between from the reference pictures using motion estimation. The inherent delay can be reduced by proper selection of the coding algorithms and coding parameters. On the other hand, coding delay associated with the hardware/software implementation of the codec can be reduced by employing processors with higher performance.

Transmission delay is the time needed to transmit the compressed bit stream to the receiver. It depends on the communication channels such as the mobile channels, phone lines, packet networks, etc. These characteristics are usually fixed and they more or less define the allowable coding delay of the system.

The implementation complexity of a codec is closely related to the data rate of the information to be compressed and the complexity of the compression algorithm. For example, more operations, storage, and internal data bandwidth will be needed to compress a (320×240) video at 30 frames/sec than a (176×144) video at 15 frames/sec, if the same compression algorithm is used. Obviously, different encoding algorithms have its own characteristics and hence different computational complexity (addition, multiplication and other operations), storage, and communications bandwidth.

9.2.4 Scalability and error resilience

Scalability of a codec refers to its ability to produce a data stream, which can be successively decoded to reconstruct the original images or videos with increasingly better quality. Scalability also simplifies progressive viewing and transmission of multimedia objects. When browsing a multimedia database for an object such as an image, it is very convenient to make available to the users a low-resolution rendition of the required objects in order to reduce the response time of the system. Similarly, in transmitting multimedia information over networks with high or variable bit error rate, such as a wireless network or a packet network, more important information, usually called the base layer, can be transmitted in higher priority packets or protected with channel codes having higher immunity to transmission errors (Error resilience) at the expense of a larger bandwidth. On the other hand, additional information in an enhancement layer can be transmitted with less priority.

Depending on the nature of the information to be encoded and the application requirements, different methods are available to support scalability. In simple applications, scalable codecs having a coarse granularity of a few layers are sufficient. For more sophisticated applications, fine granularity scalability is required. In embedded coders, say JPEG 2000, the bit stream can be truncated at any point. The more bits that are received, the better the quality of the reconstructed images or videos will be.

For image and video coding, SNR and spatial scalability are commonly used. In SNR scalability, a lower quality of the image or video is used as the base layer, while the error image or video is further encoded to a higher fidelity in additional enhancement layer(s) to improve the reconstruction quality. In spatial scalability, the original image is first decimated to a lower resolution image and it is first encoded as the base layer. The decoded low-resolution image is then interpolated to obtain an approximation of the original. Similar to SNR scalability, the coding error is further encoded in the enhancement layer(s) to a higher fidelity to improve the reconstruction quality.

In addition to SNR and spatial scalability, an additional freedom based on temporal scalability is available in video coding. The basic idea is to form the base layer by encoding the video at a lower frame rate. The in-between video frames form the enhancement layer, and these frames are encoded using bi-directional prediction from the base layer. By employing different coding and scalability techniques to form and encode the base- and enhancement layers, there are many variations of these basic schemes. Another scalability technique, which is associated with content based coding, is called object scalability. Basically, the video is described as a set of video objects (VO) and video object layers. In Chapter 12, we shall describe an object-based coding scheme for compressing simplified light fields, which offers many such desirable properties.

9.2.5 Redundancy and random access

As mentioned earlier, adjacent image pixels in traditional images and videos are usually highly correlated and can be explored to achieve data compression. As image-intensive representations are usually densely sampled higher dimensional signals, adjacent image pixels of these representations are also highly correlated. Because of their high dimensional nature, the redundancy in different dimensions can be explored to obtain a more compact representation. In traditional videos, random access at the group of picture level is usually provided, say in MPEG-2 videos, to support VCR functionalities such as fast forward, backward, and playback at selected chapters.

On the other hand, higher dimensional IBR representations such as 3D Concentric Mosaics (CMs) require random access at the line level, whereas the 4D light field and Lumigraph require random access at the pixel level. As most existing compression algorithms employ entropy coding (such as Huffman or arithmetic coding) for better compression ratio, the symbols after compression are of variable sizes. It is, therefore, very time-consuming to retrieve and decode a single line or pixel from the compressed data if there is no such provision for random access. Providing random access to the compressed data for real-time rendering is thus an important and unique problem of IBR compression. More details are given in Chapters 12 and 13.

9.3 Quantization techniques

Scalar quantization is the process of transforming a given value $x \in \Re$ into a finite set of possible output values $C = \{y_1, y_2, ..., y_L\} \subset \Re$. Analogue-to-digital (A/D) conversion of a continuous-time signal to its binary representation is an example of scalar quantization. When compressing digital signals, we are interested in representing an n-bit signal sample using fewer number of bits, say n_c. This is performed by a quantizer and it will introduce a distortion of the original value. Therefore, unlike entropy coding, quantization itself is a lossy process. Since different mappings or quantizers will result in different amount of distortion, we are interested in finding a right quantizer, say with sufficient number of bits R, such that the distortion is small enough for a desired application. It is also possible to quantize a group or vector of signal values. Such a quantizer is referred to as a vector quantizer and the process is called vector quantization (VQ).

In what follows, we shall briefly review the basic definition and concept of scalar quantization. Then, the concept of vector quantization will be introduced.

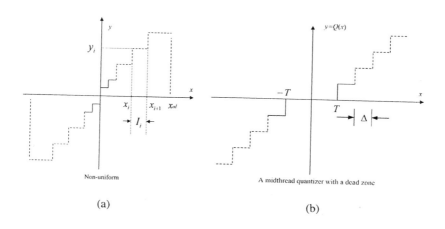

Fig. 9.2. Examples of quantizers. (a) Non-uniform quantizer, (b) uniform quantizer with a deadzone $[-T,T]$ around $x = 0$.

9.3.1 Scalar quantization

An L-level scalar quantizer Q is a mapping,

$$Q : \Re \to C, \tag{9.1}$$

which maps the real line $\Re$ to a finite set of *reconstruction values* or *representation levels*, y_k's, called the *codebook*:

$$C = \{y_1, y_2, ..., y_L\} \subset \Re. \tag{9.2}$$

In effect, the quantizer partitions the real line $\Re$ into L non-overlapping cells or partitions, I_k, $k=1, ..., L$, and maps those values of x inside I_k to the corresponding reconstruction value y_k. This is illustrated in Figure 9.2(a). More formally, we have

$$I_k = \{x \in (x_k, x_{k+1}] : Q(x) = y_k\} \qquad k = 1, 2, ..., L. \qquad (9.3)$$

x_k's are called the *decision levels*. When the signal lies in the bounded cells of the quantizer, the quantization error is usually small and it is referred to as the *granular noise*. On the other hand, for input signal that lies in the unbounded cells (with x larger than x_{ol}) of the quantizer, large distortion called *overload distortion* is experienced.

In uniform quantizers, the decision levels are regularly spaced. It is therefore very simple to determine the index k of the corresponding reconstruction level for a given input (by dividing the input x by Δ). As an example, consider the uniform midtread quantizer in Figure 9.2(b) with a dead zone [-T,T] around x=0. This quantizer is employed in the H.261 video coding standard for quantizing the DCT coefficients (AC coefficients). The purpose of including the dead zone is to avoid coding many small DCT coefficients which contributes mainly to quantization noise. The stepsize of the quantizer Δ is an even number. The index k can simply be computed as

$$k = 0, \text{if} |x| \leq T, \text{otherwise } k = round\,(x/\Delta)\,.$$

In non-uniform quantizers, x_k and y_k are not necessarily uniformly spaced, and the encoding is slightly complicated. Companding is another commonly used method to obtain non-uniform quantizers and it is useful to encode the depth maps of videos [148], because of the large dynamic range of the depth values. Also, the rendering process is usually less sensitive to errors when depth value is large, hence a larger stepsize can be used. The coding of depth map for an object-based coding algorithm for simplified light fields will be described in Chapter 13. In vector quantizer, a vector of input samples are quantized and these will be discussed later in Section 9.3.2.

Assume for simplicity that L is a power of two, i.e. $L = 2^R$, for some positive integer R. To represent the L different reconstruction levels, one can express the value k, called the *index*, as an R-bit binary number. Therefore, each input sample of x can be represented as

$$R = \log_2 L \quad \text{bits/sample}.$$

If x is sampled at f Hz (i.e. every T=1/f second), and each sample is quantized to R bits/sample, then the transmission rate required is fR bits/sec. This may be further reduced by using entropy coding. If a binary representation of k is received at the receiver, then the reconstruction level y_k will be used as an approximation of the sample x. In general, a table is needed to store all the reconstruction values y_k (except for uniform quantizers and lattice vector quantizers, where the reconstruction and decision levels can be computed quite easily). Thus, the de-quantization process amounts to a simple *table lookup operation*.

Distortion measure

As mentioned earlier, quantization is a lossy process and it introduces distortion. The quantization error of a given quantizer is the difference between its input x and its output $Q(x)$:

$$e(x) = x - Q(x). \tag{9.4}$$

Since x can assume many different values, it is usually very difficult to tell what are the optimal values for the decision and reconstruction levels of the quantizer. In order to evaluate the performance of a quantizer, x can be assumed as a random variable so that its average performance can be evaluated. The *mean squared error* (MSE) is commonly used because of its analytical simplicity:

$$\text{MSE} = E[e^2(x)],$$

where $E[\cdot]$ denotes mathematical expectation over x. Alternatively, we can collect many samples of the input x to determine the suitable parameters for the quantizer. This process is called quantizer design and more detail can be found in [82].

9.3.2 Vector quantization (VQ)

In scalar quantization, a quantizer tries to represent a range of possible input values to a small set of reconstruction levels in order to obtain a more compact representation. In many applications, successive input samples are obtained from real world signals such as images, etc, which exhibit considerable correlation. That is, adjacent samples are related or similar to each other. In these cases, it is more efficient to encode them together as a vector. This is illustrated in Figure 9.3, where a set of two-dimensional vectors are represented by five representative vectors, $\mathbf{y}_0, \mathbf{y}_1, ..., \mathbf{y}_4$. Another simple example is the generation of color palettes for displaying color images with a limited number of colors. Suppose that we are given a color image in RGB format with eight bits per color component, i.e. 24-bits/pixel. Therefore, there are altogether 65536 different colors, each corresponds to certain values of the R, G, and B components. This generates naturally a vector of three dimension in (R,G,B). The problem of designing a color palette of size L is to choose L representative colors, i.e L (R,G,B) values, so that each pixel in the original image will be mapped and displayed using one of these L colors which minimizes a certain distortion measure. In the context of quantization, the L representative colors are the reconstruction vectors and they collectively generate a codebook of L vectors. The mapping of each (R,G,B) value can be viewed as quantizing this vector to one of the reconstruction or reproduction vectors. Figure 9.4 shows an 16-color palette for the color image "Lenna" and the quantized image.

More formally, an L-level vector quantizer Q is a mapping,

$$Q : \Re^K \to C, \tag{9.5}$$

which maps $\Re^K$ to a finite set of *reconstruction vectors*, $\mathbf{y}_k$'s of K-dimension, called the *codebook*:

$$C = \{\mathbf{y}_1, \mathbf{y}_2, ..., \mathbf{y}_L\} \subset \Re^K. \tag{9.6}$$

The codebook is usually generated using the *Linde-Buzo-Gray* (LBG) algorithm or *Generalized Lloyd algorithm* (GLA) [169, 82].

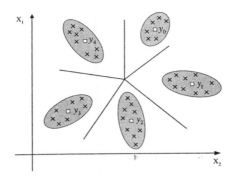

Fig. 9.3. Vector Quantization.

(a)   (b)

Fig. 9.4. Vector quantization. (a) A 16-colors palette for image "Lenna" (only grey scale values are displayed) and (b) the image "Lenna" quantized by the palette.

9.3.3 Image vector quantization

In image vector quantization, the image to be encoded is decomposed into $(N_1 \times N_2)$ nonoverlapping blocks of image pixels as shown in Figure 9.5. For color images, each pixel is represented by a vector in a certain color coordinate system such as the (R,G,B) system. By packing these components together, we obtain an K-dimensional image vector **X**. For the (R,G,B) system, $K = 3 \cdot (N_1 \times N_2)$.

For each vector, **X**, the codebook is searched for the code-vector $\mathbf{y}_k$ that *minimizes* a certain *distortion measure* $d(\mathbf{X}, \mathbf{y})$:

$$k = \arg \min_{\mathbf{y}_i \in C} d(\mathbf{X}, \mathbf{y}_i).$$

Fig. 2.14. Graph-based deghosting. ©2001 IEEE.

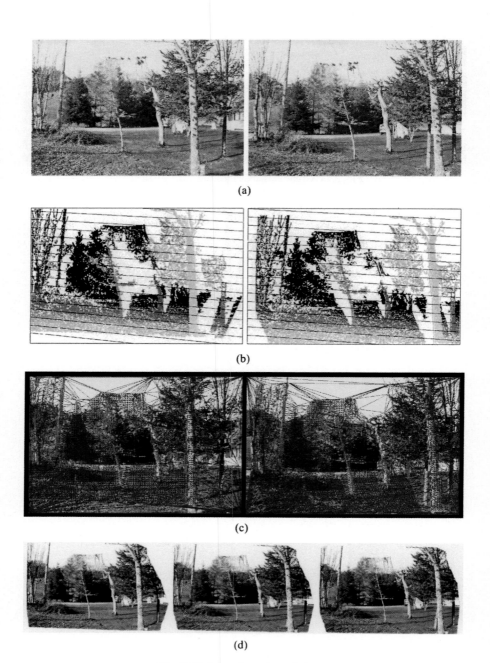

Fig. 2.19. Joint view triangulation.

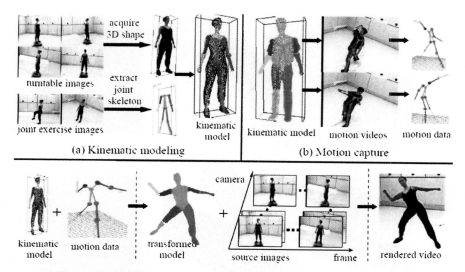

(a) Kinematic modeling   (b) Motion capture

(c) Image-based rendering (render subject #1 performing subject #2's motion)

Fig. 3.16. Overview of CMU modeling and rendering human movement, which includes movement transfer. Image courtesy of Simon Baker and German Cheung. ©2004 IEEE.

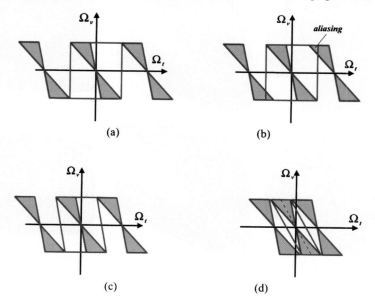

Fig. 5.4. Three reconstruction filters with different constant depths.

(a1) Scene image	(a2) EPI	(a3) Fourier transform of EPI
(b1) Scene image	(b2) EPI	(b3) Fourier transform of EPI
(c1) Scene image	(c2) EPI	(c3) Fourier transform of EPI
(d1) Scene image	(d2) EPI	(d3) Fourier transform of EPI

Fig. 5.6. Spectral support of a 2D light field.

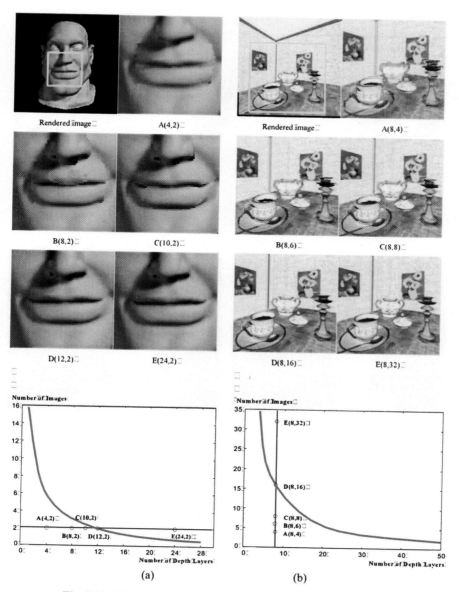

Rendered image   A(4,2)

B(8,2)   C(10,2)

D(12,2)   E(24,2)

Rendered image   A(8,4)

B(8,6)   C(8,8)

D(8,16)   E(8,32)

(a)   (b)

Fig. 5.13. Minimum sampling points in joint image-geometry space.

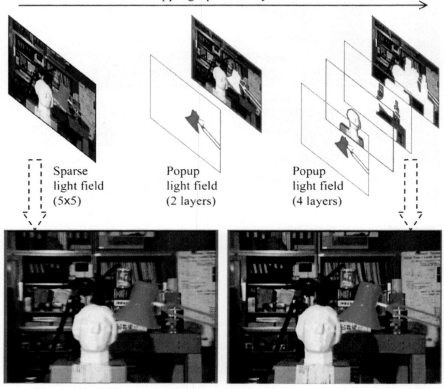

Popping up more layers

Sparse
light field
(5x5)

Popup
light field
(2 layers)

Popup
light field
(4 layers)

Rendering with sparse light field   Rendering with 4-layer popup light field

Fig. 16.1. An example of rendering with pop-up light fields.

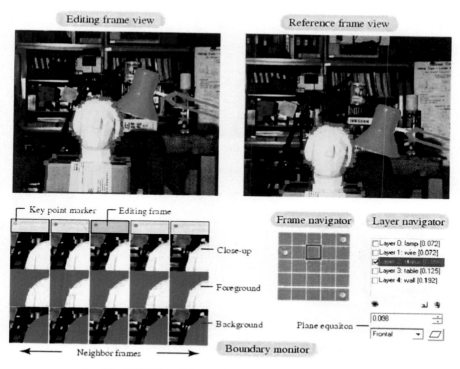

Fig. 16.7. The UI for Pop-up light field construction.

one focal plane in the front   one focal plane at the back   5 layers are popped up

Fig. 16.12. Results for *Pokemon* light field.

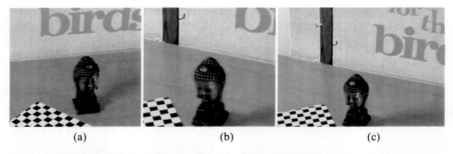

Fig. 16.13. Results for sparse images taken with unstructured camera positions.

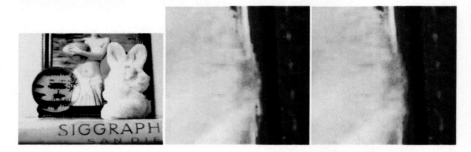

Fig. 16.14. Comparison of results for *furry rabbit*.

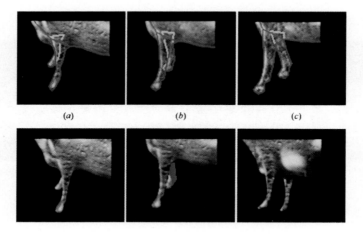

Fig. 17.5. Issues in light field warping.

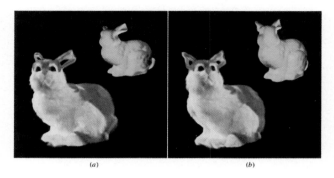

(a)　　　　　　　　　　　　　(b)

Fig. 17.10. Plenoptic texture transfer (furry cat toy to Stanford bunny).

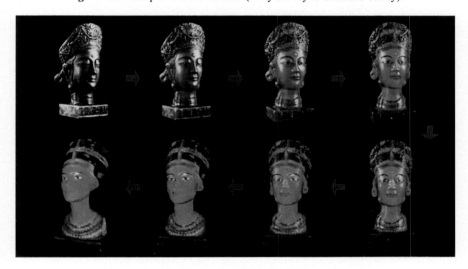

Fig. 17.11. Morphing example (antique bronze statue).

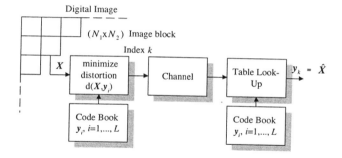

Fig. 9.5. Image Vector Quantization Block Diagram.

The index k (instead of $\mathbf{y}_k$ itself) is transmitted to the receiver, which also has a copy of the codebook C. The codebook table is then looked up to retrieve $\mathbf{y}_k = \hat{\mathbf{X}}$, the quantized value of $\mathbf{X}$. Commonly used distortion measures include the MSE and the mean absolute error (MAE) (or sometimes called mean absolute difference (MAD))

$$\textbf{MSE: } d(\mathbf{X}, \mathbf{y}_i) = \sum_{j=1}^{K} (x_j - y_j^{(i)})^2 \tag{9.7}$$

$$\textbf{MAE: } d(\mathbf{X}, \mathbf{y}_i) = \sum_{j=1}^{K} |x_j - y_j^{(i)}| \tag{9.8}$$

where K is the dimension of the vector, and $y_j^{(i)}$ is the j-th component of the i-th code vector $\mathbf{y}_i$.

In the most primitive form of image VQ, the codebook is usually generated from a training set of image vectors that are representatives of the images to be encoded. The Linde-Buzo-Gray (LBG) algorithm or Generalized Lloyd algorithm (GLA) is frequently used. Since the complexity of the encoder is much higher than that of the decoder, image VQ is said to have an asymmetric complexity. In fact, the direct computation of (9.7) or (9.8) has a complexity of:

MSE: $L \cdot (2K-1)$ addition, $L \cdot K$ multiplication, and $(L-1)$ compare operations
$$\tag{9.9a}$$

MAE: $L \cdot (2K - 1)$ addition, $L \cdot K$ absolute, and $(L - 1)$ compare operations
$$\tag{9.9b}$$

A simple trick to reduce this computation is to stop computing (9.7) or (9.8), whenever the partial sum up to an index j is greater than the minimum distortion that has been obtained so far. Another more effective solution is to employ VQ with certain attractive structures. Examples are multistage-VQ, tree-structure VQ, lattice-VQ, mean-shape-VQ, etc. More details and other aspects of VQ can be found in [82].

Although the arithmetic complexity of the decoder is very low, it also requires the storage of the L codebook vectors, which has a storage complexity of LK. For high-rate applications where L is very large, the required storage can be quite significant. The VQ structures mentioned above can also be used to alleviate this problem.

The simple decoder structure of image VQ makes it an attractive method for high-speed application such as real-time playback of compressed videos in an early version of Apple QuickTime movie. VQ is also used extensively in speech coding and the data compression of IBR such as light fields and Concentric Mosaics. Basically, the light field or mosaic images are compressed by image vector quantization. During rendering, the codebook is first loaded into the memory and the indices of the required pixels are used to retrieve the reconstruction vectors (image block in this case) for interpolation and display. The main advantage of VQ-based method is its fast rendering speed. The disadvantage, as mentioned earlier, is its limited compression ratio. Other techniques for the compression of light fields and Concentric Mosaics will be given in Chapter 12.

10

Image Compression Techniques

In this chapter, the basic principle of a commonly used technique for image compression called transform coding will be described. After a short summary of useful image formats, we shall describe two commonly used image coding standards, the JPEG and JPEG2000.

10.1 Image format

Real world images, such as color images, usually contain different components. For color images represented in the RGB color system, there will be three component images corresponding to the R, G, and B components. Since the RGB color component is relatively uniform in terms of quantization, they are frequently employed in color sensors with each component being quantized to 8 bits. From the trichromatic theory of color mixture, most colors can be represented by three properly chosen primary colors. The RGB color primary, which contains the red, green and blue colors, is most popular for illuminating sources. The CMY primary is very common for reflecting light sources and they are frequently employed in printing (the CMYK format).

Other than the RGB system, there are a number of color coordinate systems such as YIQ, YUV, XYZ, UVW, U*V*W*, L*a*b*, and L*v*v* [236, 127]. Since human visual system (HVS) is less sensitive to high-frequency chrominance information, the YCbCr color system is commonly used in image coding. The RGB image can be converted to the YCbCr color space using the following formula

$$\begin{bmatrix} Y \\ Cb \\ Cr \end{bmatrix} = \begin{bmatrix} 0.299 & 0.587 & 0.114 \\ -0.169 & -0.331 & 0.500 \\ 0.500 & -0.419 & -0.081 \end{bmatrix} \cdot \begin{bmatrix} R \\ G \\ B \end{bmatrix}. \qquad (10.1)$$

They are related to the YUV color system of the PAL and NTSC systems (see Section 11.1 for more details). The resolution of the chrominance components (Cb and Cr) is usually reduced by a factor of 2 horizontally and vertically to yield the

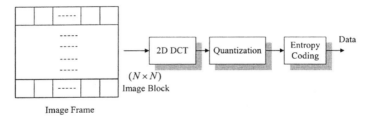

Fig. 10.1. Transform Image Coding.

YCrCb-(4:2:0) format in order to obtain a smaller data size (Figure 11.2). Hence, the Cb or the Cr components are respectively one-quarter the size of the Y component.

10.2 Transform coding of images

The concept of transform coding of images is illustrated in Figure 10.1. For simplicity, we will consider grey scale images first. For color images, the original image is usually converted to the YCrCb-(4:2:0) format and the same technique for using the Y component image is applied to the Cr and Cb component images. The image to be encoded is first divided into $(N \times N)$ non-overlapping blocks, and each block is transformed by a 2D transformation such as the 2D discrete cosine transform (DCT). The basic idea of transform coding [319] is to pack most of the energy of the image block into a few transform coefficients. This process is usually called *energy compaction*. The transform coefficients are then adaptively quantized. The quantized coefficients and other auxiliary information will be entropy coded and packed according to a certain format into a bit-stream for transmission or storage. At the decoder, the bit-stream is decoded to recover the various information. The quantized coefficients are then inverse-transformed to reconstruct an approximation of the original image.

Since the amplitudes of the transform coefficients usually differ considerably from each other, it is advantageous to use a different number of quantizer levels (i.e., bits) for each transform coefficients. This problem is called the *bit allocation problem*. The bit allocation amongst the transform coefficients can also vary over time to improve the adaptability of the coder to changing input characteristics. The bit allocation problem is discussed in [129, 266, 82]. It can be shown that the optimal transformation is called the Karhunen Loéve transform (*KLT*) [129]. Unfortunately, the *KLT*, which is constructed from the eigenvectors of the input covariance matrix, is signal dependent. Due to its high computational complexity, sub-optimal transformation with fixed coefficients such as the DCT is frequently employed. Another reason for using DCT is its good coding performance and availability of fast implementation algorithms [242]. For implementation simplicity, a *separable* transformation is frequently used. For example, the 2D separable DCT and inverse DCT (IDCT) can be written respectively as

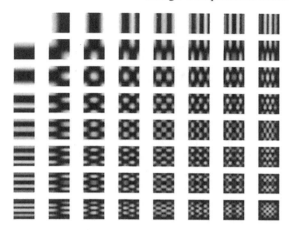

Fig. 10.2. Basis functions of (8×8) DCT.

$$X(k_1, k_2) = \frac{c(k_1)c(k_2)}{4} \sum_{n_1=0}^{N-1} \sum_{n_2=0}^{N-1} x(n_1, n_2) \cos\left(\frac{\pi(2n_1 + 1)k_1}{2N}\right) \cos\left(\frac{\pi(2n_2 + 1)k_2}{2N}\right)$$

(10.2)

where $k_1, k_2 = 0, ..., N - 1$ and $c(k) = 1/\sqrt{2}$ for $k = 0$, and 1 otherwise.

$$x(n_1, n_2) = \sum_{k_1=0}^{N-1} \sum_{k_2=0}^{N-1} \frac{c(k_1)c(k_2)}{4} X(k_1, k_2) \cos\left(\frac{\pi(2n_1 + 1)k_1}{2N}\right) \cos\left(\frac{\pi(2n_2 + 1)k_2}{2N}\right)$$

(10.3)

where $x(n_1, n_2)$ and $X(k_1, k_2)$, $n_1, n_2 = 0, ..., N - 1$, are respectively the input and transformed image blocks. For image and video coding, N is usually chosen as 8 to reduce blocking artifacts.

The above transformation can be performed by applying 1D DCT along the rows of the 2D data block, followed by 1D DCT along the columns. The order is immaterial, which means that the result is the same if column transformation is carried out first. Figure 10.2 shows the basis functions of the 2D DCT. It can be seen that they resemble the sinusoidal signals and the frequencies increase as k_1 (horizontal normalized frequency) and k_2 (vertical normalized frequency) increase. $k_1 = k_2 = 0$ corresponds to the DC component (top left corner in Figure 10.2), while the others are AC components.

As mentioned earlier, 2D-DCT is commonly used because of its good performance and availability of fast algorithms. It is adopted in a number of international coding standards including JPEG image coding standard, MPEG-1, MPEG-2, MPEG-4, H.261, and H.263 video coding algorithms. More recent standards such as H.264 uses integer approximation to DCT as well as variable image block size. Since image-based representations are usually a collection of images and videos, these techniques also form the basis for IBR compression.

Quantization

There are a number of methods to encode the transform coefficients. For example, a popular method is to employ scalar quantization followed by run-length and entropy coding. Alternatively, VQ or embedded zero-tree coding can be applied [331]. For simplicity, we only describe the first approach, which is employed in the JPEG-Baseline coding. Similar methods are also employed in other video coding standards. Consider an (8×8) luminance (Y) block, x, as shown in Figure 10.3. Most coding standards require the image pixels be preprocessed to have a mean of zero. For RGB color space, all color components have a mean value of 128 (8-bit /pixel). In YCbCr color space, the Y component has an average value of 128, while the chrominance components have an average value of zero. Therefore, we subtract 128 from each element of x before taking the 2D DCT. This gives X in Figure 10.3.

The $(0,0)$ entry of X (top left corner), $X(0,0)$, is the DC value of the block x. It can be seen that most of the signal energy is now concentrated in the low frequency transform coefficients, while most of the high frequency components are of small amplitudes. Significant compression can therefore be obtained by quantizing the elements of X. In the JPEG standard, uniform quantizers are employed and the stepsizes are specified in form of a matrix, called the quantization matrix. An example quantization matrix is:

$$Q = \begin{bmatrix} 16 & 11 & 10 & 16 & 24 & 40 & 51 & 61 \\ 12 & 12 & 14 & 19 & 26 & 58 & 60 & 55 \\ 14 & 13 & 16 & 24 & 40 & 57 & 69 & 56 \\ 14 & 17 & 22 & 29 & 51 & 87 & 80 & 62 \\ 18 & 22 & 37 & 56 & 68 & 109 & 103 & 77 \\ 24 & 35 & 55 & 64 & 81 & 104 & 113 & 92 \\ 49 & 64 & 78 & 87 & 103 & 121 & 120 & 101 \\ 72 & 92 & 95 & 98 & 112 & 100 & 103 & 99 \end{bmatrix}. \qquad (10.4)$$

The (k_1, k_2) entry of Q, $q(k_1, k_2)$, is the quantizer stepsize for the (k_1, k_2) transform coefficients (i.e., the (k_1, k_2) entry of X). In other words, the quantized (k_1, k_2) transform coefficients is

$$z(k_1, k_2) = round\left(\frac{X(k_1,k_2)}{q(k_1,k_2)}\right) = \left\lfloor \frac{X(k_1,k_2) \pm \lfloor q(k_1,k_2)/2 \rfloor}{q(k_1,k_2)} \right\rfloor,$$
$$k_1, k_2 = 0, 1, ..., 7. \qquad (10.5)$$

Note, the quantization stepsizes for high frequency components are much larger than those for low frequencies. This is because the human visual system is less sensitive to high frequency components (also called frequency sensitivity). Therefore, more quantization errors can be tolerated in these less sensitive components to yield a better compression ratio. For the transformed block X described above, the quantized DCT coefficients, Z, are shown in Figure 10.3, where the (k_1, k_2)-entry of Z is $z(k_1, k_2)$. The design of Q is usually based on psychovisual characteristics and the compression ratio required. In fact, each element of matrix Q in (10.4) is the visibility threshold of the corresponding transform coefficient, below which its spatial waveform on a plain background is usually not noticeable to human eyes [173].

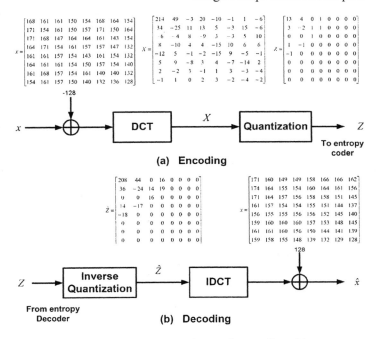

Fig. 10.3. Scalar quantization in transform coding of images.

To achieve a higher compression ratio, a quality factor q, called the q-factor, can be multiplied with Q to form a new quantization matrix with larger stepsizes.

Since the non-zero quantized coefficients are sparsely distributed, it is more efficient to represent them using *run-length coding*. The run-length code and the quantized coefficients can then be entropy coded using *Huffman* or *arithmetic coding*. This will be explained in more details when we describe the JPEG standard in the Section 10.3.

To recover the quantized coefficients, we can multiply $z(k_1, k_2)$ with the corresponding stepsize $q(k_1, k_2)$. This is called inverse quantization, which can be written as

$$\hat{z}(k_1, k_2) = z(k_1, k_2)q(k_1, k_2), \quad k_1, k_2 = 0, 1, ..., 7. \quad (10.6)$$

Taking 2D IDCT of $\hat{z}(k_1, k_2)$ and adding 128 to each element gives the reconstructed image block $\hat{x}(k_1, k_2)$. This is illustrated in Figure 10.3. Figure 10.4 shows the compression results of the JPEG Baseline algorithm which is based on DCT-based transform coding. At high to medium bit rates, the quality of the reconstructed images is very good visually. At low bit rates, significant coding artifact called "block effect" starts to appear as shown in Figure 10.4(d). This is mainly caused by the independent coding of the image blocks. At low bit rate, the coding errors will show up as discontinuities along the block boundaries. We shall focus on two commonly used image coding standards namely JPEG and JPEG2000 in this section.

(a)

(b)

(c)

(d)

(e)

(f)

Fig. 10.4. The compression results of JPEG algorithms. (a) original image "Lenna," (b) JPEG Baseline compressed to 0.93 bpp, (c) JPEG Baseline compressed to 0.38 bpp, (d) JPEG Baseline compressed to 0.23 bpp, (e) JPEG-2000 compressed to 0.38 bpp and (f) JPEG-2000 compressed to 0.23 bpp.

10.3 JPEG standard

The JPEG (Joint Photographic Experts Group) standard is an ISO/IEC international standard (10918-1) for Digital compression and coding of continuous-tone still images. It is also an ITU standard known as ITU-T Recommendation T.81. To satisfy different requirements in practical applications, the standard defines four modes of operation:

- **Sequential DCT-based**: This mode is based on DCT-based transform coding with a block size of (8×8) for each color component. The transform coefficients are runlength and entropy coded. A subset of this mode is the Baseline Mode, which is an implementation with a minimum set of requirements for a JPEG compliant decoder.
- **Progressive DCT-based**: This mode is similar to the sequential DCT-based algorithm, except that the quantized coefficients are transmitted in multiple scans. By partially decoding the transmitted data, this mode allows a rough preview of the transmitted image to be obtained at the decoder having a low transmission bandwidth.
- **Lossless**: This mode is intended for lossless coding of digital images. It uses a prediction approach, where the input image pixel is predicted from adjacent encoded pixels. The prediction residual is then entropy-coded.
- **Hierarchical**: This mode provides spatial scalability and encodes the input image into a sequence of increasing resolution. The lowest resolution image can be encoded using either the lossy or lossless techniques in other mode, while the residuals are coded using the lossy or DCT-based modes.

JPEG supports multiple component images. For color images, the input image is usually in RGB and other formats like luminance and chrominance representation (YUV, YCbCr), etc. The color space conversion process is not part of the standard, but most codecs employ the YCbCr system because the chrominance components can be decimated by a factor of two in the horizontal and vertical dimensions to achieve a better compression performance.

Either Huffman or arithmetic coding techniques can be used in the JPEG modes (except the Baseline mode, where Huffman coding is mandatory) for entropy coding. The arithmetic coding techniques usually perform better than the Huffman coding in JPEG, while the latter is simpler to implement. For Huffman coding, up to 4 AC and 2 DC tables can be specified. The input image to JPEG may have from 1 to 65,535 lines and from 1 to 65,535 pixels per line. Each pixel may have from 1 to 255 color components except for progressive mode, where at most four components are allowed. For the DCT modes, each component pixel is an 8 or 12 bits unsigned integer, except for the Baseline mode, where 8-bit precision is allowed. For the lossless mode, a range from 2 to 16 bits is supported. A brief summary of the lossless and sequential DCT-based coding modes will be summarized below. Interested readers are referred to [118] and [228] for more details.

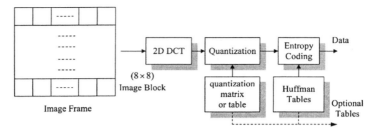

Fig. 10.5. Block diagram of a JPEG baseline encoder.

10.3.1 Lossless mode

In the lossless mode, pixels are coded in raster scan, from left to right and from top to bottom. The image pixel to be encoded is predicted from adjacent and previously encoded pixels to form a prediction residual, which is then Huffman or arithmetic coded. Consider the encoding of pixel x in the following template.

	c	b
	a	x

It can be predicted from pixels a, b, and c using one of the following predictors y:

$$y = 0, \; y = a, \; y = b, \; y = c,$$

$$y = a + b + c, \; y = a + \frac{b - c}{2}, \; y = b + \frac{a - c}{2}, \; y = \frac{a + b}{2}.$$

The prediction residual is $r = y - x$ and the predictor used is specified in the header of the data stream. The prediction residuals are encoded by either a Huffman or an arithmetic coder. The Huffman code used in lossless JPEG is a modified Huffman code, which divides the prediction residuals into categories with increasing amplitude. It consists of a pair of symbols: category and the magnitude (similar to Table 10.1). The category is Huffman coded, while the magnitude is binary coded. The compression ratio is typically around 2:1, which is much smaller than lossy modes, such as the sequential-DCT mode. The lossless mode of JPEG or JPEG2000 [123] is useful to the lossless compression of depth maps in image-based representations.

10.3.2 Sequential DCT-based coding

Since the Sequential DCT-based coding differs from the baseline mode mainly in the resolution of the image pixels, the maximum number of color components, and the number of user supplied coding tables, we focus on the baseline mode which is required in every JPEG compliant decoder.

Baseline sequential DCT

Figure 10.5 shows the block diagram of a JPEG baseline encoder. As mentioned earlier, the Baseline mode is a subset of the Sequential DCT mode. It operates on 8-bit pixels, 8-bit quantizer precision, one to four color components, and uses Huffman coding with up to 2 AC and 2 DC Huffman tables. The idea of imposing this basic requirement is to ensure interoperability between codecs from different manufacturers.

The input image is first divided into (8×8) non-overlapping blocks of component samples. The image blocks are scanned in a raster scan order, i.e., from left to right and from top to bottom. After subtracting the mean value, each block is transformed by the 2D DCT. The transform coefficients are then quantized according to a quantization table. The encoder can transmit up to 4 quantization tables with 8-bit resolution (for sequential-DCT mode, up to 8 users supplied Huffman tables can be defined). JPEG also includes a table which yields good results for CCIR-601 type images. Most applications use this table as a basis for quantizing the transform coefficients. For example, a common approach to increase the compression ratio is to multiply a q-factor to this matrix to obtain a quantization matrix with larger quantization stepsizes. The chrominance quantization table is given by (10.7).

$$
\mathbf{Q}_c = \begin{bmatrix}
16 & 18 & 24 & 47 & 99 & 99 & 99 & 99 \\
18 & 21 & 26 & 66 & 99 & 99 & 99 & 99 \\
24 & 26 & 56 & 99 & 99 & 99 & 99 & 99 \\
47 & 66 & 99 & 99 & 99 & 99 & 99 & 99 \\
99 & 99 & 99 & 99 & 99 & 99 & 99 & 99 \\
99 & 99 & 99 & 99 & 99 & 99 & 99 & 99 \\
99 & 99 & 99 & 99 & 99 & 99 & 99 & 99 \\
99 & 99 & 99 & 99 & 99 & 99 & 99 & 99
\end{bmatrix}
\tag{10.7}
$$

Table 10.1. Example JPEG Huffman code table for luminance DC difference.

Category	Code Length	Code Word	DC Coeff. Difference
0	2	00	0
1	3	010	-1,1
2	3	011	-3,-2,2,3
3	3	100	-7,$\cdots$,-4,4,$\cdots$,7
4	3	110	-15,$\cdots$,-8,8,$\cdots$,15
5	3	1110	-31,$\cdots$,-16,16,$\cdots$,31
6	4	11110	-63,$\cdots$,-32,32,$\cdots$,63
7	5	111110	-127,$\cdots$,-64,64,$\cdots$,127
8	6	1111110	-255,$\cdots$,-128,128,$\cdots$,255
9	7	11111110	-511,$\cdots$,-256,256,$\cdots$,511
10	8	111111110	-1023,$\cdots$,-512,512,$\cdots$,1023
11	9	1111111110	-2047,$\cdots$,-1024,1024,$\cdots$,2047

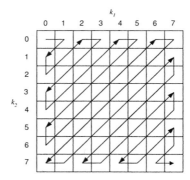

Fig. 10.6. Zig-zag scanning order for coding AC coefficients in JPEG.

Coding of DC coefficients

Since the DC coefficient corresponds to the average intensity of a block, DC coefficients of adjacent blocks are likely to have similar values. Therefore, differential coding of DC coefficients is employed and the residuals or differentials are coded using modified Huffman code as shown in Table 10.1(only the luminance component table is shown). For 8-bit-per component pixel, the range of the DC differentials is $[-2047, 2047]$, and it is divided into 12 categories. As an example, suppose that the DC differential is 195, then, from Table 10.1, it belongs to category 8. The binary part (also known as variable-length integer VLI) is 11000011 and the Huffman code of category 8 (variable-length code VLC) is 111110. The final result is the concatenation of VLC-VLI, i.e., 111110 11000011.

Coding of AC coefficients

For 8 bits per component pixel, the range of the AC coefficients is $[-1023, 1023]$. Similarly, a modified Huffman code is used and there are 10 categories and each coefficient is represented as (category, amplitude). Since AC coefficients are mostly concentrated at low frequencies and most of them are zeros, they are run-length coded following a zig-zag scanning order as shown in Figure 10.6. Each nonzero coefficient in the scan can be represented as (*run, category, amplitude*), where *run* is the number of zero AC coefficients preceding this nonzero coefficient, *category* and *amplitude* specify the quantized AC coefficient. To further improve the coding efficiency, the (*run, category*) is jointly *Huffman coded*. An example Huffman table is shown in Table 10.2. Two additional symbols are included in the Huffman code: the End-of-Block (EOB) symbol and zero-run-length (ZRL) symbol. EOB indicates that all remaining coefficients in the scan are zero. It is unnecessary to send the code if the last coefficient of the zigzag scan (i.e $k_1 = 7, k_2 = 7$) is nonzero. To avoid a lengthy Huffman table, the (*run, category*) Huffman table only specifies a run-length up to 15. For a run-length greater than 15, a ZRL symbol is sent to indicate a run of 16 zeros. As an example, suppose that there are six zeros preceding the AC coefficient

Table 10.2. Example JPEG Huffman code table for luminance AC coefficients.

Zero Run/ Category	Code Length	Code Word	Zero Run/ Category	Code Length	Code Word
0/0 (EOB)	4	1010	2/1	5	11011
0/1	2	00	2/2	8	11111000
0/2	2	01	.	.	.
0/3	3	100	2/A	16	1111111110001110
0/4	4	1011	3/1	6	111010
0/5	5	11010	3/2	9	111110111
.	.	.	.	.	.
.	.	.	3/A	.	.
0/A	16	1111111110000011	.	.	.
1/1	4	1100	.	.	.
1/2	6	111001	.	.	.
1/3	7	1111001	E/A	16	1111111111110110
1/4	9	111110110	F/0 (ZRL)	11	11111111001
.	.	.	.	.	.
.	.	.	.	.	.
1/A	16	1111111110000011	F/A	16	1111111111111110

and the value of the nonzero coefficient is -18. 18 belongs to category 5 and its VLI value is 0110 1, the run/category code for 6/5 is 1101. The final result is 1101 01101.

10.4 The JPEG-2000 standard

The JPEG-2000 standard [123] defines a set of lossless (bit-preserving) and lossy compression methods for coding continuous-tone, bi-level, grey-scale, or color digital still images. It not only provides better compression of images over the JPEG standard, but also provides greater flexibility in extracting the compressed data for editing, processing and targeting particular devices and applications. Also, the compressed codestream can be arranged so that the encoded image can be reconstructed at a lower resolution or bit-rate, and at specific regions of interest (ROI). This allows the matching of a codestream to the transmission channel, storage device, or display device, regardless of the size, number of components, and sample precision of the original image. The packet and layer structures of the standard provide certain degrees of random access to the compressed bit stream. This considerably speeds up the rendering of image-based representations compressed by wavelet type coder. More details on the application of the related wavelet coding methods for compressing Concentric Mosaics can be found in [175, 328].

JPEG-2000 is divided into 12 Parts. Part-1 was published as an international standard [123] and it specifies the minimum compliant decoder, which should be used to provide maximum interchange. Part-II is optional, and it includes a valued-added extension not required in all implementations. Notable differences with Part-I in-

Table 10.3. Features in Part-I of the JPEG-2000 standard.

Bitstream	Fixed and variable length markers
Arithmetic coder	MQ-coder
Coefficient modelling	Independent coding of fixed size blocks
Quantization	Scalar quantizer with dead-zone and truncation of code-blocks
Transformation	Low complexity La Gall (5,3) integer wavelet, or high performance Daubechies (9,7) floating point wavelet.
Component transformation	Reversible component transforms (RCT), YCbCr transformation.
Error resilience	Resynchronization markers
Bit-stream ordering	Progressive transmission by tile-part, then SNR, or resolution, or component. Random (spatial) access to the bitstream. Region of interest coding.
Compressed domain processing	Examples: rotation and cropping

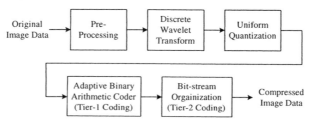

Fig. 10.7. Block diagram of JPEG-2000 encoder.

clude the use of Trellis Coded Quantization (TCQ), new markers, user defined filters, arbitrary point transform or reversible wavelet transform across components, fixed length entropy coder and repeated headers for error resilience, and more flexibility in incorporating metadata. Main features in Part-I of the standard are summarized in Table 10.3. Readers are referred to [184, 132] for information on the JPEG-2000 development process and other aspects of JPEG-2000. A comprehensive reference on JPEG-2000 is [294].

10.4.1 JPEG-2000 compression engine

Figure 10.7 shows the general block diagram of the JPEG 2000 encoder. After pre-processing for different component images, the discrete wavelet transform is applied to decorrelate the source image. The transform coefficients are then uniformly quantized and entropy coded before packing into an output bitstream for storage or transmission. In the decoder, the bitstream is first entropy decoded, dequantized, and inverse discrete wavelet transformed to construct the image data. The functional blocks will be briefly described below.

Preprocessing

JPEG-2000 supports multiple-component images (e.g., RGB), each possibly sub-sampled by a different factor. The input image is divided into tiles, which are rectangular and non-overlapping arrays of component images making up the same relative portion of the image. Each tile of a component image must be of the same size, except those around the border of the image. Tiling of the image creates tile-components that can be decoded, reconstructed and extracted independently. It also provides one of the methods for extracting a region of the image called tile.

The unsigned sample values in each component are level shifted by subtracting a fixed value from each sample to make its value symmetric around zero. When encoding multiple component images in say RGB format, a point-wise decorrelating transform may be used to decorrelate the color components for better compression. Part-I of the standard supports two component transformations: an irreversible component transformation (ICT), based on the YCbCr transform commonly used in JPEG for lossy coding, and a reversible component transformation (RCT) that may be used for both lossless or lossy coding.

Forward RCT:

$$Y = \lfloor \tfrac{1}{4}(R + 2G + B) \rfloor$$
$$Cb = B - G$$
$$Cr = R - G$$

Inverse RCT:

$$G = Y - \lfloor \tfrac{1}{4}(Cb + Cr) \rfloor$$
$$R = Cr + G$$
$$B = Cb + G$$

Discrete wavelet transform (DWT)

Given a tile, an L-level dyadic DWT is performed on each component image. The DWT involves a bank of filters, or filter banks, which decompose the image into four sub-images, or subbands, corresponding to the horizontal low frequency and vertical low frequency (LL), horizontal high frequency and vertical low frequency (HL), horizontal low frequency and vertical high frequency (LH), and horizontal high frequency and vertical high frequency (HH) components of the image. This composition can be applied successively to the LL subbands to obtain a series of subband signal at different resolutions. A two-level decomposition of the image "Lenna" is illustrated in Figure 10.8(a). The DWT can be irreversible or reversible. The default irreversible transform defined in Part-I is the Daubechies (9,7) DWT [5], while the default reversible transformation is implemented by Le Gall 5-tap/3-tap filter bank [78, 1]. Multiple wavelets including "user defined" are allowed in Part II. A lower reconstructed resolution image can be generated by decoding a selected subset of these sub-bands up to a certain level.

Quantization and partitions

After DWT, most of the coefficients are very small and they can be quantized to yield a more compact representation. This operation can be made lossless, if a quantization step of 1 is used together with the reversible integer 5/3 wavelet, because the output of the latter are integers. For lossy compression, all the wavelet coefficients will undergo uniform scalar quantization with a fixed deadzone. The quantized outputs are in sign-magnitude representation. One quantization step size, which can be selected by the user to achieve a given level of quality, is allowed to quantize all the coefficients for each subband. To achieve a fixed rate, the default behavior of the verification models (VM), which are reference implementations of the standard, is to quantize each coefficient rather finely, and makes use of subsequent truncation of embedded bitstreams to achieve the desired rate.

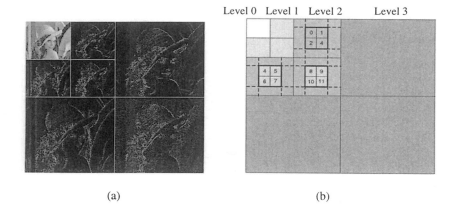

(a)   (b)

Fig. 10.8. (a) 2-level wavelet decomposition of the image "Lenna". (b) Example packet partition location and code-blocks showing twelve code-blocks of one packet partition location at resolution level 2 subband in a three level dyadic wavelet decomposition. The packet partition location is marked in dark thick lines.

To provide a finer granularity than tiles in supporting medium-grain level of spatial locality in the bitstream for efficient memory implementation, streaming, and random access to the bitstream, each subband can be further divided into regular non-overlapping rectangles called "packet partition" as shown in Figure 10.8(b). In this example, three spatially consistent rectangles, one from each subband at the same resolution level, comprise a packet partition location. Each packet partition location is further divided into regular non-overlapping rectangles called code-blocks, which is the basic unit of entropy coding. These rectangular arrays of coefficients can be extracted independently.

Block coding

Each code-block will be independently entropy coded using a context-dependent, binary, arithmetic coding of bitplanes. The MQ-coder from JBIG2 standard [122], which is an ISO Standard for coding bi-level images, is employed. A bit plane of a code-block is obtained by extracting a given binary digit of all its quantized coefficients. The first bitplane consists of the MSB of all the magnitudes, followed by the next bit plane until the least significant digit. Starting from the first bitplane which has a nonzero coefficient, each bit-plane is coded in three coding passes:

1. significant propagation pass : a location is coded if it is not significant, but at least one of its eight-connected neighbors is significant.
2. refinement pass : all significant locations detected in previous bitplanes are coded.
3. clean-up pass : take care of any bits not coded in the first two passes (say a "1" without significant neighbors).

A location is significant if a "1" has been coded for that location in the current or previous bitplanes. The coding in the first and the third passes is identical, except that run length coding will sometimes be employed in the third pass. Each of these coding passes collects contextual information about the bit-plane data, which will be used by an MQ-arithmetic coder to encode the bit-stream. Unlike JBIG and JBIG2, JPEG-2000 uses no more than nine contexts to code any given type of bit in order to speed up the probability adaptation. (For more details of context based arithmetic coding, see the binary shape coding of MPEG-4 standard in Section 11.4.5.)

10.4.2 Bit-stream organization

The collection of certain bitplanes, called sub-bitplanes, from the code-blocks in a given packet partition location forms the body of a "packet". The packet structure corresponding to the packet partition location in Figure 10.8(b) is shown in Figure 10.9. A packet consists of a packet header, which contains the following information: block inclusion information, which indicates whether data is presence for a given code-block, the number of completely zero bitplanes for each block, the number of sub-bitplanes included in the packet for each block, and the number of bytes used to store the coded sub-bitplanes of each block. More details on the header structure and code method can be found in [294]. Each code-block can contribute a different number of sub-bitplanes (it can be zero) to the packet. Hence, a packet can be viewed as one quality increment for one resolution level at a certain spatial location.

A collection of packets, one from each resolution level and packet partition location, constitutes a "layer". Thus, it can be viewed as one quality increment for the entire image at full resolution. All the packets from a tile are interleaved in one of several orders and placed in one, or more, tile-parts. The tile-parts have a descriptive tile-part header and can be interleaved in any order. The codestream has a main

Packet Header	n_0 sub-bitplanes from code-block 0	n_1 sub-bitplanes from code-block 1	– – – – – –	n_{11} sub-bitplanes from code-block 11

Fig. 10.9. Packet structure for the packet partition location in Figure 10.8(b).

header at the beginning that describes the original image and the various decomposition and coding styles that shall be used to locate, extract, decode, and reconstruct the image with the desired resolution, fidelity, region of interest, and other characteristics.

10.4.3 Progression

While JPEG offers several methods for generating progressive bitstreams for progressive transmission, progression in JPEG-2000 is simply achieved by ordering the packets of compressed information within the bitstream, which are coming from different resolutions, quality, spatial locations, and components. For example, to achieve SNR scalability or progression by SNR, the bitstream can be ordered in layers, one after the other. Inside each layer, the compressed information for each component image can be transmitted successively. Likewise, inside a particular component image, the compressed information can be transmitted by resolution levels. Finally, inside one resolution level, one can send the compressed data for different partition locations. In other words, the bitplanes of the subband coefficients of the image are transmitted successively. The progression type can also be changed at various parts within the bitstream. One interesting feature of JPEG-2000 is that the progression type of the bitstream is defined by markers in the bitstream. A parser can be used to read all the markers and change the type of progression, by changing the markers and other auxiliary information, without having to run the MQ-coder, the context model, or even decode the block inclusion information. The flexible bitstream structure of JPEG-2000 therefore simplifies considerably the editing and processing of the compressed information. Other important features of JPEG 2000 include region of interest (ROI), error resilience, and visual frequency weighting [123].

10.4.4 Performance

JPEG-2000 provides better rate-distortion performance than the JPEG standard at any given bit rate. The greatest improvement, however, is observed at very high and very low bitrates. Some compression results of JPEG-2000 for the image "Lenna" are shown and compared with JPEG in Figure 10.4. The key advantage of JPEG-2000 is its flexibility in progressive transmission over progressive JPEG. Basically, only the order of the compressed data, but not the data itself, is changed and the performance is almost identical to the performance of using a single layer optimized at the same rate. The loss is mainly due to the increased overheads for the additional layers. JPEG-2000 produces much less blocking artifact and hence a better visual quality at low bit rate than the JPEG Baseline algorithm.

On the other hand, JPEG-2000 requires considerably higher computational complexity than JPEG. This is mainly associated with the multi-pass bitplane context model and arithmetic entropy coder. JPEG-2000 also requires more memory than sequential JPEG.

For lossless compression of continuous-tone images, the performance of JPEG-2000 is similar to JPEG-LS [121], but substantially better than JPEG lossless. For images with text and graphics, JPEG-LS is significantly better than JPEG lossless and JPEG-2000. The performance of the (9,7) wavelet is usually much better than the (5,3) wavelet, while the latter is simpler to implement.

10.5 Appendix: VQ structures

In this appendix, we describe multistage VQ, tree structure VQ, and lattice VQ.

10.5.1 Multistage VQ

In multistage VQ, the input vector is successively quantized (also called successive approximation) by a set of codebooks. This is illustrated in Figure 10.10, where the input vector X is first quantized by a codebook C1 of size L_1 to obtain a reproduction vector X_0. The residual or error vector $R_0 = X - X_0$ is then quantized by another codebook C2 of size L_2 to give X_1. The indices I_0 and I_1 for X_0 and X_1, respectively, are sent to the receiver. At the receiver, X is reconstructed as $X_0 + X_1$, and the quantization error is R_1. More stages can be used to reduce the quantization error to a sufficiently small value. It can be seen that the number of effective codevectors is $L_1 \cdot L_2$, although only $L_1 + L_2$ different codevectors need to be stored. This gives a significant saving in storage as compared with a codebook with size $L_1 \cdot L_2$. The encoding complexity is also drastically reduced. According to (9.9a), the encoding complexity of this 2-stage MS-VQ is

$$\textbf{MSE:} (L_1 + L_2) \cdot (2K - 1) \text{ addition}, (L_1 + L_2) \cdot K \text{ multiplication},$$
$$\text{and } (L_1 + L_2 - 2) \text{ compare operations;} \tag{10.8}$$

The encoding complexity for a codebook of the same size will require

$$\textbf{MSE:} (L_1 \cdot L_2) \cdot (2K - 1) \text{ addition}, (L_1 \cdot L_2) \cdot K \text{ multiplication},$$
$$\text{and } (L_1 \cdot L_2 - 2) \text{ compare operations;} \tag{10.9}$$

Also, the number of stages can be made adaptive to provide more flexibility in achieving the given rate or distortion constraints.

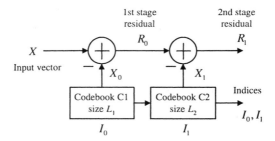

Fig. 10.10. Two-stage VQ

10.5.2 Tree structure VQ (TSVQ)

In designing a codebook using the LBG algorithm, the computational time is usually very long and good initial values of the code vectors are required in order to converge to a good local minimum. Usually, the mean of the training vectors is first computed and it is then split successively to more codevectors. Tree structure VQ is originally proposed for reducing the complexity in searching a codebook for the optimal code vector by arranging the codevectors as a binary tree (or more generally a K-d tree with K instead of two branches per node) as shown in Figure 10.11. Later, it was found that the concept of tree-structure VQ can also be used to generate efficiently codebooks with good performance.

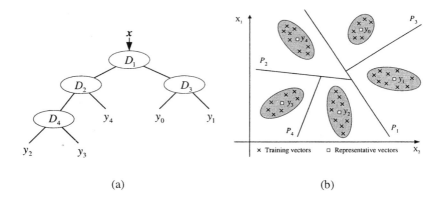

(a)   (b)

Fig. 10.11. (a) Binary decision tree of Tree structure VQ (b) Partitioning of training vectors in the arbitrary hyperplane codebook generation algorithm.

First, the training data is divided using a hyperplane P_1 given by

$$P_1 : \mathbf{h}_1^T \mathbf{x} + \beta_1 = 0$$

into two partitions corresponding to those vectors respectively on the left and right of the plane. This division can also be viewed as designing a codebook with two elements. Then, one of these partitions will be chosen according to a certain criterion for further subdivision, e.g. P_1 as shown in Figure 10.11(b). The process repeats at each stage by choosing an appropriate partition for subdivision, until the desired number of partitions (i.e number of codevectors) is obtained. In Figure 10.11(b), five code vectors are employed. According to the centroid condition, the mean of each partition yields one code vector of the codebook. This partitioning can also be represented by a binary decision tree as in Figure 10.11(a). Moreover, it can be used to perform codebook searching. Given an input vector x to be quantized, the decision tree is transversed and at the i-th tree node, the following decision is made:

$$D_i : \begin{matrix} \mathbf{h}_i^T \mathbf{x} + \beta_i \leq 0 \\ \mathbf{h}_i^T \mathbf{x} + \beta_i > 0 \end{matrix},$$

to determine which branch of the tree should be transversed. The codevector associated with the leaf node where the search terminates is the quantized value of this vector. The worse case encoding complexity of a TSVQ is only

MSE: $D \cdot (K - 1)$ addition, $D \cdot K$ multiplication, and D compare operations,
$$(10.10)$$

where D is the depth of the tree. If the tree is balanced, then $D = \log_2 L$. Interested readers are referred to [34] for determining the hyperplanes and related references.

10.5.3 Lattice VQ

In uniform scalar quantization (SQ), the real line is partitioned into intervals with equal length. The regularity of uniform SQ considerably simplifies its encoding and decoding. A generalization of uniform SQ to multidimensional space is called a lattice vector quantization (LVQ). The code vectors are obtained from the lattice points of a structure called lattice. An n-dimensional lattice, Λ_n, is defined as

$$\Lambda_n = \{\mathbf{x} \in R^m | \mathbf{x} = \mathbf{G} \cdot \mathbf{z}, \mathbf{z} \in Z^n\}, \tag{10.11}$$

where $\mathbf{G}$ is the $(m \times n)$ generator matrix of the lattice $(m \geq n)$. Thus, the lattice points are generated by integer combination of the column of $\mathbf{G}$, $\{\mathbf{g}_i's\}$:

$$\mathbf{x} = \mathbf{G} \cdot \mathbf{z} = \sum_{i=1}^n \mathbf{g}_i z_i, \ z_i \in Z. \tag{10.12}$$

Examples of lattices include D_n, E_8, and A_n:

$$D_n = \{ x \in Z^n \mid \sum_{i=1}^n x_i = \text{even number}\}. \tag{10.13}$$

The lattice points are those points in the Z^n with coordinate sum being an even number. E_8 is obtained by the union of the D_8 lattice and its coset $[D_8 + (\frac{1}{2}, \frac{1}{2}, \frac{1}{2}, \frac{1}{2}, \frac{1}{2}, \frac{1}{2}, \frac{1}{2}, \frac{1}{2},)]$:

$$E_8 = D_8 \cup [D_8 + (\tfrac{1}{2}, \tfrac{1}{2}, \tfrac{1}{2}, \tfrac{1}{2}, \tfrac{1}{2}, \tfrac{1}{2}, \tfrac{1}{2}, \tfrac{1}{2},)].$$ (10.14)

A coset is obtained from a lattice by adding a fixed vector to its lattice points. It can be shown that for a smooth pdf and sufficiently fine quantization, a LVQ derived from the densest sphere packing can approach the minimum squared quantization error for a given entropy [333]. In 2- and 3-dimensions, A_2 (hexagonal lattice) and A_3 are known to yield densest sphere packing. In higher dimensions, only E_8 and the leech lattice (24 dimensions) are known to give densest sphere packing. E_8 and other lattices have been proposed for coding subband coefficients [86]. Interested readers are referred to [52, 53, 278] for fast encoding and decoding algorithms of several lattices and other aspects of sphere packings.

11

Video Compression Techniques

In this chapter, we turn to video compression techniques. First of all, the basic concept and techniques of video compression will be briefly reviewed in Sections 11.2 to 11.3. A brief survey of various video coding standards is given in Section 11.4.

11.1 Video formats

In this section, we describe formats for analog and digital videos.

11.1.1 Analog videos

In composite analog video (CAM) such as NTSC (National Television Systems Committee), PAL (Phase Alternation Line), and SECAM (Systeme Electronique Color Avec Memoire), each primary is considered as a separate monochromatic video signal.

In the NTSC receiver primary system (R_N, G_N, B_N), which was developed as a standard for television receivers, three phosphor primaries that glow in the red, green, and blue regions of the visible spectrum were adopted. R_N, G_N, B_N range from 0 to 1 and they indicate the excitation of the phosphor, 0 indicates no excitation and 1 indicates maximum excitation. The reference white corresponds to $R_N = G_N = B_N = 1$. The luminance is the sum of the luminance of the red, green, and blue phosphors and is given by:

$$Y = 0.299 \cdot R_N + 0.587 \cdot G_N + 0.114 \cdot B_N. \qquad (11.1)$$

It has a scale from 0 to 1 and the weighting indicates the relative contributions of the primaries to the total luminance. The color information can be expressed as the chrominance components. The advantage of the luminance-chrominance representation is that the human visual system is less sensitivity to high frequency chrominance components, and their bandwidth can be reduced to save transmission bandwidth. For instance, in the NTSC transmission (Y,I,Q) system, the I and Q chrominance signals defined below

$$
\begin{bmatrix} Y \\ I \\ Q \end{bmatrix} = \begin{bmatrix} 0.299 & 0.587 & 0.114 \\ 0.596 & -0.274 & -0.322 \\ 0.211 & -0.523 & 0.312 \end{bmatrix} \begin{bmatrix} R_N \\ G_N \\ B_N \end{bmatrix} \tag{11.2}
$$

occupy respectively 1.6 and 0.6 MHz bandwidth, while the Y component has a bandwidth of 4.2MHz. A related chrominance representation called the YUV color difference coordinate system is used in the PAL color TV systems. The U and V signals are expressed as a set of color differences:

$$
U = 0.493(B_N - Y); \tag{11.3}
$$

$$
V = 0.877(R_N - Y). \tag{11.4}
$$

The SECAM system uses the YDbDr coordinate, where Db and Dr are related to U and V by Db = 3.059U, Dr = -2.169V.

The U and V signals are related to the I and Q signals by a simple rotation in the color space.

$$
I = -U \sin(33^o) + V \cos(33^o), \tag{11.5}
$$

$$
Q = U \cos(33^o) + V \sin(33^o). \tag{11.6}
$$

11.1.2 Digital videos

Digital videos can be obtained directly from digital cameras or by digitizing analog video signals. In digital cameras (e.g., using CCD sensors), video signals are digitized into image frames at regular interval T, as shown in Figure 11.1(a). The frame rate is equal to 1/T. Each frame is represented as a rectangular array of pixels. The number of horizontal line and pixel per line defines the resolution of the image or video. In Figure 11.1(a), the resolution of the video is $N_1 \times N_2$. For color videos, each pixel is represented by its color components.

To save transmission bandwidth, a traditional analog video system employs an interlaced scanning method, where the even and odd lines of an image frame are transmitted alternately as shown in Figure 11.1(b). In progressive scanning, each image frame contains both the odd and even lines. Early video coding standards usually work with the progressive scanning format and hence conversion from interlaced to progressive scanning is required. On the other hand, more recent standards such as MPEG-2 and H.264 are able to handle both progressive and interlaced scanning.

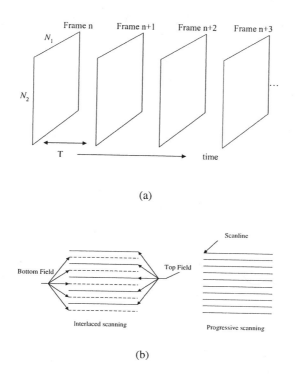

(a)

(b)

Fig. 11.1. Digital videos (a) as sequence of images (b) interlace and progressive videos.

11.1.3 ITU-T BT.601 (formerly CCIR601)

In order to exchange digital videos between different applications and products, different standardized formats have been developed for digital videos. For example, to permit international exchange of production-quality programs digitalized from various analog video systems, the International Telecommunications Union - Radio Sector (ITU-R) (formerly known as International Radio Consultative Committee (CCIR)) developed a digital video format for TV studios with 4:3 and 16:9 aspect ratios, known as the BT.601 recommendation. The 4:3 aspect ratio version is formerly known as the CCIR601 format, which we shall briefly discuss.

In order to match the horizontal and vertical sampling resolution for the 525-line NTSC, and 625-PAL/SECAM systems, the sampling frequency is chosen as 13.5MHz, which is an integer multiple of the horizontal sweep frequencies of the two systems. The number of pixels per line for NTSC and PAL/SECAM are thus 858 and 864, respectively, and the two formats are called 525/60 and 625/50 signals. Taking into account the samples during horizontal and vertical retraces, the active pixels for the 525- and 625-line systems are 720×480 and 720×576.

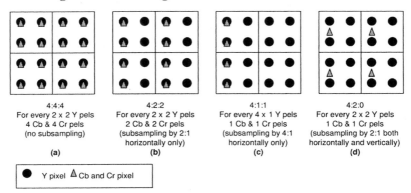

4:4:4	4:2:2	4:1:1	4:2:0
For every 2 x 2 Y pels	For every 2 x 2 Y pels	For every 4 x 1 Y pels	For every 2 x 2 Y pels
4 Cb & 4 Cr pels	2 Cb & 2 Cr pels	1 Cb & 1 Cr pels	1 Cb & 1 Cr pels
(no subsampling)	(subsampling by 2:1 horizontally only)	(subsampling by 4:1 horizontally only)	(subsampling by 2:1 both horizontally and vertically)
(a)	(b)	(c)	(d)

● Y pixel △ Cb and Cr pixel

Fig. 11.2. BT.601 color component formats: YCbCr (a) (4:4:4), (b) (4:2:2), (c) (4:1:1), (d) (4:2:0) formats.

The color components are represented by the YCbCr color coordinate system. It uses the same Y coordinate as the YUV system, where U and V are scaled and zero-shifted to obtain Cb and Cr respectively as follows:

$$Cb = (U/2) + 0.5; \tag{11.7}$$

$$Cr = (V/1.6) + 0.5. \tag{11.8}$$

In practice, these values are multiplied by 255 so that they can be represented by an 8-bit integer. For example, the integer YCbCr values are related to the integer RBG values (in the range 0 to 255) by

$$\begin{bmatrix} Y \\ Cb \\ Cr \end{bmatrix} = \begin{bmatrix} 0.257 & 0.504 & 0.098 \\ 0.148 & -0.291 & 0.439 \\ 0.439 & -0.368 & -0.071 \end{bmatrix} \begin{bmatrix} R \\ G \\ B \end{bmatrix} + \begin{bmatrix} 16 \\ 128 \\ 128 \end{bmatrix} \tag{11.9}$$

where R = $255R_N$, B= $255B_N$ and G = $255G_N$.

Table 11.1. Picture size for the H.263 picture formats.

Picture Format	Image size for Luminance (pixel × lines)	Image size for Chrominance (pixel × lines)
sub-QCIF	128 × 96	64 × 48
QCIF	176 × 144	88 × 72
CIF	352 × 288	176 × 144
4CIF	704 × 576	352 × 288
16CIF	1408 × 1152	704 × 576

Because of the limited sensitivity of the human visual system, the chrominance components can be decimated. There are four different formats called YCbCr (4:4:4),

(4:2:2), (4:1:1), and (4:2:0) formats as shown in Figure 11.2. No decimation of chrominance components is performed in the (4:4:4) format and it is intended for applications requiring very high resolution. In the (4:2:2) format, the chrominance components are decimated by a factor of two horizontally. For four luminance pixels, there are two Cb and two Cr pixels, hence the name (4:2:2). In the (4:2:0) format, the chrominance components are decimated by a factor of two in the horizontal and vertical directions. The (4:1:1) format differs from the (4:2:0) formats in that the chrominance signals are decimated by a factor of 4 horizontally. Although the number of chrominance samples is the same, it yields a very asymmetric resolution in the horizontal and vertical directions, which is undesirable. Table 11.1 summarized some of the digital video formats.

In additional to spectral and spatial redundancies in digital images, digital videos also exhibit considerable redundancy in the temporal domain. This suggests the use of motion estimation to predict one video frame from others. This is usually referred to as motion compensation/prediction, which is a very efficient method in video coding. They can be combined appropriately with other waveform coding methods to form a wide range of coders to meet different complexity/performance tradeoffs. One very successful method, called motion compensated hybrid DCT/DPCM coding, is to combine motion compensation/prediction with transform coding. Because of its good performance and reasonable implementation complexity, they form the basis for most video coding standards.

11.2 Motion compensation/prediction

Motion compensation (MC) plays a very important role in video compression especially in motion compensated hybrid DCT/DPCM codecs. MC is an effective method in exploiting the temporal correlation of pixels between adjacent video frames arising from motion of objects or camera. The basic idea is to predict a group of pixels in a video frame from nearby pixels in previously encoded frames called reference frames. A certain motion model is needed to form the predictor and its parameters have to be estimated, which is called the motion estimation problem. By entropy coding these motion parameters and the quantized prediction residuals, significant compression can be achieved.

Although more sophisticated motion models are available, most video codecs employ the simple linear translation model to reduce the arithmetic complexity in estimating the motion parameters and to avoid the transmission of too many motion parameters. In the linear translation model, objects are assumed to undergo linear translation motion. If the object motion is slow in successive video frames, we can further assume that the object shape remains unchanged. As a result, we can simply compare the current pixels to those in the previous frame, which are linearly shifted in the x- and y-dimension by a certain amount relative to the current pixels. The displacement that minimizes a certain distortion measure such as the MSE is an estimate of the motion vector of the pixels.

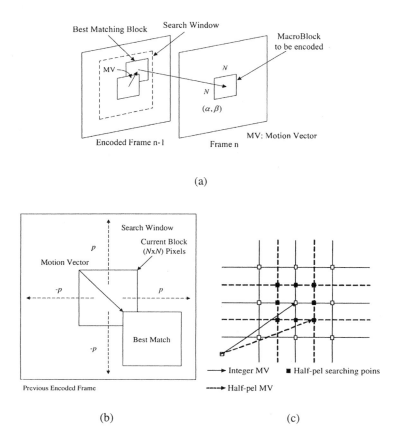

(a)

(b)   (c)

Fig. 11.3. (a) Motion compensation/prediction (b) Full search motion estimation (c) Half-pel motion estimation.

In order to combine MC with block-based DCT transform coding, hence the name motion compensated hybrid DCT/DPCM coding, motion estimation is usually performed on a square block of pixels. This is usually called block matching motion estimation and it is illustrated in Figure 11.3. Here, a predictor is searched in a nearby searching window of the reference frame (frame n-1 in this case) for the current block of pixels at frame n. Since the matching process is computationally very intensive, the mean absolute difference (MAD) is commonly employed to avoid the squaring operation in computing the MSE. Denote the intensity images of the current and reference frames by $I(x, y)$ and $\hat{I}(x, y)$, respectively. The motion vector for a block with its lower corner at (α, β) computed using the MAD distortion measure is given by

$$MV = \arg\min_{(i,j)} MAD(i,j)$$

$$MAD(i,j) = \sum_{n_1=0}^{N-1} \sum_{n_2=0}^{N-1} \left| I(\alpha + n_1, \beta + n_2) - \hat{I}(\alpha + n_1 + i, \beta + n_2 + j) \right|$$

$$-p \le i, j \le p,$$

$$(11.10)$$

where $(2p + 1)$ is the x- and y-dimensions of the search window as shown in Figure 11.3(b). The prediction residual is then $e(n_1, n_2) = I(\alpha + n_1, \beta + n_2) - \hat{I}(\alpha + n_1 + MV_x, \beta + n_2 + MV_y)$, where MV_x and MV_y are respectively the x- and y- components of the motion vector.

Apart from the motion activities in the video, the quality of block-based motion estimation also depends on the choice of 1) the size and precision of the search window, and 2) the block size used for motion estimation.

11.2.1 Size and precision of search window

In principle, the MV can be real-valued. However, because of the high complexity of motion estimation and the diminishing return in coding high precision MV, they are chosen as integers (full or integer pel motion estimation) or half-integers (half-pel motion estimation) in most video coders. Integer MVs can lead to considerable prediction residuals, which produce annoying artifacts at low bit rate. In an earlier standard such as H.261, a loop filter has to be used in order to smooth out undesirable artifact in the reconstructed video. More recent standards, such as H.263 and MPEG, use half-pel motion estimation for better motion prediction and the loop filter is unnecessary. Instead of searching over all the half integers inside the search grid, the integer grid is first searched for the best integer MV. The 8 half-pel locations around this integer vector are then examined. In so doing, the reference frame has to be interpolated, usually using bilinear interpolation, for computing the MAD. This is illustrated in Figure 11.3(c).

Research has shown that very little can be gained if the search grid is smaller than a quarter pel [76]. For video conferencing applications, where object motion is usually not so severe, a search window of (-15,15) in both dimensions is commonly used. In MPEG-1/2 video coding, the reference frames are not necessarily adjacent to the frame to be encoded, a larger search window is thus necessary.

11.2.2 Block size

For scenes with moderate motion, block-based motion estimation usually performs quite well except at object boundaries, where pixels from the background or other objects are rather difficult to predict from blocks in the reference frame. The amplitude of the residuals will increase and more bits are thus required. This problem is less serious when the block size is small say (8×8). However, more motion vectors will be required. Therefore, there is a tradeoff between coding of motion parameters and residuals. In video coding standards such as H.261 and MPEG-1/2, a block size of (16×16), called a macro-block (MB), is employed as a tradeoff between the two components. In H.263, an optional advanced prediction mode is defined to offer the flexibility of having one MV for each (8×8) luminance block in a MB. The four MVs are differentially coded and it is up to the encoder to decide whether one MV or 4 MVs are employed for a particular MB (usually based on the number of bits generated from these two coding modes). Another method in the advance prediction mode for improving the performance of block-based method is overlapped block motion compensation (OBMC). The basic idea of OBMC is to predict the pixels in the current block, not only by the MV of the current block, but also the MV of its neighboring blocks. Experimental results show that OBMC lead to better coding performance. Interested readers are referred to [219] and [126] for more details. In H.264 [125], variable block size motion estimation is employed. Interested readers are referred to [125] for additional information.

To reduce implementation complexity, motion estimation is usually performed on the luminance component only. The motion vector so obtained will be appropriately scaled for carrying out motion compensation for the chrominance components. If motion compensation is satisfactory, which means that the residual is small enough, then the residual will be transformed, quantized and entropy coded. On the other hand, if motion compensation fails, the original image blocks will be transform coded.

There are several approaches in reducing the arithmetic complexity of motion estimation at the expense of slightly degraded performance. These algorithms reduce the number of computations by either reducing the number of locations searched [146, 128, 281] or the number of arithmetic operations in each comparison by pixel decimation [172]. Other techniques include the hierarchical search [16] and many of its variations [283].

11.3 Motion compensated hybrid DCT/DPCM coding

Motion-compensated hybrid DPCM/DCT coding is a commonly used and efficient compression technique for digital videos. International video coding standards such as H.261 [30], H.263 [126], MPEG-1/2 [119, 120] and MPEG-4 [117] are based on this scheme. Figure 11.4 shows the block diagram of a motion compensated hybrid DCT/DPCM encoder.

11.3.1 Coding of *I*-frames

The first image frame is usually encoded using DCT-based coding, where the image frame is segmented into macroblocks of size usually (16×16) for motion estimation. For color images using the (4:2:0) YCbCr format, each macroblock will contain 4 (8×8) luminance (Y) and two (8×8) chrominance blocks, one for Cb and one for Cr. Each block will undergo 2D DCT and the DCT coefficients are quantized and entropy encoded. Since the picture is coded, and hence decoded, without reference to other pictures, it is usually called an *I*-picture and the coding method is called INTRA-frame coding. *I*-pictures can serve as reference frames for predicting other

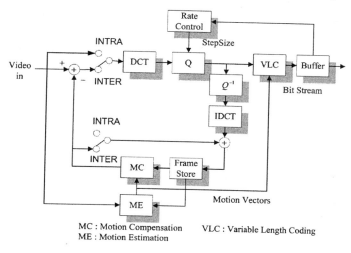

Fig. 11.4. Block diagram of Motion compensated hybrid DCT/DPCM encoder. Q: quantization, Q^{-1}: inverse quantization.

pictures. They also serve as random access point for decoding a group of pictures in the MPEG videos and many MPEG-like algorithms for coding image-based representations. The latter will be discussed in Chapters 12 and 13.

11.3.2 Coding of *P*-frames

To improve coding efficiency, motion compensation is usually applied to subsequent video frames using for example the block matching motion estimation algorithm described in Section 11.2. As mentioned earlier, motion estimation is usually performed at the macroblock level and is based on the Y component. The macroblocks are usually coded from left to right, and from top to bottom. For each macroblock, if motion estimation is successful (i.e., the prediction residuals are small enough to justify the use of the prediction) then a motion vector (MV) is used to specify the prediction of the current macroblock from previous reference pictures. The motion vector is then differentially encoded, using adjacent motion vectors as predictors, and the residuals are entropy coded. This coding mode is usually referred to as "INTER", because it involves inter-frame prediction. For chrominance blocks, the motion vector is divided by two in forming the prediction from the chrominance components of the previous reference frame. After subtracting the prediction, each of the (8×8) blocks, i.e., the 4 Y-, Cb-, and Cr- blocks will undergo 2D-DCT. The DCT coefficients are then quantized and entropy encoded (usually in a zig-zag order using run-length and Huffman codes). The entropy coded MV and transform coefficients, and other coding information will be packed in a certain format to produce an output bit stream at a variable rate. The number of bits generated depends on the stepsize of the quantizer Q (and in more advanced coders, the choice of possible coding modes).

For macroblocks associated with uncovered objects in a scene, motion estimation might fail since a good match may not be found from previous reference frame. In this case, the prediction residuals will become very large and it is preferred to encode the blocks inside the macroblock directly using DCT-based coding without prediction. This coding mode is usually referred to as "INTRA", as no inter-frame prediction is employed.

After encoding the current macroblock, the quantized coefficients are locally reconstructed through inverse quantization, Q^{-1}, and then inverse transformed and motion compensated to obtain the quantized version of the current macroblock. It is then saved in the frame store so that it can be used for predicting future video frames through motion compensation.

11.3.3 Rate control

Since the compressed video is usually transmitted over a constant rate channel (say 32-, 64-, and 128kbps, etc), a buffer is needed to smooth out the variable output rate from the VLC. This buffer is filled at a variable rate from the VLC and is reading out at a constant rate equal to the channel rate. When the buffer is being filled faster than it is being emptied, buffer overflow occurs. On the other hand, if the buffer is being emptied faster than it is being filled, buffer underflow occurs. To avoid buffer overflow and underflow, a rate-control mechanism is needed to determine the quantization stepsizes and other necessary action such as frame skipping for proper operation. For example, when there is little activity in the input video, motion compensation will be very efficient and the transform coefficients are usually of small amplitudes. Under this situation, the quantizer stepsize needs to be decreased in order to generate more bits to avoid the buffer from underflowing.

On the other hand, when there are lots of activities in the input video, the quantizer stepsize needs to be increased in order to prevent excessive number of bits being generated by the VLC from overflowing the buffer. The determination of the various control parameters, such as quantizer stepsize and the choice of coding modes, to minimize the distortion subject to certain rate constraints is generally referred to as the rate control or buffer control problem. Traditional buffer control algorithms are usually based on feedback control approach in which the buffer status is used to control the quantizer stepsize, and therefore the rate. The maximum size of the buffer is usually constrained by the maximum allowable coding delay. For delay sensitive applications, rate control is very difficult due to the limited buffer size. For more information on various rate control algorithm, see [45, 220, 286, 246, 310, 154, 103].

Figure 11.5 shows the block diagram of the decoder. The motion vectors, transform coefficients, and other information are decoded from the bitstream in order to reconstruct the image blocks.

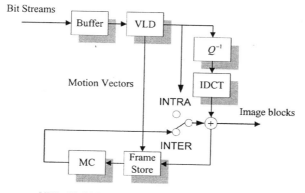

VLD : Variable Length Codes Decoding

Fig. 11.5. Block diagram of Motion compensated hybrid DCT/DPCM decoder.

11.4 Video coding standards

Over the last two decades, different video coding standards have been developed to address a wide range of applications including video communications, production, distribution and storage. The two major organizations responsible for the standardization processes are the ITU-T Video Coding Experts Groups (VCEG) and the ISO/IEC Moving Picture Experts Group (MPEG). VCEG has traditionally focussed on low bitrate video coding applications for communications, while MPEG groups are targeting for higher bitrates entertainment and broadcasting applications. The ITU-T H.261 recommendation [30] is the first coding standard used, which is based on motion compensated hybrid DCT/DPCM coding.

Motivated by the need for more efficient digital storage of videos and audios at the VCR quality, the MPEG-1 standard was developed later, which operates at a higher bit of 1.5Mbps. It was soon recognized that MPEG-1 is unable to efficiently compress interlace digital videos at broadcast quality. A joint effort between VCEG and MPEG is then set up to develop the MPEG/H.262 [120] standard and its main purpose is to enable MPEG-1 like functionality for interlaced pictures, primarily using the ITU-R BT.601 (formerly CCIR601) 4:2:0 format. The target was to produce TV-quality pictures at data rates of 4-8 Mbps and high quality pictures at 10-15 Mbps for high definition TV (HDTV) applications. In 1995, H.263 [126], which is a more efficient coder based on H.261, was developed to address low bitrate applications over say a public switched telephone network (PSTN). Later standards, such as MPEG-4 [117], MPEG-7 and MPEG-21 [203], on the other hand, focus respectively on the development of new coding functionalities, new description tools, and a new multimedia framework.

At the same time, the ITU-T group continued with the extension of the H.263 standard, H.263+ in 1997 and H.263++ in 2000, and it results in a set of options for further improving the compression efficiency of the basic H.263. The newest standard in the family, H.264 (also known as MPEG-4 part 10, Advanced Video Coding (AVC)) is a joint effort between ITU-T and ISO/IEC, and the primary focus is to provide efficient compression solution for a wide range of applications. Important features of H.264 include a simplification back-to basics approach, high compression performance, improved network friendliness, and enhanced error and packet loss resilience tools for real-time applications over error-prone channels. A brief introduction to these standards will be given below. To make our discussion more concise, interested readers are referred to [125, 247, 90]. Some of their applications to IBR compression will be further elaborated in Chapters 12 and 13.

11.4.1 H.261

H.261 is an ITU-T recommendation for coding and decoding of the video component of audiovisual services at rates $p \times 64$ kbits/s over integrated services digital network (ISDN), where p is between 1 to 30. The standard was published in 1990. It specifies a set of protocols that every compressed video bit stream has to follow. It also specifies a set of operations that every standard compatible decoder must be able to perform. Like many international standards introduced later, only the bit stream format and capability of the decoder are defined and the actual implementation of the encoder is left to the innovations of individual vendors.

Figure 11.6 is a functional block diagram of the H.261 video codec. It is a motion compensated hybrid DCT/DPCM coder, which employs motion compensated inter-picture prediction to explore temporal redundancy and transform coding to reduce spatial redundancy. The multiplexer combines the compressed data and side information to indicate alternative modes of operation in the encoder. A transmission buffer is employed to smooth out the variable bit rate from the source encoder for transmission over a fixed rate communication channel.

Picture formats

Two picture scanning formats are specified: common intermediate format (CIF) and quarter-CIF format (QCIF). CIF is a non-interlaced format with 352 pixels per line and 288 lines per picture. These values are chosen to facilitate the conversion from BT.601 (formerly CCIR601) 625- and 525-lines signals. One only needs to perform a picture rate conversion in 625 - line 25 fps PAL/SECAM systems, and a line-number conversion in 525-line NTSC systems. QCIF is for lower bit rate applications such as videophones and its resolution is one-quarter of CIF, i.e., (176×144). A H.320 terminal compliance decoder must be able to decode QCIF format at a rate of 7.5 Hz. Supporting CIF coding and decoding is optional. A color picture is represented as the YCbCr (4:2:0) format. The Cb and Cr components are subsampled by a factor of two in both the horizontal and vertical directions. Their sizes are 176 pixels per line and 144 lines per frame for CIF and 88×72 for QCIF.

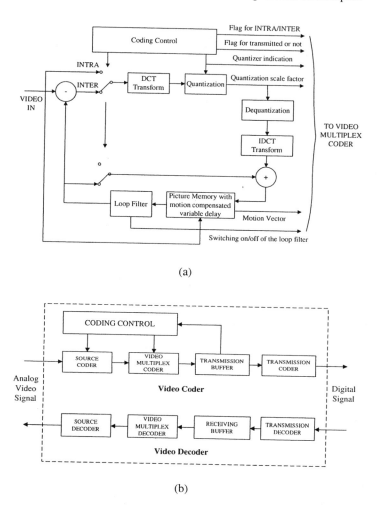

(a)

(b)

Fig. 11.6. (a) Block diagram of an H.261 source coder. (b) Block diagram of the H.261 video codec.

H.261 source coder

From Figure 11.6(a), it can be seen that the main elements of the coder are frame prediction, DCT transformation and quantization. Since H.261 is designed for real-time communications, it uses only the closest previous frame for inter-frame prediction in order to reduce the encoding delay. Image frames are divided into macroblock (MB) of size (16×16), each consists of 4 (8×8) luminance blocks, and two (8×8) chrominance blocks, one for Cb and one for Cr.

Frame prediction

Since only *I*-pictures and *P*-pictures are used, frame prediction is done in *P*-pictures using integer-pel forward motion estimation. One vector is used for all four luminance blocks in the MB. The motion vector for both color difference blocks is obtained by halving the component value of the MB motion vector and truncating it towards zero. Both horizontal and vertical components of these motion vectors are integer valued not exceeding a search range of ± 15. The MB is then classified as intra-type or inter-type, depending on the results of motion estimation and the coding strategies. Their respective inputs to the DCT coder will be the input MBs for the INTRA mode or the differential MBs between the current frame and the previous frame for the INTER mode. Not all MBs need to be coded and transmitted. For example, at low bit rates, MBs with low activities (and up to three full frames) can be skipped. The choices of transmitting or skipping MBs, using intra- or inter-frame coding, and the rate control mechanism are not part of the standard. It may vary dynamically depending on the complexity of the input signal and the output data rate constraints.

To tell the decoder whether a MB will be coded and which coding modes will be used, auxiliary information called macroblock type (MTYPE) and coded block pattern (CBP) are sent to the decoder. These combinations are Huffman coded.

In the prediction path, an optional spatial loop filter can be employed to reduce the artifacts generated by large prediction error in the reconstructed frame due to inaccurate integer-pel motion estimation. The loop filter is a separable 2D filter that operates on (8×8) blocks. The loop filter is particular useful to the suppression of coding artifact at low bit rates, such as 64kbps.

DCT transformation and quantization

Like JPEG and many other video coding standards, the non-zero DCT coefficients are scanned in a zigzag manner and the (run-level) values of non-zero coefficients are Huffman coded. Unlike JPEG, which uses the same set of quantizer stepsizes for all blocks, a different quantizer structure is used which allows stepsizes to be changed at the MB level to facilitate rate control. The INTRA DC coefficient is linearly quantized with a stepsize of 8 and no dead-zone. Within a MB, the same quantizer is used for all the coefficients, except the INTRA DC coefficient. There are 31 quantizers, and each of them is linear but with a central dead-zone around zero to reduce coding noise. (Section 9.3). The stepsize is an even integer in the range 2 to 62.

Rate control

Because the coder is operating at a constant output rate, rate control has to be performed. Methods for controlling the generation of coded video data include: processing prior to the source coder, changing the quantizer stepsize, and block and frame skipping. The overall control strategy is not specified in the H.261 standard. In the informative part of the H.261 document [30], the quantizer parameter MQUANT is updated by the following formula

$$stepsize = 2 \times \text{INT}(buffer\_content/[200 \times q]) + 2, \qquad (11.11)$$

where INT denotes the truncation of the fraction and the bitrate is $q \times 64$ kbit/s.

In the forced updating mode, the INTRA mode will be employed. The update pattern is not defined. However, to control the accumulation of error arising from mismatch in IDCT implementation for the encoders and decoders, a MB should be forcibly updated at least once per every 132 times it is transmitted. The encoder must control its output bitstream to comply with the requirements of the Hypothetical Reference Decoder defined in the Recommendation. When operating with CIF, the number of bits created by coding any single picture must not exceed 256 kbits. When operating with QCIF the number of bits created by coding any single picture must not exceed 64 kbits.

11.4.2 H.263

This recommendation specifies a coded representation that can be used for compressing the moving picture component of audio-visual services at low bit-rates, say over public switched telephone network (PSTN). The basic configuration of the video source coding algorithm is based on ITU-T Recommendation H.261. Four negotiable coding options are included for improved performance.

The five standardized image formats are 16CIF, 4CIF, CIF, QCIF and sub-QCIF (SQCIF) (Table 11.1). In addition to CIF and QCIF defined in H.261, a smaller resolution called the SQCIF format and higher resolutions called 4CIF and 16CIF formats are defined in H.263. Each picture is divided into groups of blocks (GOBs). A group of blocks (GOB) comprises of $k \times 16$ lines, depending on the picture format (k = 1 for sub-QCIF, QCIF and CIF; k = 2 for 4CIF; k = 4 for 16CIF). The number of GOBs per picture is 6 for sub-QCIF, 9 for QCIF, and 18 for CIF, 4CIF and 16CIF. The GOB numbering is done using the vertical scan of the GOBs, starting with the upper GOB (number 0) and ending with the lower GOB. Data for each GOB consists of a GOB header (may be empty) followed by data for macroblocks. Data for GOBs is transmitted per GOB in increasing GOB number.

Source coding algorithm

The H.263 coding algorithm is an extension of H.261, which is also based on hybrid DPCM/DCT video coding. Both standards use techniques such as DCT, motion compensation, variable length coding and scalar quantization and both use the well-known macroblock structure. Differences between H.263 and H.261 include:

- Availability of a number of options. In the original H.263 standard, there are four options : PB-frames; Unrestricted motion vectors, Syntax-based arithmetic coding; Advanced prediction mode using overlapped block motion compensation and 8 pel × 8 pel motion vectors;
- H.263 has an optional GOB level;
- H.263 uses different VLC tables at the MB and block levels;
- In H.263, there is no error detection/correction included like the BCH in H.261;
- H.263 uses a different form of MB addressing;
- H.263 does not use the end of block marker.

The general form of the source coder is similar to that in Figure 11.6(a). However, half-pixel precision is used for the motion compensation, as opposed to H.261 where full pixel precision and an optional loop-filter are used. In addition to the core H.263 coding algorithm, four negotiable coding options are included that will be briefly described. All these options can be used together or separately.

Unrestricted motion vector mode

In this optional mode, motion vectors are allowed to point outside the picture. In forming the predictor, the pixels, which lie outside the image, are replaced by the pixels at the image boundaries. A significant gain is achieved if there is movement across the boundary of the picture, especially for smaller picture formats. Additionally, this mode extends the range of the MV so that larger MVs can be used. This is especially useful in case of camera movement.

Syntax-based arithmetic coding mode

In this optional mode, arithmetic coding is used instead of variable length coding. By removing the restriction of fixed integral number of bits for VLC code, higher compression ratio can be achieved at the expense of higher encoder/decoder complexity.

Advanced prediction mode

In this optional mode, overlapped block motion compensation (OBMC) is used for the luminance part of P-pictures in order to reduce the blocking artifact. In OBMC, each pixel in an (8×8) luminance prediction block is a weighted sum of three predicted values computed from three motion vectors - the MV of the current and those from two MBs that are closest to the current (8×8) block. A MB can also use four vectors (8×8) instead of one MV for the entire (16×16) block and the coding mode is decided by the encoder. Due to the increased flexibility in motion estimation, this generally gives a considerable improvement. The subjective quality is also improved because OBMC results in less blocking artifact.

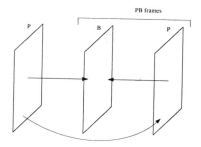

Fig. 11.7. Prediction in PB-frames mode.

PB-frames mode

A PB-frame consists of two pictures being coded as one unit. The name PB comes from the name of picture types in ITU-T Recommendation H.262 (MPEG-2 video) where there are *P*-pictures and *B*-pictures. Thus a PB-frame consists of one *P*-picture, which is predicted from the previous decoded *P*-picture, and one *B*-picture, which is predicted from both the previous decoded *P*-picture and the encoded *P*-picture in the same PB-frame (using backward motion prediction). The name *B*-picture was chosen because parts of *B*-pictures may be bidirectionally predicted using motion compensation from the past and future pictures. With this coding option, the picture rate can be increased considerably without increasing too much the bit-rate. The prediction process is illustrated in Figure 11.7.

Video multiplex arrangement

Similar to H.261, the video multiplex of H.263 is arranged in a hierarchical structure with four layers. From top to bottom the layers are: Picture, Group of Blocks, Macroblock and Block. Due to page limitation, the details are omitted, interested readers are referred to [126] for more information.

The development of H.263 had three phrases. An initial standard was finished in 1995. Extensions of H.263, nicknamed H.263+ and H.263++, were incorporated later at 1997 and 1999. The description above is based on the initial standard [126]. Some of the preferred modes not mentioned above included: 1) Advanced intra coding (Annex I) where intra-blocks are coded using the block to the left and above as predictors, provided that block is also coded in intra-mode. It increases coding efficiency for *I*-pictures. 2) Deblocking filter (Annex J) where an adaptive filter is applied across the boundaries of the decoded (8×8) blocks to reduce blocking artifacts. 3) Improved PB-frame Mode (Annex M) where backward motion compensation, like the *B*-frames in MPEG, is allowed.

11.4.3 MPEG-1 video coding standard

As mentioned earlier, the MPEG-1 standard was designed for coding of moving pictures and associated audio for digital storage media at up to about 1.5 Mbits/s. The

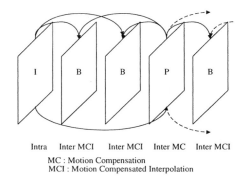

Intra   Inter MCI   Inter MCI   Inter MC   Inter MCI

MC : Motion Compensation
MCI : Motion Compensated Interpolation

Fig. 11.8. Group of pictures in MPEG video.

target of MPEG-1 video was to provide near-VHS quality video at about 1.2 Mbits/s with VCR-like interactivity, such as fast forward, fast reverse, and random access into compressed data at every half-second [198]. The video is assumed to be progressively scanned using the YCrCb (4:2:0) color component format. A source intermediate format (SIF), which is a quarter the size of the active area in the BT.601 signal (formerly CCIR601), was defined.

There are two SIF formats: 360 × 240 pixels at 30 frames/s or 360 × 288 pixels at 25 frames/s. SIF pictures can be obtained from BT.601 frame using filtering and subsampling. The MPEG-1 video is based on motion compensated hybrid DCT/DPCM coding and it bears many similarities with H.261. However, to provide random access capability and better prediction, a new picture type called the bi-directionally predictive-coded frame (*B*-frame) is employed in additional to intra-coded frames (*I*-frames) and predictive-coded frames (*P*-frames). A DC-coded frame (*D*-frame) is also defined for quick browsing of the video. Details of the codec will be described below.

Source coder

As mentioned earlier, each picture is encoded as *I*-, *P*-, or *B*-frames. Like H.261, the entire image is divided into (16 × 16) macroblocks, each containing four luminance, one Cr, and one Cb (8 × 8) blocks. Figure 11.8 shows a typical Group of Picture (GOP) structure of MPEG video with arrows indicating the dependencies among the video frames. *I*-frames are coded using DCT-based transform coding without reference to other pictures and they provides moderate compression. One important function of *I*-frames is to provide potential random access points into the compressed bitstream in order to support features like: fast forward, fast reverse, random search and editing to the video, since they can be decoded independently.

On the other hand, *P*-frames are predicted by motion compensation using the preceding reference frame, which can be a *P*- or *I*-frame as shown in Figure 11.9(a). The prediction residuals or the original pixels (if motion-compensation fails) are encoded using transform coding. Like *I*-frames, *P*-frames can also serve as prediction

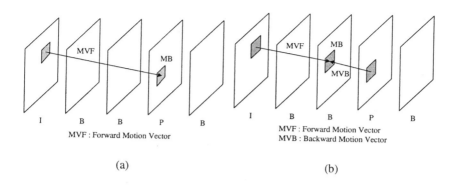

(a)   (b)

Fig. 11.9. (a) Forward motion prediction in *P*-frames and (b) Bidirectional motion prediction in *B*-frames.

references for nearby *B*-frames and future *P*-frames. As *I*- and *P*-pictures are served as reference frames for predicting other pictures, their quantization stepsizes are usually smaller than *B*-frames to prevent excessive propagation of coding errors.

Bidirectional pictures, or *B*-frames, are predicted using motion compensation/ interpolation from preceding and next reference pictures, as shown in Figure 11.9(b). For each MB in a *B*-frame, two motion vectors can be estimated from the reference pictures, one from the preceding and the other from the next reference pictures. Four coding methods are possible for each MB in a *B*-frame: (1) intra coding when motion compensation fails, (2) forward prediction where the predictor is derived from the preceding reference picture using forward motion compensation, (3) backward prediction where the predictor is derived from the next reference picture using backward motion compensation, and (4) bidirectional prediction where the predictor is obtained by averaging the predictors from the preceding and next reference pictures. In the last case, two motion vectors will be required for encoding the MB.

The prediction residuals or original pixels (intra) are then transform coded using DCT. Backward prediction is effective in predicting uncovered areas that might appear in the future, but not the previous reference pictures. *B*-frames offer high compression ratio due to the use of non-causal prediction. In addition, their quantization stepsizes can be increased relative to *I*- and *P*-frames to obtain a higher compression because the quantization errors in *B*-frames do not propagate to other pictures. A simplified block diagram of the MPEG-1 encoder is shown in Figure 11.10.

As we shall see later in Chapters 12 and 13, *I*-frames also serve as random access point for coding Concentric Mosaics, light fields, and other dynamic IBR representations. However, in order not to complicate the decoding, *P*-pictures are usually not employed as reference pictures in coding light fields.

Since MPEG-1 employs motion compensation with half-pel accuracy, the loop filter in H.261 is unnecessary. Another difference between MPEG-1 and H.261 is that the default Quantization matrix (Qmatrix) for intra-coded block (Table 11.2) has

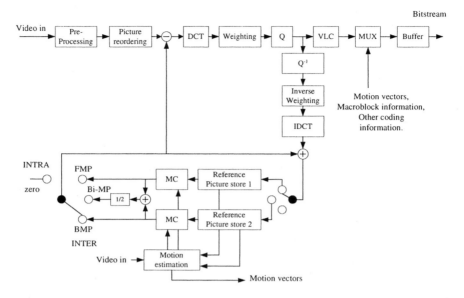

Fig. 11.10. Block diagram of the MPEG-1 encoder.

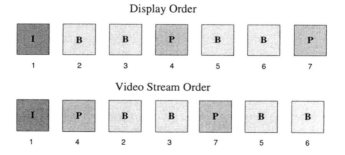

Fig. 11.11. Display and Transmission Order.

larger stepsizes for high frequency DCT coefficients because they are less sensitive to the human visual system. In contrast, all entries in the default Qmatrix for inter-coded blocks are all set to 16. To perform rate control, the Qmatrix is scaled by multiplying each element by a constant, called MQUANT, with values between 1 and 31. A higher value of MQUANT increases the quantization stepsize and reduces the bits generated by the encoder. To achieve a better compression performance, the DC coefficient of an intra-coded block may be predicted from the DCT coefficient of its left neighbor. This concept is different from H.261 and was also employed in H.263 and MPEG-4.

Figure 11.11 shows the transmission and display order of the *I*-, *P*- and *B*- frames in MPEG-1. It can be seen that the transmission and display order of image frames

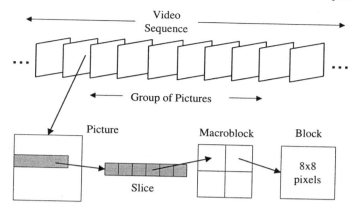

Fig. 11.12. Layer structure of MPEG-1 Video.

are different. The next reference frames, say frame number 4 and 7 in Figure 11.11, are transmitted before the B-pictures in between to reduce the storage and delay at the encoder and decoder. More precisely, since the reference frames are received before the B-frames in between, the decoder can start decoding the B-frames once they are received.

Table 11.2. Qmatrix for intracoded blocks.

8	16	19	22	26	27	29	34
16	16	22	24	27	29	34	47
19	22	26	27	29	34	34	38
22	22	26	27	29	34	37	40
22	26	27	29	32	35	40	48
26	27	29	34	38	46	56	69
26	27	29	34	38	46	56	69
27	29	35	38	46	56	69	83

Layer structure

A MPEG video stream is organized as a hierarchy of layers as shown in Figure 11.12 The first layer is the Sequence Layer which defines the overall video sequence. Context information for the stream, such as the image size and picture rate, is contained in this layer. The Constrained Parameters Bitstream (CPB) is designed for CD-ROM playback. These settings bound a sequence to no less than CCIR size (720×576 pixels), 30 fps with a bit rate of about 1.5 Mbps.

Below the Sequence layer is the Group of Pictures (GOP) Layer. It includes a header and a series of one or more pictures intended to allow random access into the sequence. The decoder can always begin decoding at the start of a GOP, without referencing the pictures that come before or after.

The next layer is the Picture Layer, which is the primary coding unit of a video sequence containing actual image data. Each picture can be divided into slices which contain consecutive macroblocks in raster scan manner. It allows decoding of the next slice, when transmission errors are encountered in the current slice. Motion compensation in P-and B-frames are performed at macroblock level which is chosen to be (16×16). A macroblock contains six (8×8) subblocks: four luminance blocks and two chrominance blocks. As mentioned earlier, the prediction residuals or the original pixels (if motion estimation fails) will be encoded by transform coding. The motion vectors and the encoded transform coefficients will be packed in the macroblock layer. At the block level, the transform coefficients are zigzag scanned and the run-level values of non-zero coefficients are entropy coded and transmitted. The end of transmission of each block is indicated by the end_of_block codeword.

11.4.4 MPEG-2 video coding standard

As mentioned earlier, the main purpose of MPEG-2 is to enable MPEG-1 like functionality for interlaced pictures, primarily using the ITU-R BT.601 (4:2:0) format. The target was to produce TV-quality pictures at data rates of 4-8 Mbps and high quality pictures at 10-15 Mbps (HDTV). A wide range of bit rates, applications, resolutions, signal qualities and services are also addressed.

The following are some major differences between MPEG-1 and MPEG-2 videos:

1. The chroma samples in the (4:2:0) format are located horizontally shifted by 0.5 pels relative to MPEG-1, H.261, and H.263.
2. MPEG-2 is able to code interlaced sequences in the 4:2:0 format and allow for much higher bit rates.
3. MPEG-2 allows additional scan patterns for DCT coefficients and motion compensation with blocks of size (16×8) pels.
4. MPEG-2 allows 10-bit quantization for the DC coefficient of the DCT. It also employs nonlinear quantization and better VLC tables.
5. It support various modes of scalability: spatial scalability which enables different decoders to extract videos of different picture sizes from the same bit stream; temporal scalability where a bitstream can be decoded into video sequences of different frame rates; SNR scalability where different amplitude resolutions are supported from the same bit stream.

Extending the concept of a constrained parameter set in MPEG-1, MPEG-2 defines profiles that describe the tools required for decoding a bit stream and levels that describe the parameter ranges for these tools. They constitute subsets of the MPEG-2 features and parameter ranges. For example, the Main profile supports I-, P-, and B-frames. The Main profile at Main level (MP@ML) is widely used, especially for TV broadcasting. The Multiview profile enables transmission of several video streams in parallel to support stereo presentations. This functionality is implemented using temporal scalability, which allows Main profile decoders to decode one of the video streams. The coding of Concentric Mosaics, light fields and other image-based representation is closely related to multiview coding of videos, except that the pixels or lines in the pictures have to be randomly accessed during rendering as we shall see later in Chapters 12 and 13.

11.4.5 MPEG-4 standard

As defined in ISO/IEC 14496, MPEG-4 is designed to address the requirements of a new generation of highly interactive multimedia applications over the various types of network, which range from digital television, streaming video, to mobile multimedia, games and so on. Apart from efficient coding of the multimedia objects, such applications also require other content-based functionalities, such as interactivity with individual objects, scalability of contents, and a high degree of error resilience.

The main difference between MPEG-4 from other coding standards is that it provides a rich set of tools to support content-based functionalities and object-based coding of natural and synthetic audios and videos, and graphics. MPEG-4 consists of a number of parts, and the main parts are the systems, audio, and visual. Due to page limitation, only the visual parts will be briefly described in this section. Interested readers are referred to [117, 314] for more details.

Since MPEG-4 allows video objects to be coded and manipulated separately, they are well-suited to the coding of pop-up light fields (Chapter 14) and other layered based image-based representation. In these applications, objects in the image-based representation need to be segmented and rendered using different depth information in order to reduce rendering artifacts. This calls for efficient tools for coding the shape, texture, and depth maps of the representation, which can be handled effectively using the framework of the MPEG-4 standard. An example system for light fields and the plenoptic videos will be discussed in more details later in Chapters 12 and 13.

MPEG-4 video coding standard

The visual part of MPEG-4 consists of a set of efficient coding tools that enables several classes of functionalities. The applications could range from 5 kbit/s (e.g., mobile multimedia) up to 1 Gbit/s (e.g., HDTV). The video formats supported include both progressive and interlaced scans, and the resolutions supported cover a broad range, typically from sub-QCIF for low bitrate video communications to 'Studio' resolutions at 4k × 4k pixels.

In order to provide functionalities such as content-based interactivity, MPEG-4 allows separate decoding and reconstruction of arbitrarily shaped video objects in a sequence, probably with certain semantic meaning. Coding and representing Video Objects (VOs) rather than video frames enables applications to access, manipulate, and interact with the content more flexibly. Consider the "Dance" sequence in Figure 11.13(a), where a lady is dancing in a gymnasium. To avoid rendering artifacts and provide efficient rendering in pop-up light fields, we need to segment the scene

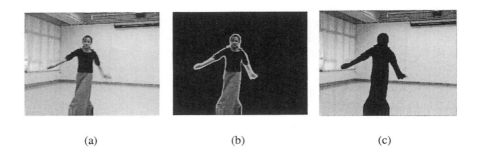

(a)   (b)   (c)

Fig. 11.13. (a) An image frame of the "Dance" sequence. Segmented VOPs for the (b) "Dancer" and (c) "Background" VOs.

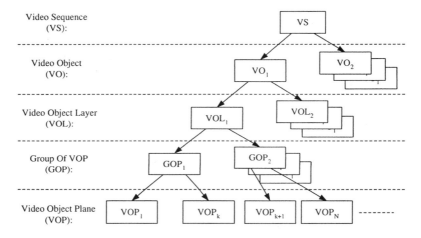

Fig. 11.14. Hierarchical bitstream structure of MPEG-4 video object coding.

into different depth layers. For the "Dance" sequence, one possibility is to segment the sequence into two objects - the lady and the background. In MPEG-4 terminology, the sequences of object constitute a Video Object (VO). Each VO can be further divided into different Video Object Layers (VOLs). For example, the wall and curtain of the room can be further segmented from the background. Generally, a VOL can represent different layers of a scalable bit stream or different parts of a VO. A time instant of a VOL is called a Video Object Plane (VOP). Figures 11.13(b) and 11.13(c) show the VOPs of the "Dancer" and the "Background" VOs segmented from the scene.

The hierarchical bitstream structure of MPEG-4 video object coding is shown in Figure 11.14 and the various terminologies are briefly summarized below:

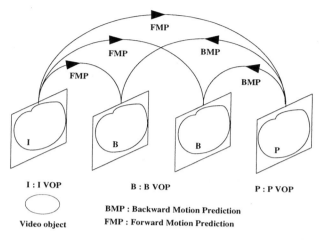

Fig. 11.15. Predictive structure using *I*-, *P*-, and *B*-VOPs.

- *Visual Object Sequence (VS):* As the highest syntactic structure of the coded visual bitstream, a VS can be considered as a conventional frame-based video to be coded. The scene of a VS may contain multiple 2D or 3D natural or synthetic objects and their enhancement layers.
- *Video Object (VO):* A VO corresponds to a particular object in the scene of video with three dimensions (2D plus time).
- *Video Object Layer (VOL):* Each VO can be encoded in scalable (multi-layer) or non-scalable form (single layer), depending on the application, represented by the Video Object Layer (VOL). Hence, the VOL provides support for scalable coding. One of the layers is called base layer, which can be decoded independently. Other layers are called enhancement layers, which can only be decoded together with the lower layers.
- *Group Of Video Object Planes (GOV):* A GOV consists of a group of successive VOPs, similar to the definition of Group Of Picture (GOP) in MPEG-2. GOVs can provide the random access points within the bitstream, thereby enabling random access functionality.
- *Video Object Plane (VOP):* A VOP is an instance of VO at a given time. A VOP is a rectangular video frame or a part thereof with an arbitrary shape. A VOP includes two different types of information, its texture (luminance signal and chrominance signal) and its shape (binary shape or gray-scale shape).

As mentioned earlier, VOP is a rectangular video frame or a part of it, and is defined by its texture (luminance and chrominance values) and its shape. Like MPEG-2, each VOP can be encoded by the motion, texture, and shape coding tools as *I*-, *P*-, and *B*-VOPs. Consecutive VOPs can be grouped into a Group Of Video Object Planes (GVOP). This is illustrated in Figure 11.15 where there are two *B*-VOPs be-

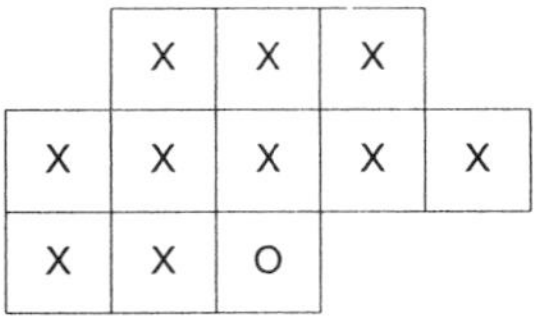

	X	X	X		
X	X	X	X	X	
X	X	O			

Fig. 11.16. Template for predicting the pixel "o" from pixels "x".

tween successively referenced VOPs (*I*- and *P*-VOPs). At the decoder, different VOs are decoded and composed into a scene and displayed. We now describe the coding of these VOs.

Object-based video coding

The shape of a VOP is described using alpha-maps. The alpha map of a VOP has the same resolution as its luminance component and the value at a particular position defines the transparency of the image pixel at the same location. Alpha maps can be binary or grey-scale valued. Binary alpha-maps of a VOP define pels that belong to it, while grey-scale alpha-maps further define the transparency of the image pixels in a VOP. The latter is very useful when one object is pasted on another background. By making the values of the alpha map gradually decreasing to zero along the VOP boundary, a more natural "matting" or "mixing" of the VOs can be achieved. Furthermore, excessive artifacts around image boundaries due to imperfect segmentation can be greatly reduced. Because of these reasons, alpha maps have also been extensively employed in the rendering of image-based representations such as the pop-up light fields and plenoptic videos, which will be described later in Chapter 13.

Once the shape of a VOP is known, we only need to encode the image pixels or texture inside the VOP. We now briefly describe the methods for shape and texture coding of VOPs.

Binary shape coding

The (16×16) binary alpha-map of an MB is called a binary alpha block (BAB). The rectangular alpha map bounding a VOP is first divided into non-overlapping blocks of size (16×16). The blocks that do not lie inside the VOP are called transparent. Those completely inside the VOP are called opaque. The blocks containing the boundary of the VOP are called boundary or binary alpha block (BAB) and they contain both transparent and opaque pels. For nonboundary blocks, the encoder can just signal whether the block is part of the VOP or not. For BABs at object boundaries, a method for coding binary image called context-based arithmetic coding (CAE) [142, 21] is employed.

In this method, pels are coded using arithmetic coding in scan-line order and row by row. A template of n pels is used to define the causal context (i.e., the value of the pels in the template) for predicting the value of the current pel. Figure 11.16 shows a template that uses ten pels to define the context for coding the current pel. Since the current pel is binary-valued and there are 10 binary pels in the template, we need to store the probabilities of the white or black symbol (because the other can be computed easily). There are altogether 1024 different combinations, with each corresponding to a combination of the context. By so doing, the arithmetic code can adapt efficiently to the spatially-varying statistics of the binary image.

There are also two different coding modes for boundary binary shape MBs, intra-CAE and inter-CAE. Intra-CAE is used to code shape MBs of I-VOP independently, while inter-CAE makes use of motion prediction from a binary shape mask reference, and they are used in P-VOPs and B-VOPs. Because the context in inter-CAE is derived from the current encoded as well as the reference pels, improved performance over intra-CAE is usually observed in coding highly correlated video sequences.

Texture coding

Texture coding in MPEG-4 is similar to hybrid motion/compensated DCT/DPCM coding in other video coding standards, except that special object-based tools are employed to handle the arbitrary shape VOPs. For transparent MBs, which lie completely inside the VOP, the coding is analogous to conventional video coding. For boundary MBs, two methods are available:

- Padding: The basic idea of padding is to extend the boundary MBs with arbitrarily shaped texture data into (16×16) MBs so that the traditional block based technique can be applied (Figure 11.17). The pels in the transparent region of a boundary MB are filled by extrapolating the pels from the opaque region in a way that not many high-frequency components are created as a result of padding. When a boundary MB is encoded in intra-mode (MB of I-VOP), the pels in the transparent region are first filled with the mean value of the texture pixels in the opaque region. Low-pass filtering is then applied to the transparent pixels by averaging the pels of its four neighbors. For inter-coded boundary MBs, they are padded with zero values.
- Shape-Adaptive DCT (SA-DCT): In SA-DCT as illustrated in Figure 11.17, the opaque pels are shifted vertically to form vertical sequences with different lengths. 1D DCTs with appropriate size are then applied to the opaque pels. Then, a similar operation is performed horizontally. SA-DCT also belongs to a part of Advanced Video Coding (AVC) and interested readers are referred to [275] for more details.

For P-VOPs and B-VOPs, the reference VOPs are usually padded to a larger size and motion estimation is performed using the current VO to be encoded with arbitrary size. If motion estimation fails, the VO will be intra-coded. Otherwise, the residuals will be padded to form a MB and inter-coded similar to traditional MPEG video coding standards. The block diagram of the overall coder is summarized in Figure 11.18.

In addition to the object-based coding tools, a number of efficient coding tools have been incorporated in the texture coding of the standard. Some of the major features are briefly summarized below:

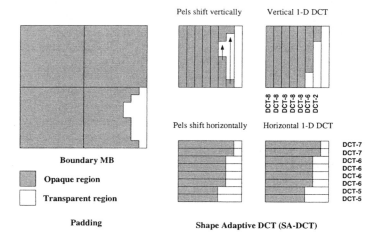

Fig. 11.17. Padding and Shape Adaptive DCT.

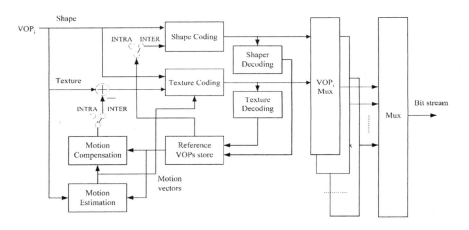

Fig. 11.18. Block diagram of the object-based coder in MPEG-4.

- Transformation: Similar to MPEG-1 and MPEG-2, DCT (Discrete Cosine Transform) is used as the major component to encode 2D (8×8) blocks of pixels.
- DC Coefficient Prediction: The DC coefficient value of a block after performing DCT can be predicted with reference to either the previous block or block above the current block.
- AC Coefficients Prediction: MPEG-4 employs AC prediction of DCT coefficients. The first column or row of AC coefficients of the current block is predicted by choosing the previous block or the above block as the predictors.

- Direct prediction mode: In this prediction mode, the MVs in recently coded P-VOPs, are appropriate scaled to yield forward and backward MVs for MBs in B-VOPs. As a result, much fewer bits are used for coding motion vectors.

- Selective Scan: MPEG-4 scans the AC coefficients by selecting a better one from the alternate vertical scan and alternate horizontal scan when the AC prediction is decided.

- 3D VLC: An improved 3D VLC is used to encode the DCT coefficients, similar to H.263.

- Unrestricted Motion Vectors: A much wider motion vector range up to 2048 pixels may be applied as in H.263.

- Four Motion Vectors: Like H.263, four Motion Vectors can be used in a Micro-Block (MB), one for each block.

- Sprite: A sprite is similar to a panorama and it efficiently represents the background of a VO throughout the video segment. The original sprite image can be transmitted once at the beginning of the transmission. In subsequent encoding, only the mapping parameters involved in warping the sprite, instead of the background, are transmitted. A very high compression efficiency can thus be achieved.

- Global Motion Compensation: MPEG-4 supports Global Motion Compensation (GMC). The main goal is to better represent and encode the global motion of a VOP resulting from camera motion, camera zoom, or large moving object. GMC coding supports the five transformation modes for the warping process: stationary, translational, isotropic, affine, and perspective. This tool is a part of *Advanced Video Coding* (AVC).

- Quarter-Pixel Motion Compensation: Compared to half-pixel motion compensation, quarter-pixel motion compensation gives better prediction with small syntactical and computational overheads. This tool is also a part of *Advanced Video Coding* (AVC).

Other functionalities and tools

Apart from the coding tools mentioned above, MPEG-4 also supports many other functionalities such as scalability, still texture coding, error resilience, rate control, etc. Interested readers are referred to [117] for more details.

12

Compression of Static Image-based Representations

12.1 The problem of IBR compression

In this chapter, the compression of image-based representations is discussed. Compression is an essential part of any practical IBR system because of the large data size of these representations. Unlike traditional image and video compression that we have surveyed in Chapters 10 and 11, there are specific requirements in IBR that need to be addressed. These requirements and different compression approaches in IBR will be explained in Section 12.1.1. The compression of several important static image-based representations including Concentric Mosaics and light fields are discussed in this chapter. Compression of dynamic representations including panoramic videos and plenoptic videos will be discussed later in Chapter 13.

12.1.1 IBR requirements

Image-intensive representations are usually densely sampled higher dimensional signals. The data sizes and dimensions of several image-based representations are summarized in Table 9.1. It can be seen that their data sizes are huge, but their samples are highly correlated. Direct application of traditional compression algorithms described in Chapters 10 and 11, however, usually results in sub-optimal performance. Providing random access to the compressed data for real-time rendering is another important and unique problem of IBR compression. Unlike conventional video coding, which supports random access at the picture or group of picture (GOP) level, higher dimensional image-based representations such as 3D Concentric Mosaics (CMs) require random access at the line level, whereas the 4D light field and Lumigraph require random access at the pixel level. As most existing compression algorithms employ entropy coding (such as Huffman or arithmetic coding) for better compression ratio, the symbols after compression are of variable sizes. It is, therefore, very time consuming to retrieve and decode a single line or pixel from the compressed data if there is no such provision for random access.

In addition, it is often impossible to decode the complete bit stream of a high dimensional representation in main memory for rendering due to its large data

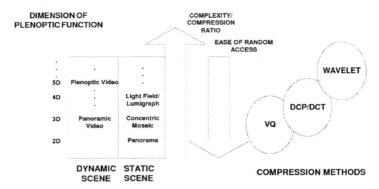

Fig. 12.1. Comparison of different image-based representation and compression methods in terms of their complexity. The ease of random access increases as the dimension of plenoptic function decreases, while the complexity and potential for compression both increase with the dimension. DCP refers to disparity compensation/prediction methods (see Section 12.1.2).

sizes. For example, the 3D CMs of the "Lobby" scene (Figure 2.9) require 297 MB of random access memory (RAM). To overcome this problem, VQ [267, 268] or just-in-time (JIT) decoding [164, 341] is usually used. Only those lines required for rendering are decoded online from the compressed images. Random access mechanisms at the "line level"are, therefore, needed to locate and decode individual compressed line image. These problems are even more pronounced in higher dimensional representations such as the light field and Lumigraph. Consider the 4D light field of the Buddha statue [160], which consists of a 32×32 array of images, each having a resolution of 256×256 with 24-bit per pixel. The total amount of storage is 192 MB. Decoding the entire light field into the main memory is, thus, prohibitive, especially when the resolution gets increasingly higher. Similar problems exist in the transmission of image-based representations. Techniques to support selective transmission/reception and a scalability data stream are, thus, of paramount importance. A simple comparison of different image-based representations and compression methods in terms of their complexities, compression ratios, and ease of random access is shown in Figure 12.1.

12.1.2 Different compression approaches

In general, there are two approaches to reduce the data size of image-based representations. The first one is to reduce their dimensionality, often by limiting viewpoints or sacrificing some realism. Light fields and CMs are such examples. The second approach is more classical, namely, to exploit the high correlation (i.e., redundancy) within the representation using waveform coding or other model-based techniques. The scene geometry may be used explicitly or implicitly. The second approach can further be classified into four broad categories, which are: pixel-based methods, disparity compensation/prediction (DCP) methods, model-based/model-aided methods and object-based approach.

Fig. 12.2. Snapshots of "Dance" sequence.

In pixel-based methods, the correlation between adjacent image pixels is exploited using traditional techniques such as VQ and transform coding. Very little geometry information, however, is used. In the DCP methods, scene geometry is utilized implicitly by exploiting the disparity of image pixels, resulting in better compression performance. (Disparity refers to the relative displacement of pixels in images taken in adjacent physical locations.) It is somewhat similar to motion of objects in video coding and they have been used in coding stereoscopic and multiview images [7, 158, 174, 204, 213, 214, 238, 285, 334]. As an illustration, Figure 12.2 shows a simplified light field called "Dance" constructed by two horizontally placed camera arrays. It can be seen that the positions of the same object in the light field images are shifted relative to each other. Since the disparity is related to the object depth, as well as the viewing geometries, these methods also implicitly use the scene geometry to improve their coding performances. In contrast, model-based/model-aided approaches [180, 181] recover the geometry of the objects or scene in coding the observed images. The models and other information such as prediction residuals [180] or view-dependent texture maps [181] are then encoded. It is clear that an image-geometry tradeoff also exists in IBR compression. In the object-based approach, the image-based representations are segmented into IBR objects, each with its image sequences, depth maps and other relevant information such as shape information. The main advantages of this approach is that it helps to reduce the rendering artifacts and hence the required sampling rate. It can also be viewed as an extension of the DCP method to individual IBR objects.

Pixel-based methods are easy to implement and, in some cases, the random access problem is usually less complicated. However, their compression performance is limited compared with the other approaches. The model-based/model-aided methods have the potential to offer higher compression ratios and other functionalities such as model deformation. On the other hand, it requires the acquisition of 3D models, and the encoding and decoding algorithms are more complicated. In this book, we only focus on the compression techniques of image-based representations. Details on geometry compression [65] and model acquisition are omitted.

We first review techniques for encoding image-based representations of static scenes. We start with compression techniques for CMs since its random access problem is the easiest to illustrate. We then deal with the compression of light fields in later sections.

12.2 Compression of Concentric Mosaics (CMs)

As described in Section 2.1.3, CMs are constructed from images captured using a forward-displaced rotating camera. A novel view is reconstructed by retrieving appropriate vertical lines from these images. It can be seen from Figure 2.5 that the CM has large spatial resolution, and thus it has to be compressed for efficient digital storage and transmission. It is natural to apply standard image compression techniques like transform coding, vector quantization and wavelet transform to compress these images because the images are highly correlated. Most of these techniques are based on a pixel-based method or DCP. The main problem is to provide a special mechanism to support random access at the line level.

12.2.1 Random access problem

As mentioned earlier, entropy coding usually complicates the rendering of the Concentric Mosaic because the symbols after compression are of variable size. In fact, it is very time-consuming to retrieve the slit images if the bit stream does not support any mechanism for randomly or efficiently accessing the compressed slit images. In the original work on CMs [267], image vector quantization (VQ) with a fixed vector size is used to simplify the random access problem. The compression ratio reported was 12 : 1. This can be viewed as a pixel-based method where correlation between adjacent image pixels are explored. Like the VQ method first proposed by Levoy and Hanrahan [160] to overcome the random access problem in light fields compression, the fixed size of the VQ index allows quick access to the required pixel data from the compressed light field or CMs for rendering. It also makes real-time decoding possible because VQ decoding involves only a simple look-up table. A compression ratio of 6 : 1 to 23 : 1 was reported in [160] for light fields at good reconstruction quality. However, the compression ratio of simple VQ is rather limited; it will also be unable to cope with future generations of image-based representations with extensive synthetic, as well as real world scenes.

Fortunately, since the mosaic images are retrieved column by column to reconstruct a novel view, the access pattern of the Concentric Mosaic, in contrast to the light field and Lumigraph, is relatively regular. This considerably simplifies the random access problem mentioned earlier. In fact, an entire image line instead of individual pixels is retrieved at the same time to reconstruct a novel view. Therefore, it becomes feasible to use a set of pointers for indexing the starting locations of each line in the compressed data. On the other hand, the storage overhead due to the pointers is still at an acceptable level. Moreover, several adjacent lines can be

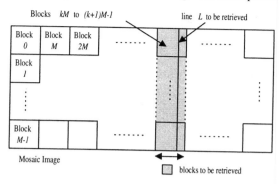

Fig. 12.3. Accessing a line L in a mosaic image.

grouped together to further reduce the storage overhead. This opens up new possibilities of using more advanced coding techniques for compressing the Concentric Mosaic. Of course, it requires more computational power to cope with the increased complexities of the decoder. But with proper choice of the coding algorithms and the processing power of today's computers, this is a promising and indeed possible direction.

Apart from this practical issue, successive mosaic images CM_k's also exhibit significant amounts of redundancy, similar to that of conventional video sequences. Therefore, DCP similar to motion compensation techniques can be used to reduce such redundancy making the representation more compact. Based on these observations, Shum *et al.* [268, 269] proposed an MPEG-like compression algorithm for CMs, which supports random access at the line level. We shall base our discussion on this algorithm and summarize other approaches later at Section 12.2.5.

12.2.2 Pointers structure

Let us consider the mosaic image in Figure 12.3. For simplicity, the image is assumed to be compressed by some block-based techniques, such as transform coding using the discrete cosine transform (DCT) (Section 10.2). The reasons for choosing a DCT-based codec are: i) its reasonably good performance at medium to high bit-rate, and ii) the availability of efficient software and hardware implementations. Other coding schemes such as the wavelet transform can also be used after appropriate modifications, as suggested in the following, to achieve fast decoding. In Figure 12.3, the image is divided into non-overlapping blocks of size 8×8 (or 16×16 if MPEG-2 algorithm is used). Here, the blocks are scanned vertically so that pixel data of each vertical line are contained in a group of consecutive blocks. In order to retrieve the pixel data of line L, the compressed data of blocks kM to $(k+1)M-1$ have to be located and decoded. Due to the use of variable length coding mentioned earlier, each compressed group of blocks (GOB) is of variable length. Locating the required data by searching the headers of the blocks can be very time consuming, especially for real-time rendering. To overcome this problem, a set of pointers to the

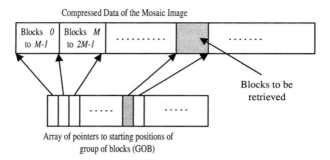

Compressed Data of the Mosaic Image

Fig. 12.4. Accessing a line L in a mosaic image.

starting locations of the vertical GOB in the compressed data is first determined and stored in an array prior to rendering. Alternatively, the pointers can be embedded in the compressed bit streams to avoid creating the pointer arrays when new mosaic images are loaded into the memory at the expense of slightly lower compression ratio. During rendering, the compressed data for the required GOB can be located very quickly. Figure 12.4 illustrates how the pointers can be used to locate the required compressed data. In the next section, a coding scheme similar to the MPEG-2 standard [120] is used to reduce the spatial redundancy of the mosaic images.

12.2.3 Predicting mosaic images using DCP

As mentioned earlier, successive mosaic images have a significant amount of spatial redundancy. Thus, DCP similar to motion estimation in video coding can be applied to exploit this redundancy. In [268, 269], a modified MPEG-2 video codec is used to compress the CM. Recall from Sections 11.4.3 and 11.4.4 that in the MPEG-2 algorithm, the *I*-pictures are coded using DCT-based coding without reference to other image frames. The *P*-pictures are predicted by motion compensation using the nearest encoded *P*- or *I*-pictures as references. The prediction residuals or the original pixels (if motion-compensation fails) are encoded using transform coding. Like *I*-pictures, *P*-pictures serve as prediction references for *B*-pictures and future *P*-pictures. *B*-pictures, which offer much higher compression ratio, are coded by bi-directional motion estimation and transform coding. By first transmitting and decoding the reference pictures, it is possible to decode each of the *B*-pictures in the group of pictures (GOP) independently so as to provide efficient access to individual pictures.

If each of the mosaic images is treated as a video frame, then it is possible to apply the MPEG-2 algorithm to compress and decompress the Concentric Mosaic. Generally, the mosaic images can be encoded in two different representations: multiperspective panoramas, and image sequences obtained in the normal setup (shown in Figure 12.5). The first approach normally leads to faster rendering speed, while the latter usually yields higher compression ratio.

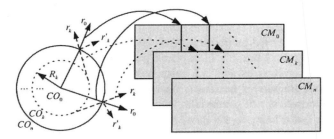

Fig. 12.5. Construction of the Concentric Mosaic from one circle, with the camera facing the normal direction.

Fig. 12.6. The data volume of the Concentric Mosaics: (top) "Kids" and (bottom) "Lobby."

Multiperspective panoramas

Multiperspective panoramas are obtained by putting together the lines $L_i(\theta)$ at the same horizontal position, i, for successive images captured in the normal setup (see Figures 12.5 and 12.8). This is illustrated in Figure 12.6 for the CMs "Lobby" and "Kids." Figures 12.7 and 12.8 show how the MPEG-2 algorithm can be used to compress the mosaic images and its equivalent multiperspective panoramas, respectively. The main reason for choosing multiperspective panoramas is their larger spatial redundancy between successive panoramas. The correlation between successive mosaic images, on the other hand, is slightly lower because the rays are further apart when the radius of the circles decreases (see lines I_{I_1}, I_{B_1}, and I_{I_2} in Figure 12.7).

For an I-picture, the pointer structure mentioned earlier can be used to access the compressed data of a group of blocks (GOB). These pointers can either be embedded in the compressed bit stream or created during decoding. If B-pictures are added to

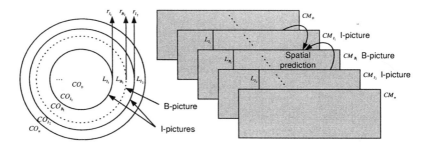

Fig. 12.7. Rays captured in the Concentric Mosaic and spatial prediction of mosaic images.

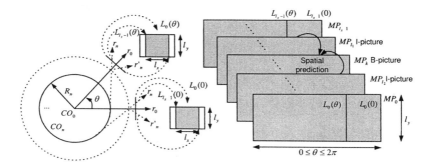

Fig. 12.8. Spatial prediction of multiperspective panoramas (MP).

achieve a higher compression ratio, the pointer structure would only allow us to efficiently decode the motion vectors and the prediction residuals of that GOB. Their predictors in the reference I-pictures still need to be retrieved. However, since the predictors in the I-pictures are in general located in different GOBs, these blocks in the I-pictures also need to be decoded. The situation is even worse if P-pictures are involved because they are in turns predicted from previous P-pictures. One solution to this problem is to decode all the I- and P-pictures and save them in memory for later use, or implement a cache to store previously decoded pictures, at the expense of higher memory requirement and complexity. In other words, there is a tradeoff between rendering speed and the amount of compression that can be achieved.

Fortunately, it was found in [269] that using the multiperspective panoramas, the number of I- and P-pictures required can be significantly reduced. Out of the 320 multiperspective panoramas of the Concentric Mosaic "Lobby," only two I-pictures at the beginning and the end, and 7 P-pictures in between are needed. Therefore, there are approximately 39 B-pictures between two P- or I-pictures. If the P-pictures are decoded when the Concentric Mosaic is loaded into the memory, then the complicated interdependence of the P-pictures mentioned above can be avoided. Consequently, faster rendering speed can be achieved. Alternatively, the P-pictures can be

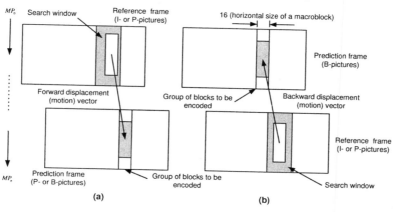

Fig. 12.9. Global motion estimation in multiperspective panoramas: (a) Forward prediction; (b) Backward prediction.

replaced with *I*-pictures to reduce the loading time, at the expense of a slightly lower compression ratio.

Another reason for the increased separation between the *P*- and *I*-pictures is that global motion estimation can be applied to the whole GOB in the multiperspective panoramas, before carrying out the motion estimation for the macroblocks in the MPEG-2 algorithm. This reduces the search range in the motion estimation of the MPEG-2 algorithm and hence the data required to represent the motion vectors. More precisely, the block size used in the global motion estimation is 16×208, which consists of a vertical stripe as shown in Figure 12.9. Because of the differences in field of view of the multiperspective panoramas, the upper and lower boundaries of a given multiperspective panorama do not, in general, appear in its previous neighbour, Figure 12.9(a). Therefore, only the middle portion is used to estimate the global displacement (motion) vectors for a group of blocks in "forward prediction," Figure 12.9(a). "Backward prediction" in the multiperspective panoramas is done similarly as shown in Figure 12.9(b). The global motion vectors are used as initial positions for carrying out the motion estimation of the macroblocks in the MPEG-2 algorithm. The differential motion vectors of the macroblocks in the vertical slice are coded using the MPEG-2 algorithm while the global motion vectors are included by modifying the GOB headers of the standard.

Normal setup sequence

The normal setup sequence as shown in Figure 12.5 can be viewed as a video sequence with a constant camera panning motion. As a result, there is more spatial correlation within the normal setup sequence than that of using multiperspective panoramas. Consequently, the compression ratio when using the normal setup sequence is also higher. The separation of the *I*-pictures, on the other hand, is much smaller than the former. This is due to the significant amount of uncovered scenes occurring at each image frame, which cannot be predicted by bi-directional motion estimation when the *I*-pictures are set further apart. Therefore, it would require a significant amount of storage to decode all the *I*-pictures and load them into memory for fast rendering. Instead, the required GOBs together with their predictors in the reference frames are selectively decoded. Therefore, the rendering speed is slightly slower than using multiperspective panoramas. For efficient rendering, *P*-pictures are usually not employed in the algorithm, due to their inter-dependencies.

Table 12.1. Coding results of the CMs using multiperspective panoramas (Y component).

"Lobby" (Q = 16)			"Kids" (Q = 24)		
GOP structure	C.R.	PSNR (dB)	GOP structure	C.R.	PSNR (dB)
I only	24.28	33.67	I only	22.05	28.87
I 5B	43.22	34.52	I 5B	33.67	29.94
I 10B	46.50	34.53	I 10B	35.64	30.02
I 20B	45.03	34.54	I 20B	35.28	30.07
I 39B	42.01	34.50	I 43B	33.14	30.06

C.R. stands for compression ratio.

Table 12.2. Coding results of the CMs using normal setup sequence (Y component).

GOP structure	"Lobby" (Q = 16)				"Kids" (Q = 24)			
	Q = 16		Q = 24		Q = 24		Q = 32	
	C.R.	PSNR (dB)	C.R.	PSNR (dB)	C.R.	PSNR (dB)	C.R.	PSNR (dB)
I only	30.49	35.15	39.76	33.15	29.48	31.07	36.19	29.67
I 1B	47.04	35.59	63.82	33.60	43.73	31.66	54.97	30.22
I 2B	56.90	35.72	79.25	33.74	50.70	31.79	64.94	30.33
I 3B	60.89	35.70	86.88	33.69	54.22	31.84	70.46	30.36
I 4B	63.92	35.74	92.80	33.74	55.61	31.83	73.09	30.34
I 5B	64.83	35.73	95.38	33.73	55.39	31.80	73.33	30.30
I 6B	65.28	35.72	97.08	33.71	54.24	31.77	72.22	30.25
I 7B	64.68	35.69	96.98	33.65	52.60	31.74	70.40	30.20
I 8B	63.63	35.66	96.11	33.61	50.91	31.71	68.36	30.16
I 9B	62.14	35.62	94.42	33.55	49.03	31.69	65.93	30.13

C.R. stands for compression ratio.

12.2.4 Compression results

We now illustrate the performance of the coding algorithms using the CMs "Lobby" and "Kids" described in Section 12.2.3 as examples. For simplicity, no rate control

Fig. 12.10. Rendered views of the Concentric Mosaic "Lobby" after decompression: (Left) using the normal setup sequence (compression ratio: 65.28; mean PSNR: 35.72 dB); (Right) using multiperspective panoramas (compression ratio: 42.01; mean PSNR: 34.50 dB)

algorithm is applied and a uniform quantization scale factor (Q) is used for all the I- and B-pictures.

For the CMs "Lobby", the multiperspective panoramas are represented as 320 24-bit true color panoramas, each having a resolution of 1350×240. Table 12.1 shows the compression results of the multiperspective panoramas. It can be seen that the coding performance is reasonably good even when the separation between successive I-pictures is increased to 39. Since there are only 9 reference I-pictures, they can be decoded and stored in the main memory for fast decoding of all the 320 image frames. The memory requirement for these reference I-pictures is only 8.34 MB ($1350 \times 240 \times 3 \times 9$ Bytes). It can also be seen that the use of global motion compensation considerably outperforms direct application of the MPEG-2 algorithm. A Peak Signal to Noise Ratio (PSNR) of 34.50 dB can be achieved at a compression ratio of 42.01 (or 0.571 bpp). On the other hand, the compression ratio for the Concentric Mosaic "Kids" is much lower than that of "Lobby" at a given PSNR. It is because the former contains significantly more details than the later.

The normal setup sequence of the CMs "Lobby" consists of 1350 (320×240) images with 24-bit true color. Table 12.2 shows its compression results with Q equal to 16 and 24. If only I-pictures are used, a Peak Signal to Noise Ratio (PSNR) of 35.15 dB can be achieved at a compression ratio of 30.49 (or 0.787 bpp). The performances of using different combinations of I- and B-pictures are also given. It can be seen that using more B-pictures improves the coding performance when the separation between successive I-pictures is less than 6 for Q equal to 16 and 24. When the compression ratio is increased to 65.28 (or 0.368 bpp), using 6 B-pictures between two consecutive I-pictures improves the PSNR to 35.72 dB. This shows that there is a significant amount of inter-frame redundancy in the normal setup sequence. Also shown in Table 12.2 are the coding results for the Concentric Mosaic "Kids" with Q equal to 24 and 32. As a comparison, a compression ratio of about 20 can be achieved using vector quantization and entropy coding [267]. Therefore, the MPEG-2 based compression algorithms can provide much higher compression ratios than simple vector quantization while preserving the random access capability.

Next, the rendering speed of the compression algorithms are briefly compared. In the experiments, there are respectively 39 and 6 B-pictures between successive I-pictures in coding the multiperspective panoramas and the normal setup sequence.

Table 12.3. Rendering speed comparison on Pentium II 300 MHz PC (frames per second).

Algorithm	352×168	800×372
VQ [267]	40.00	10.53
Multiperspective Panorama (I 39B)	41.67	10.87
Normal Setup Sequence (I 6B)	27.03	9.62

One thousand novel views at the center of the Concentric circles are continuously rendered with an angular spacing of 0.006 radians. The averaged rendering speed of the three algorithms on a Pentium II 300 MHz PC with 64 MB memory are shown in Table 12.3. It can be seen that the rendering speed of using multiperspective panoramas is comparable to that of using VQ. The algorithm using the normal setup sequence is about 35% slower than the formers but it can still achieve a rendering speed of 27 frames per second at a resolution of 352×168. Figure 12.10 shows the rendered views of the Concentric Mosaic "Lobby" obtained respectively from the compressed normal setup sequence and the multiperspective panoramas. They show good quality reconstruction with a compression ratio of 65 and 42, respectively.

12.2.5 Other approaches

A similar MPEG-like algorithm, called the reference block coder (RBC), was also proposed in [340]. The mosaic images are classified as anchor (A) and predicted (P) frames. A-frames are independently encoded in a similar manner as the *I*-pictures in MPEG-2, while the *P*-frames are encoded using DCP with reference to the surrounding A-frames. The *P*-frame in RBC differs from the *P*-pictures of MPEG-2 in that it refers only to the A-frames to facilitate random access. In addition, a two-level hierarchical table is embedded in RBC for indexed bit stream access. The compression ratio is slightly better than direct application of MPEG-2 after taking into account the regular panning nature of the image sequence. An interesting feature of RBC is the extensive use of data caches to reuse previously decoded macroblocks, which improves rendering speed. The rendering system is able to run smoothly on a Pentium II 300 desktop PC. The RBC was also the first algorithm that enabled the online streaming of CMs [342].

The application of wavelet transform to the compression of CMs was studied in [175, 328, 329]. Potential advantages of wavelet transform are its higher coding performance and ability to provide resolution and quality scalabilities. Direct 3D wavelet transform coding [151], however, yields a performance only comparable to that of MPEG-2. By using a smart rebinning approach to align successive images in a CM, the wavelet-based approach produces very encouraging results, which outperforms the MPEG-2 based algorithm by 3.7 dB on average. The success of the rebinning method is due to its ability to exploit the redundancy of multiple mosaic images arising from the disparity of image pixels.

The rendering operation is, however, complicated by the long filter support of the wavelet transform (compared with block transforms). In fact, decoding a given pixel involves decoding other adjacent pixels. To overcome this problem, the progressive inverse wavelet synthesis (PIWS) method [328] only performs the necessary inverse calculations to reconstruct the coefficient used in the current view. With extensive cache usage, PIWS was able to perform real-time rendering. A multiresolution subband coder using nonlinear filter bank [209] has also been proposed to overcome the long filter support of wavelet transform for progressive transmission.

12.3 Compression of light field

12.3.1 Conventional techniques

The light field and Lumigraph sample the plenoptic function in a 2D plane and generate a 2D array of images of the scene. Since adjacent light field images appear to be shifted relative to each other, there is considerably redundancy in the 4D data set. In additional to conventional pixel- and disparity-based methods, a number of model-based/model-aided algorithms that explicitly explore the scene geometry were proposed. We now briefly describe the various methods for light field compression under the first three categories mentioned previously. The object-based approach will be separately treated in Section 12.3.2.

Pixel-based methods

Earlier approaches on light field or Lumigraph compression were mostly based on conventional pixel-based methods. The original work of Levoy and Hanrahan [160] used VQ to provide random access in light fields; DCT coding [196] and wavelet coding [150, 230] were subsequently used. More recently, DCP and model-based/model-aided methods were proposed to achieve a higher compression ratio for storage and transmission.

DCP methods

Disparity compensated prediction, as with CMs, can be applied to predict one light field image from the others. This is illustrated in Figure 12.11, where the array of light field images is divided into I- and P-pictures. The P-pictures can be predicted by disparity compensation from the nearest encoded I-pictures, which are evenly distributed. An example is the V-coder described in [177, 179], which is based on the H.263 video coding algorithm [126]. Like conventional video coders, the P-images are divided into 16×16 blocks. Eight different coding modes are incorporated to efficiently exploit the characteristic of the light field. Mode selection was determined using a rate-constrained approach and was solved using the method of Lagrange multipliers. Prior to rendering, the I-images are decoded and kept in local memory to provide instantaneous access to a low-resolution version of the light field. However, rendering speed may be adversely affected if the compressed light field is decoded online. This is because random access of light rays (pixel) is unavailable.

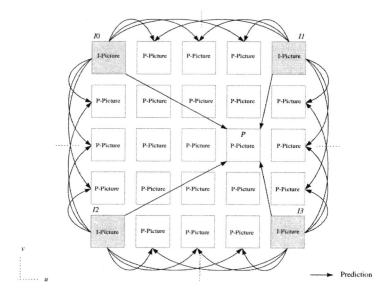

Fig. 12.11. DCP in light field compression.

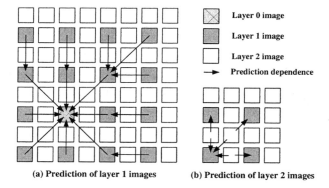

(a) Prediction of layer 1 images   (b) Prediction of layer 2 images

Fig. 12.12. Prediction order of an 8×8 light field in [298]. (a) Each layer 1 image is predicted from a nearest layer 0 or layer 1 image. (b) Each layer 2 image is predicted from two nearest layer 1 images. Only a quarter of the light field is shown.

Recently, Tong and Gray [298] combined disparity compensation prediction (DCP) and VQ (HDCP) and proposed a hierarchical light field coder as shown in Figure 12.12. The 2D array of light field images is divided into layers, with the lowest layer being vector quantized without any prediction. Images in higher layers are predicted from images in the lower layers using DCP. The prediction residuals are again vector quantized and different coding modes are incorporated to improve coding efficiency. In Figure 12.12, each layer 1 image is predicted from a nearest layer 0 or layer 1 image, whereas each layer 2 image is predicted from two nearest layer

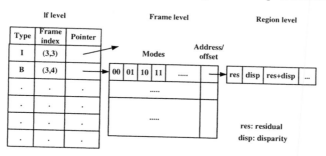

Fig. 12.13. The three levels of light field data structure proposed in [298].

1 images. To facilitate random access, the residuals and disparities are not entropy encoded. Moreover, the predictive coded images are divided into regions, and each is associated with a 4-Byte offset to support random access. Significantly better compression rates were obtained for the "Buddha," "Dragon," and "Lion" light fields, compared with using simple tree-structure VQ (TSVQ) (Appendix 10.5), due to the use of DCP. Figure 12.13 shows the three levels of light field data structures. The highest level contains three fields: the type, frame index, and an address pointer for each light field image or frame. The type field indicates whether the frame is coded as *I*-, *P*- or *B*-frames. The frame index is the (u, v) coordinate of the light field while the address pointer locates the corresponding compressed data. At the frame level, the compressed data are either arranged as fixed-length VQ indices for an *I*-frame or disparity and residual data for *P*- and *B*-frames coded with different modes. For example, the first macroblock in Figure 12.13 is coded with mode 00, which has no disparity and residual. Other macroblocks may be coded with residual only (01), disparity only (10) and both (11). They are referenced using the address/offset field. Finally, the compressed data are contained in the next level, the region level. In [298], one codebook is used for each layer.

The *D-coder*, which was also proposed in [178, 179], relies on disparity compensation of light field images. The four corner images in the image array are first encoded as *I*-images. Their disparity maps are then estimated and Huffman coded. From the encoded corner images and their disparity maps, the center image, and then the images midway between any two corner images, are predicted. The residuals, if any, are DCT coded. These nine encoded images are then used to divide the image array into four quadrants, each of which is recursively encoded using a similar method. Due to the hierarchical nature of the D-coder, the decoding of the image pixels is very time consuming. This slows down the rendering speed if the compressed data is decoded online.

Zhang and Li [341] have also extended the reference block coding to the encoding of Lumigraph using multiple reference frame (MRF) prediction. Disparity compensation is applied to the 2D light field array instead of the one-dimensional (1D) image sequence in CMs. As with *I*-images in [179, 298], certain images in the light field array are chosen as the anchor frames (A frames), which serve as references for predicting the remaining P-images. A two-level index table is incorporated into the bit stream for quick access to individual picture and macroblocks. Like CMs, this reduces the compression ratio. At a compression ratio of 100 : 1, the overhead incurred is 10%. The overhead increases to 30% when the compression ratio reaches 160 : 1. A caching scheme is also incorporated to speedup the rendering.

Model-based/model-aided methods

It has been shown that 3D scene geometry can improve coding efficiency and rendering quality considerably [42, 323]. The model-based coding (also known as texture-based coding) proposed in [181] makes use of the scene geometry to convert the images from a spherical light field to view-dependent texture maps. These maps exhibit greater inter-map correlation than the original images and are more effectively encoded using a modified set partitioning in hierarchical trees (SPIHTs) 4D wavelet codec [250]. On the other hand, model-aided predictive coding [180] makes use of geometry information to morph and predict new views from already encoded images. The prediction residuals are encoded using DCT-based coding. Like the hierarchical light field coder in [298], a decimated version of the spherical light field array are encoded as intra- or I-pictures, and they serve as references for predicting images at the next layer. By arranging the images in a hierarchical manner, a multiresolution representation of the image data is obtained which facilitates progressive rendering and decoding. Both algorithms encode the geometry of the objects using the embedded mesh coding (EMC) in which the vertex coordinates and mesh connectivity are jointly encoded to provide better scalability and improved performance. Experimental results showed that the model-aided approach is more robust to variations of the geometric models. Readers are referred to [182] for more details.

12.3.2 Object-based light field compression

A difficult problem of rendering light fields is the excessive artifacts due to depth variations. If the scene is free of occlusions, then the concept of plenoptic sampling [33] (Part II) can be applied to determine the sampling rate in the camera plane. Unfortunately, because of depth discontinuities around object boundaries, the sampling rate is usually insufficient around object boundaries and significant rendering artifacts due to occlusion are observed. Moreover, appropriate mean depths for objects have to be determined to avoid blurring within the objects and ghosting at the boundaries. Thus, depth segmentation or some kind of depth information is necessary in order to improve the rendering quality. Motivated by Gortler *et al.*'s work on Lumigraph [91] and the layered depth images of Shade [264], [80] proposed to augment each image pixel in a static and dynamic light field with a depth value. Due to the limited amount of information that we can gather from images and videos, a very high-resolution depth map is usually unavailable. Besides, the data rate of these detailed depth maps sequences is very high. Fortunately, plenoptic sampling tells us that dense sampling of image-based representation will tolerate this variation within the segments by interpolating the plenoptic function. In other words, it is highly desirable to focus on objects with large depth discontinuities. By properly segmenting the light field images into objects at different depths, the rendering quality in a large environment can be considerably improved using mean depth values [270].

These observations motivated Shum *et al.* [270] to propose a light field representation called pop-up light fields. Later, Gan *et al.* [79] developed an object-based approach to simplified dynamic light fields called the plenoptic videos. In pop-up light fields, to be described in Chapter 14, and the object-based plenoptic videos, the light field images are segmented into IBR objects, each with its image sequences, depth maps and other relevant information such as shape information. Moreover, this allows other content-based functionalities such as scalability of contents, error resilience, and interactivity with individual IBR objects to be incorporated similar to the MPEG-4 standard. For instance, IBR objects can be processed, rendered and transmitted separately to meet different requirements of the channel, processing speed, and presentation styles. In wireless transmission, different IBR objects might be given different number of bits (and different amounts of channel coding) and hence different reconstruction qualities (error resilience). They might also be transmitted at different frame rates to achieve object scalability. In this chapter, we shall describe how these video objects are segmented, rendered, and inpainted. The compression aspects of the object-based plenoptic videos are described later in Section 13.4.

Using the object-based representation, it is relatively simple to detect possible occlusions during rendering and estimate the rendered pixels. In the framework of plenoptic sampling, the operation can be viewed as a spatially varying reconstruction filter in the frequency domain. Basically, our rendering algorithm explores and observes the physical model and constraints of image formation so that the rendering quality can be improved at lower sampling rate. In [81], a portable plenoptic video system, which consists of two linear arrays each carrying 6 video cameras, for large and dynamic environment scenes was constructed to demonstrate the usefulness of the object-based approach.

Here, we shall review some important aspects of the object-based approach proposed in [81]. In Section 12.3.3, an interpretation of the object-based approach in the context of plenoptic sampling is given. Then, the object tracking, rendering and matting algorithms will be described respectively in Sections 12.3.4 and 12.3.5.

12.3.3 Sampling and reconstruction of light fields

In Part II, we described the concept of plenoptic sampling [33], where the number of pictures required to render a given scene or the sampling density was studied. Here, we shall further consider its applications and interpretation in the context of object-based representation of IBR. For the standard two-plane ray space parameterization, the camera plane and the focal plane are respectively parameterized by the parameters (s, t) and (u, v). Each ray in the parameterization is uniquely determined by the quadruple (u, v, s, t). For fixed values of s and t, we obtain an image taken at a location indexed by (u, v). Interested readers are referred to [91, 160] for more details.

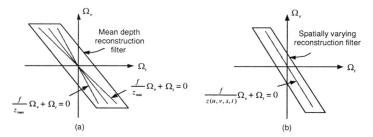

Fig. 12.14. Left to right: (a) Mean depth reconstruction filter. (b) Spatially varying reconstruction filter with local mean depth.

Assuming a pinhole camera model, the pixel value observed is the convolution of the plenoptic function $l(v,t)$ with the point spread function. If we know the spectral support of the Fourier transform of $l(v,t) : L(\Omega_v, \Omega_t)$, then it is possible to apply the sampling theorem to predict the required sampling density. Assuming that there are no occlusions or depth discontinuities, it was found that the spectral support of $L(\Omega_v, \Omega_t)$ is dependent on the depth of the objects as shown in Figure 12.14 (a) for a 2D light field. From the figure, it can be seen that objects at a certain depth z will appear as a line in the frequency domain. Thus, if the maximum and minimum depth values are known, a reconstruction filter using the mean depth of the scene can be used to reconstruct the light field from its samples and it also defines the sampling density for a given sampling geometry. This result allows us to determine the sampling rate for proper reconstruction of the light field and to avoid undesirable aliasing effects.

Since the Fourier transform of a light field is a global frequency description of the entire light field, it only gives us the frequency components or spectrum in the entire light field, but not its local behavior. For scenes with large depth variations, objects with different depth values will contribute to the entire spectrum. If we window the light field at a particular location (u, v, s, t), and compute its Fourier transform at this location, it will give us its local frequency content. For regions with less depth variations, we would expect a spectrum similar to that shown in Figure 12.14 (b) with an orientation predicted approximately by plenoptic sampling. A reconstruction filter tailored for this particular mean depth can be used to reconstruct the light field locally. Therefore, the reconstruction filter should be spatially varying and it should depend on the local depth image of the light field. Ideal reconstruction filters with support shown in Figure 12.14 (a) usually have long filter length and they will cause ringing artifacts in reconstruction. The spatially varying reconstruction allows simpler reconstruction filter such as bilinear interpolation to be used, if the local mean depth of the region is known.

Although the quality of rendering will be improved with the amount of depth information or geometric information we have, very accurate depth values are generally not required inside regions with limited depth variations according to plenoptic sampling. At object boundaries, image pixels cannot be interpolated simply because of a discontinuity generated by the occlusion. Fortunately, the light field looks like a piecewise continuous 2D signal with pixels from foreground covering those from the background. If the objects are nearly coplanar and if they are at a large distance from the cameras, the magnitude of the discontinuity will be smaller and fewer artifacts will be generated. Therefore, the previous analysis using locally adaptive reconstruction filter and plenoptic sampling can be applied to individual objects except around the boundaries.

In sampling theorem, the band-limited signal is reconstructed by interpolation of the data samples. On the other hand, in recovering piecewise continuous signals, the discontinuity has to be identified and interpolation/extrapolation are performed independently on each side of the discontinuity in order to avoid excessive artifacts. In other words, by exploring the structure of the light field or the physical geometry, it is possible to reduce the sampling rate in order to get an acceptable reconstruction, provided that depth information, especially the location of the depth discontinuity, can be identified. Therefore, methods for detecting and handling occlusion are important issues. Rendering algorithms using this concept and the pixel-depth representation that we have mentioned earlier will be described in the next section. It is shown that the rendering quality can be significantly improved with additional depth information.

12.3.4 Tracking of IBR objects

As mentioned earlier, objects at large depth differences are segmented into layers and are compressed and rendered separately. This helps to avoid the artifacts at object boundaries due to depth discontinuities. In the method in [81], an initial segmentation of the objects is first obtained using a semi-automatic approach. Tracking techniques are then employed to segment the objects at other video streams and subsequent time instants. The method in [81] is based on the level-set method or geometric partial differential equations (PDE). The use of PDE and curvature-driven flows in tracking, segmentation and image analysis has received great attention over the last few years [263, 224, 252, 188, 221]. The basic idea is to deform a given curve, surface, or image according to the PDE, and arrive at the desired result as the steady state solution of this PDE. The problem can also be viewed as minimizing a certain energy function:

$$U_I(C) = \int_I F(C, x) dx \qquad (12.1)$$

as a function of a curve or surface C. The subscript indicates that the energy is computed from the given images I. Usually, $F(C, x)$ is designed to measure the deviation of the desired curve from C at point x. To minimize the functional in (12.1), the variational approach can be employed to convert it to a PDF. A necessary condition for C to be a local minimum of the functional is $U_I'(C(t)) = 0$. A general numerical approach is to start with an initial curve C_0 and let it evolve over a fictitious time variable t according to a PDE, which depends on the derivative $U_I'(C(t))$ as follows:

$$\frac{\partial C(t)}{\partial t} = U_I'(C(t)). \tag{12.2}$$

However, conventionally finite difference methods are unsuitable to solve (12.2), because the PDE might be singular at certain points. A major breakthrough in solving (12.2) is due to Osher and Sethian [222], and the method is commonly referred to as the level-set method. The basic idea behind the level-set method is to represent a curve or surface in "implicit form" such as the zero level sets or isophone of a higher dimensional function. More formally, the time evolution of curves $C(x, t)$ is represented as the level set of an embedding function $\phi(x, t)$:

$$L_c(x, t) := \{(x, t) \in R^2 : \phi(x, t) = c\} \tag{12.3}$$

where c is a given real constant. (12.2) can be rewritten as a PDE of $\phi(x, t)$ as follows:

$$\frac{\partial \phi(t)}{\partial t} = \beta \|\nabla \phi\| \tag{12.4}$$

where β is the velocity of the flow in the normal direction and it is derived from above. The initial curve C_0 is associated with the level set with $c = 0$, i.e., zero level set, and its time evolution is computed numerically by solving the following equation for $\phi(t)$, after discretizing at a sufficiently small time interval or step Δt:

$$\phi((n + 1)\Delta t) = \phi(n\Delta t) + \Delta t \cdot G(\phi, x) \tag{12.5}$$

where $G(\phi, x)$ is an appropriate approximation of the right hand side of (12.4). The desired solution is obtained when the PDE converges at sufficiently large value of n. For the object tracking problem in [81], the following energy function for curve C is defined:

$$U_I(C) = \alpha \int_I C_{inside}(x, y) dx dy - \beta \int_I C_{outside}(x, y) dx dy + \lambda Length(C) \tag{12.6}$$

where $C_{inside}(x, y)$ and $C_{outside}(x, y)$ are two functions designed respectively to control the expansion and contraction of the curve C at location (x, y), and $Length(C)$ measures the length of the curve. If we assume that the pixel values are independent and Gaussian distributed with means c_{in} and c_{out} respectively inside and outside the curve, then it can be shown that the PDE so obtained can be written as:

$$\left.\frac{\partial \phi}{\partial t}\right|_{(x,y)} = \alpha(u_{(x,y)} - c_{in})^2 - \beta(u_{(x,y)} - c_{out})^2 + \lambda \cdot div(\frac{\nabla \phi}{|\nabla \phi|}) \tag{12.7}$$

where α, β and γ are positive parameters, $u_{(x,y)}$ is the value of pixel (x, y), c_{in} denotes the driving force inside the curve C, and c_{out} represents the driving force

Fig. 12.15. (a) Tracking result of global-based method. (b)-(d) Tracking results of our method.

outside the curve C. The third term, which is derived from $Length(C)$, makes the curve smooth and continuous.

There are two different methods for determining c_{in} and c_{out}: global-based and local-based methods. The global-based method which is adopted in [38] utilizes all the pixels to drive the curve C, where c_{in} denotes the mean of all pixels inside the curve C, and c_{out} is the mean of all pixels outside the curve C. There are many advantages associated with a global-based method, e.g., fast evolution speed and insensitive to noise. However, some fine features along the boundary of the objects to be tracked might be lost. Figure 12.15(a) shows an example tracking result using the global-based method. It can be seen that the girl's right hand is outside the curve, because its mean is more similar to the background than to its body. On the contrary, local-based method uses local mean value inside a window instead of all the image pixels. In [188, 337], a local-based method is exploited, where c_{in} and c_{out} are set as follows: $c_{in} = u_{(x+i,y+j)}$, where $(u_{(x,y)} - u_{(x+i,y+j)})^2$ is the minimum value over all integer pairs (i, j) such that $|i| \leq m$ and $|j| \leq m$ and pixel $(x + i, y + j)$ is inside the curve C; $c_{out} = u_{(x+i,y+j)}$, where $(u_{(x,y)} - u_{(x+i,y+j)})^2$ is the minimum value over all integer pairs (i, j) such that $|i| \leq m$ and $|j| \leq m$ and pixel $(x + i, y + j)$ is outside the curve C. Obviously, this method utilizes local features of the image to cope with objects having a non-uniform energy distribution. Unfortunately, this method is rather sensitive to image noise, because only one pixel is chosen for determining both c_{in} and c_{out}. In [81], combining the advantages of both the global-based and local-based methods is proposed by employing the following c_{in} and c_{out}:

$$\begin{cases} c_{in} = \text{average}(u_{(x+i,y+j)}), \text{ where } |i| \leq m, |j| \leq m \\ \qquad\qquad \text{and pixel } (x + i, y + j) \text{ is inside the curve } C \\ c_{out} = \text{average}(u_{(x+i,y+j)}), \text{ where } |i| \leq m, |j| \leq m \\ \qquad\qquad \text{and pixel } (x + i, y + j) \text{ is outside the curve } C \end{cases} \qquad (12.8)$$

12.3.5 Rendering and matting of IBR objects

The depth information is estimated for each IBR object after it has been segmented using Lazy snapping [165]. In [80], a depth matching algorithm for rendering and post-processing of plenoptic video with depth information was proposed. This algorithm brought satisfactory rendering results, but the arithmetic complexity of this algorithm is very high. An improved rendering algorithm with a much low computational complexity is described here.

More precisely, instead of finding the depth value of the image pixel to be rendered from adjacent light field images, the two images are projected using the depth values of each pixel to the current viewing position. Consider the reconstruction of a pixel V in the viewing grid. If the two pixels obtained from projecting the left and right images to the position of pixel V have the sample depth values, then there is no occlusion and the value of V can be interpolated from these pixels according to bilinear interpolation. On the other hand, if their depth values differ considerably (say larger than a threshold), then occlusion is said to be occurred. The projected pixel with a small depth value will then occlude the other. Therefore, the value of pixel V should be equal to the one with a smaller depth value. Furthermore, if multiple pixels are projected to the location of pixel V, the intensity of pixel V is assigned to the one with the smallest depth value. If only one pixel from the left or right image is projected to the position of pixel V, the intensity of pixel V is set to the intensity of this pixel. Finally, due to occlusion, pixel V might not have any projected pixels from adjacent light field images. In this case, the image consistency concept is used to "guess" the intensity of these pixels [80] from neighboring rendered pixels using interpolation. A linear interpolation from the two image pixels just before and after this occlusion region is performed to cover all undetermined pixels. Image inpainting techniques [56] can also be employed to fill in holes that result from occlusion or change of viewpoints.

As mentioned earlier, due to possible segmentation errors around boundaries and finite sampling at depth discontinuities, it is preferred to calculate a soft, instead of a hard, membership function between the image-based objects and the background. In other words, the boundary pixels are assumed to be a linear combination of the corresponding pixels from the foreground and background:

$$I = \alpha F + (1 - \alpha)B \qquad (12.9)$$

where I, F and B are the pixel's composite, foreground and background colors, and α is the pixel's opacity component or the alpha map. Using this model, it is possible to matte a given object with the original background at different views and other backgrounds. The digital analog of the matte (the α-map) is introduced by Porter and Duff [235] in 1984. In natural matting, all variables α, F and B need to be estimated and the problem is to find the most likely estimates for α, F and B, given the observation I. This can be formulated as the maximization of the posteriori probability $P(F, B, \alpha | I)$. Using the Bayesian rule, we have:

$$\max_{F,B,\alpha} P(F, B, \alpha | I) = \max_{F,B,\alpha} P(I | F, B, \alpha) P(F, B, \alpha) / P(I). \qquad (12.10)$$

Since the optimization parameters are independent of $P(I)$, the latter can be dropped. Further, if F, B, α are assumed to be independent, then (12.10) can be written as:

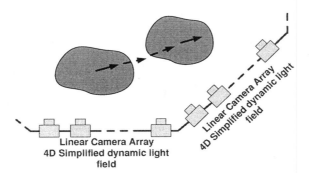

Fig. 12.16. Multiple linear camera array of 4D simplified dynamic light fields with viewpoints constrained along line segments.

$$\arg\max_{F,B,\alpha} P(F, B, \alpha | I) = \arg\max_{F,B,\alpha} P(I | F, B, \alpha) P(F) P(B) P(\alpha)$$
$$= \arg\max_{F,B,\alpha} \{ \ln P(I | F, B, \alpha) + \ln P(F) + \ln P(B) + \ln P(\alpha) \}. \quad (12.11)$$

Taking the derivatives of (12.11), one gets a set of equations in the estimates of α, F and B. Interested readers are referred to [47] for more information.

It can be seen from the above discussion that for the proper rendering of an image-based object, we also need an alpha map and additional geometrical information in the form of a depth map, apart from its conventional texture pictures or maps. The alpha map is produced through the natural matting method discussed above, and they are used in the composition and rendering of the image-based objects. This information needs to be compressed for efficient storage and transmission of the plenoptic videos. More details will be discussed in Sections 13.3 and 13.4. We now present some experimental results of the object-based light field system.

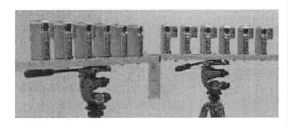

Fig. 12.17. Two linear camera arrays, each consists of 6 JVC video cameras.

12.3.6 Experimental results

The performance of the above tracking method is evaluated using the "Dance" and the "Pingpong" sequences captured by the IBR system in [79]. The system is designed to capture both static and dynamic simplified light fields. The simplified dynamic light field is also called the plenoptic video and it is a 4D plenoptic function. It is obtained by capturing videos, which are regularly placed along a series of line segments, instead of a 2D plane in the static light fields [36, 79], as shown in Figure 12.16. The main motivation is to reduce the large dimensionality and excessive hardware cost in capturing dynamic representations. Despite the simplification employed, plenoptic videos can still provide a continuum of viewpoints, significant parallax and lighting changes along line segments joining the camera arrays. More details of plenoptic videos will be given in Section 13.3. Figure 12.17 shows the plenoptic video system constructed in [79] for capturing dynamic scenes. This system consists of two linear arrays of cameras, each hosting 6 JVC DR-DVP9ah video cameras. The spacing between successive cameras in the two linear arrays is 15cm and the angle between the arrays can be flexibly adjusted. More arrays can be connected together to form longer segments. Because the videos are recorded on tapes, the system is also more portable for capturing outdoor dynamic scenes.

Along each linear camera array, a dynamic 4D or a static 3D simplified light field can be captured, and the user's viewpoints are constrained along the linear arrays of video cameras. The use of multiple linear arrays allows the user to have more viewing freedom in sports events and other live performances. It also represents a design tradeoff between simplicity and viewing freedom. Other configurations can also be employed. The cameras are calibrated using the method in [346]. In order to use this method to calibrate the camera array, a large reference grid was designed so that it can be seen simultaneously by all the cameras. Using the extracted intrinsic and extrinsic parameters of the cameras, the videos of the cameras can be rectified for rendering. After capturing, the video data stored on the tapes can be transmitted to computers through FireWire interface. All these components are relatively inexpensive and they can readily be extended to include more cameras. For each frame, the initial curve C_0 is the tracking result of the previous frame, and the object curve of the first frame is obtained manually by lazy snapping, which is a semi-automatic method. The level-set contour evolution is implemented using the narrow band method, where (12.7) is used as the speed function. The window size m for the local energy calculation is fixed to 6.

Figure 12.15(a) shows typical tracking result of the global-based method for the "Dance" sequence as shown in Figure 12.2. The tracking results are shown in Figures 12.15(b)-(c), where the boundaries of the objects are well delineated. It can be seen from the results that the method in [80], [81] gives more reasonable result for objects with non-uniform energy distribution. Although this method is capable of tracking the objects satisfactorily for a number of frames (such as 30), the performance will start to deteriorate due to accumulation of tracking errors. This is illustrated in Figure 12.15(d) where parts of the girl's head and right hand are not well delineated. This problem can be alleviated by incorporating the motion information of the objects.

Fig. 12.18. Layered depth map of "Dance" sequence.

Fig. 12.19. Rendering results of "Dance" obtained by the algorithm proposed in [81].

Fig. 12.20. (a) Input image (left). (b) alpha map. (c)-(d) New images of compositing extracted foreground over other background scenes.

Figure 12.18 shows some of the layered depth maps of the "Dance" sequence, and the rendering results are shown in Figure 12.18. The results of natural matting the image-based object are illustrated in Figure 12.20. Figures 12.20(a) and (b) show an example snapshot of a segmented image-based object called "Dancer" and its associated alpha map computed. Figures 12.20(c) and (d) show example renderings of the image-based object, after matting with two different backgrounds or scenes.

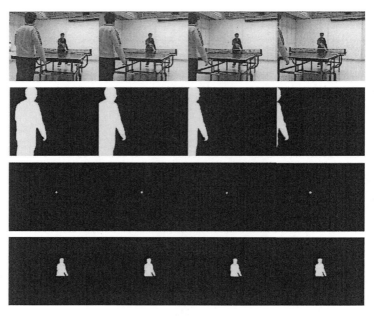

Fig. 12.21. Segmentation results of the "Pingpong" sequence from the 1st linear camera array. Top: Snapshots of original simplified light field images. Second to forth rows: alpha maps of the image objects "Player 1", "Ball", and "Player 2."

Figures 12.21 and 12.22 show additional segmentation results of another simplified light field sequence called the "Pingpong" sequence. Some of the rendering results at different viewing angles are shown in Figure 12.23. It can be seen that the object-based approach yields high quality renderings and it is effective to suppress the ghosting and blurring artifacts in a conventional approach with a single mean depth.

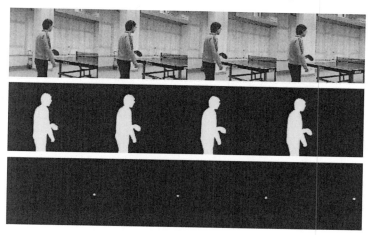

Fig. 12.22. Segmentation results of the "Pingpong" sequence from the 2rd linear camera array. Top: Snapshots of the original simplified light field images. Second and third rows: alpha maps of the image objects "Player 1" and "Ball."

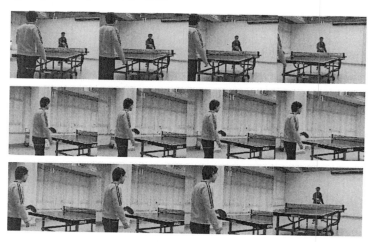

Fig. 12.23. Renderings of the "Pingpong" sequence.

13

Compression of Dynamic Image-based Representations

13.1 The problem of dynamic IBR compression

The compression algorithms of image-based representations discussed so far are associated with static scenes. The compression and transmission of general dynamic image-based representations are not well studied. This is mostly due to the difficulties in the capturing, processing and rendering of dynamic image-based representations, which are usually of high dimension. In fact, there are several practical as well as theoretical problems in capturing and rendering dynamic image-based representations. First of all, to provide users with a good immersive viewing experience, the viewing freedom has to be sufficiently large. This calls for a considerable number of synchronized video cameras. Secondly, the simultaneous recording of these video streams might require sophisticated compression hardware, which can be very expensive. Thirdly, the processing and rendering of these video streams to provide real-time of nearly real-time performance can be prohibitively large. Finally, the calibration of the multiple video cameras is also very complicated and its quality will affect significantly the rendering quality of the representations. Nevertheless, the ability of image-based techniques in creating photorealistic images of real scenes has stimulated a lot of interest in constructing sensor systems for capturing dynamic environments from multiple viewpoints. We have seen some example systems in Chapter 3. These include notably the early examples of the Stanford University, Stanford, CA, Multicamera Project[1] and the Carnegie-Mellon University, Pittsburgh, PA, Virtualized Reality Project [133]. The goal of the Multicamera Project is to build an array of 128 video cameras using low-cost CMOS camera, inexpensive lens, and other processing and compression hardware. A prototype system with six cameras was reported [317]. The Virtualized Reality Project uses a set of synchronized cameras, and allows the viewer to virtually fly around and watch the event from new positions. This is made possible by reconstructing 3D (octree) models at every frame offline.

---

[1] [Online]. Available: http://graphics.stanford.edu/projects/array/

In this chapter, we shall describe the construction and compression of two dynamic image-based representations called panoramic videos [210] and plenoptic videos [36], which are 3D and 4D plenoptic functions. We first start with the panoramic video in Section 13.2, which has a much lower data requirement. Then, we proceed to a simplified dynamic light field called plenoptic videos in Section 13.3. Other related approaches will be described. Finally, an object-based approach for the compression and rendering of plenoptic videos are described. The main advantages of using the object-based representation are: 1) by properly segmenting IBR into objects at different depths, the rendering quality in large environment can be significantly improved (Section 12.3.2); 2) by coding the plenoptic video at the object level, desirable functionalities such as scalability of contents, error resilience, and interactivity with individual IBR objects (including random access at the object level), etc, can be achieved. More information on the segmentation or object-based approach can be found in [270] and Chapter 14 on pop-up light fields.

13.2 Compression of panoramic videos

A panoramic video [20, 10, 77] is a sequence of panoramas taken at different time instants. It can be used to capture dynamic scenes at a stationary location or in general along a path, which is also known as a dynamic or time-varying environment map. It is basically a video with 360 degrees of viewing freedom. Another application of panoramic videos is to implement virtual walkthrough applications where a series of panoramas of a static scene along a given path is captured. Therefore, it is a static environment map where one can freely navigate along predefined paths and freely change their viewpoints. Much emphasis has been put on the construction of panoramic videos and how they can be constructed and rendered [288, 20, 271, 10, 77, 114, 4, 83].

Although the amount of data associated with panoramic videos is significantly reduced when comparing to other possible dynamic image-based representations, it can still be very high, thereby posing a number of practical problems when good resolution and interactive response are required. To illustrate the severity of this problem, let us consider a 2048 × 768 panoramic image without compression. It will occupy about 4.5 MB of storage. A 25 fps video at this resolution would require 112.5 MB/s of digital storage or transmission bandwidth. Another problem of high-resolution panoramic videos is the high computational complexity in software-only real-time decoding.

In this section, efficient methods for the compression and transmission of high-resolution panoramic videos for both dynamic environment maps and virtual walkthrough applications will be introduced. Our discussion is based on the work in [210]. For dynamic environment map applications, a high-performance MPEG-like compression algorithm, which takes into account the random access requirement in changing one's viewing angle and the redundancy of panoramic videos, was presented in [210]. For virtual walkthrough applications, the indexing structure proposed in [268] is employed to support random access for individual panoramic im-

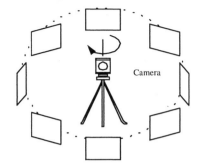

Fig. 13.1. Construction of a panoramic mosaic.

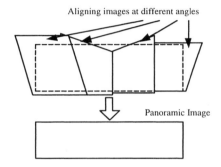

Fig. 13.2. Mapping of images onto a cylinder to generate a panoramic image.

ages so that the user can freely change the viewing position and angle along pre-defined paths. The transmission of panoramic videos over cable networks, local area networks (LANs) and the Internet are also briefly discussed.

13.2.1 Construction of panoramic videos

A panoramic mosaic can be obtained by projecting a series of images (after registration and stitching) on a cylindrical or spherical surface. Figures 13.1 and 13.2 show the construction of a panoramic mosaic. Since it is obtained by stitching several images together, its resolution is usually very large (e.g., 2048 × 768). Several algorithms for constructing such mosaics or panoramas were previously reported in [41, 288, 291, 92]. Using the panorama, it is possible to emulate "virtual camera panning and zooming" by projecting appropriate portions of the panorama onto the user's screen [41]. Different projections can be used to map the environment map to 2D planar coordinates. The cylindrical projection is the most popular for general applications since it is very easy to be captured. A drawback of the cylindrical projection, however, is the limited vertical field of view as compared to the spherical projection. The cubic projection [93] is another efficient representation of environment maps.

Fig. 13.3. Frame 8 of the Cafeteria panoramic video sequence.

The captured environment map is projected onto the sides of the cube. Therefore, each environment map consists of 6 images each associated with one face of the cube, making it very simple to manipulate. By capturing a sequence of panoramas at different time instants, a panoramic video can be constructed.

Capturing panoramic videos

A time-varying environment map can be obtained by taking panoramas at regular time intervals either at a given location or along a trajectory. Such time-varying environment map or panoramic video closely resembles a video sequence with very high resolution. There are different methods to capture a panoramic video [10, 77, 114, 135, 83]. For example, in the FlyCam system [77], multiple cameras are mounted on the faces of an octagon with each side equal to 10 cm. In the system reported in [10], the camera is fitted with a mirror to produce panoramic videos. Specialized hardware for capturing panoramic videos has also been reported in [114], where six closely spaced CCDs are assembled together to minimize parallax. Each CCD is used to capture an image pointing at one of the six faces of a cube. Their outputs are synchronized and streamed directly to disks for storage.

In [210], the compression of real-world and synthetic panoramic videos are considered. For real-world scenes, panoramic videos captured by the omni-directional setup proposed in [135] was used. It comprises a catadioptric omni-directional imaging system [207] with a 1300×1100 pixel camera, all placed on a movable cart. To capture a panoramic video, four video streams of the omni-directional video are taken at different camera orientations (front, left, back, right) along the same path. This arrangement is used because each omni-directional image has blind spots in the middle, and has only about 200 degrees field of view from side to side. The resulting panoramic video (with a frame resolution of 2048×768) is created by stitching these four video streams frame by frame. The panoramic video consists of 381 panoramic images. Figure 13.3 shows a typical panorama of the *Cafeteria* panoramic video sequence.

For the synthetic scene, the mosaic images of the environment map were rendered using 3D Studio Max®. Cubic projection is used for storing the panoramic video. Each panorama has six input images with a resolution of 256×256 and there are

Fig. 13.4. A typical cubic environment map of the synthetic environment.

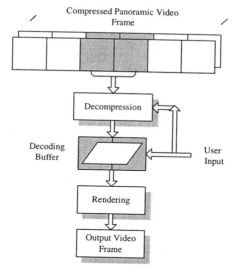

Fig. 13.5. Rendering of panoramic video.

altogether 2910 images. Figure 13.4 shows a typical cubic environment map of the synthetic panoramic video sequence *Village*.

Rendering novel video view

Figure 13.5 is a flow chart showing the decoding of panoramic videos. At the viewer side, the compressed videos are decoded and rendered to create a scene at a given viewing angle. As the resolution of the panoramic video is usually very large, the decoding or transmission of the whole panoramic video is often very time-consuming. This problem can be remedied by reducing the resolution of the decoded video and/or decoding only a given portion of the whole video frame. In virtual walkthrough applications, it is unnecessary to decode the entire video frame because only a fraction of the panorama will be used for rendering the novel view. Because of this reason, the panorama is usually divided into tiles to simplify decoding and data transfer from slower devices such as CD ROM [41].

For a panoramic video sequence with 2D planar images, like the real panoramic video *Cafeteria*, each panoramic video frame can be divided into six vertical tiles as shown in Figure 13.5. If the whole panorama has a view of 360 degrees, the maximum viewing angle of each tile is 360 / 6 = 60 degrees, which is sufficient for most applications. It is therefore only necessary to concurrently decode at most two tiles at a time. Based on the current viewing angle, the tiles involved (the shaped ones) are decoded and placed in the decoding buffer. Appropriate portion of the panorama inside the buffer is used to render the novel view. Tile switching might happen when the user changes his/her viewpoint during the playback of the panoramic video. Therefore, additional mechanism must be provided in the compressed data stream to provide fast tile seeking. This issue is discussed in the following section on the compression of panoramic videos.

13.2.2 Compression and rendering of panoramic videos

As mentioned earlier, a panoramic video can be used to capture dynamic scenes at a stationary location or along a given path. It can also be used to provide seamless walkthrough by constraining the virtual camera location to a predefined path for image acquisition. Both of these applications are discussed below.

MPEG-2 video coding of sub-tiles for dynamic environmental map

Similar to traditional videos, successive panoramic images have significant amount of temporal and spatial redundancies. These can be exploited for data compression by video coding techniques such as motion compensation. As mentioned in the previous section, each mosaic image is usually divided into smaller tiles to avoid decoding the whole panoramic video and to reduce the data transfer requirement when slower secondary devices are used. It is therefore natural to treat each of these tiles as a video sequence and compress these tiles individually. If a panoramic video with a resolution of 2048 × 768 is divided into six non-overlapping tiles, it yields six video sequences each of which has a resolution of 352 × 768. To provide functionalities such as fast forward/backward and to make the panoramic video compatible to most decoders, one can employ the commonly used MPEG-2 video coding standard [120] to compress each of these video streams. Another advantage of MPEG-2

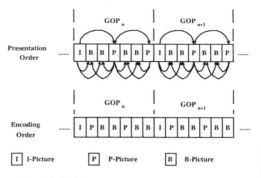

Fig. 13.6. GOP setting in MPEG-2 bitstream.

(as shown by the results later) is that it is very efficient in compressing high resolution panoramic videos with a compression ratio of more than 100 times, yet with reasonably good reconstruction quality. For applications involving frequent editing of the videos, separate coding of the mosaic images might be desirable. Under these circumstances, the use of still image coding techniques such as JPEG 2000 (Section 10.4) are desirable. Next, we shall consider the organization of the compressed video streams in order to provide efficient access to individual tile during decoding.

The selective decoding problem (tile seeking)

For transmission and storage of panoramic videos, individual tiles must be organized in an efficient manner in order to support fast switching between tiles during decoding. Figure 13.6 shows the format of a tile or video stream encoded using the MPEG-2 standard. Consecutive image frames of a given tile are arranged in groups called Group of Pictures (GOP). In each GOP, the image frames are encoded as *I*-, *P*-, or *B*-pictures. The arrows in Figure 13.6 show the inter-dependency of various pictures in a GOP due to motion prediction. As shown in the figure, the coder has seven pictures in each GOP, which consists of one *I*-picture, two *P*-pictures and four *B*-pictures. Also shown in Figure 13.6 is the sequence order of the compressed image frames to be transmitted. Note that the reference pictures are transmitted before the *B*-pictures because they must be decoded before the *B*-pictures. They serve as references for reconstructing the *B*-pictures in between.

Figure 13.7 illustrates how the six tiles (video streams) of the panoramic video are multiplexed in the method in [210]. Each tile is encoded by the MPEG-2 standard with the same GOP structure shown in Figure 13.6. The compressed data of the tiles in the same panoramic video frame are packed together. This allows the decoder to locate very quickly the corresponding *I*-pictures when decoding the required tiles. An individual picture in each tile can be accessed randomly by searching for the appropriate picture header. During decoding, the viewer can selectively decode the tiles required by the user, for example, streams 1 and 2 in Figure 13.7. The novel view can then be generated by re-mapping appropriate pixels in the tiles onto the user's screen.

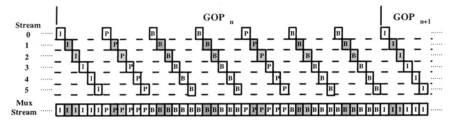

Fig. 13.7. Multiplexing of the tiles (streams) in the MPEG2 compressed panoramic video.

When the viewing angle is changed in such a way that some of the required pixels are no longer in the tiles currently being decoded, switching to the new tile(s) has to be performed. If this happens during the decoding of *P*- and *B*-pictures in a GOP, tile switching can only begin in the next GOP. It is because the *I*-pictures of the new tiles in the current GOP might not be available. (In practice, previously decoded data is usually not buffered.) Hence, the separation of *I*-pictures in panoramic video streams should not be very large. Otherwise, it would introduce an unacceptable delay in switching from one stream to another. As mentioned earlier, there are seven images in each GOP. At a frame rate of 25 fps, the maximum delay during tile switching is therefore 0.28 second, which is quite acceptable. Other values can be chosen according to one's tradeoff between the compression performance and the delay in response time. The synchronized *I*-pictures also allow us to preserve the fast forward and fast reverse capability in the MPEG-2 standard. Notice that the number of *P*- and *B*-pictures in GOPs from different tiles can be different (as well as GOP from the same tile), provided that their *I*-pictures are synchronized. It helps to improve the compression performance, but at the expense of more complicated encoding and decoding processes.

Modified MPEG-2 video coding for virtual walkthrough over static scenes

For virtual walkthrough applications in a static scene, users are allowed to move along a given path and change freely their viewpoints. The compressed panoramic video bitstream is usually stored in local storage or downloaded from the network before decoding. The image frames of the panoramic video are then accessed on demand for rendering according to the specified viewing position. The situation is somewhat similar to the coding of Concentric Mosaics, except that the MPEG algorithm has to be modified in order to support random access to individual image frames. In Figure 13.8, a set of pointers to the starting locations of each image frame in the compressed data is first determined and stored in an array prior to rendering. During rendering, the compressed data for the required image can be located quickly. Alternatively, the pointers can be embedded in the compressed bitstreams. For an *I*-picture, the pointer structure mentioned earlier can be used to access the compressed data. If *B*-pictures are added, the pointer structure only enables us to efficiently decode the motion vectors and the prediction residuals and we still need to decode the two reference *I*-pictures. The dependence will be very complicated if *P*-pictures are

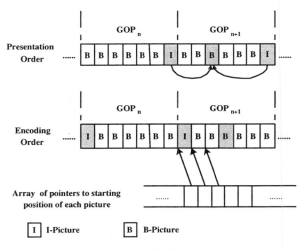

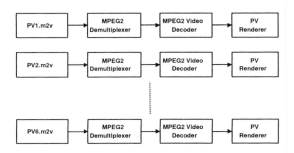

Fig. 13.8. GOP setting in MPEG-2 bitstream for static scene.

Fig. 13.9. Filter graph of panoramic video viewer.

employed which, as mentioned earlier, is very similar to the situation in coding Concentric Mosaics and light fields in Sections 12.2.3 and 12.3, respectively. Because of these reasons, P-pictures are usually not employed in this compression algorithm due to their inter-dependencies. In [210], no rate control algorithm is applied and a uniform quantizer is used for the I- and B-pictures for simplicity.

Rendering of panoramic videos

For dynamic environment maps, the panoramic videos are usually streamed from some servers. In [210], the panoramic video viewer of the system is implemented using the Microsoft(R) DirectShow(R) and Direct3D(R) application programming interfaces (APIs) [66]. The DirectShow API is a media-streaming architecture for the Microsoft(R) Windows(R) platform, which provides high-quality capture and playback of multimedia streams. The basic building block of DirectShow is a software component called a *filter*. A filter generally accepts a multimedia stream as its input and performs a single operation on it to produce the output. For example, a filter

Fig. 13.10. Frame 8 of the decompressed panoramic video sequence Cafeteria at the bit rate of 1.5 Mb/s per tile.

for decoding MPEG-2 videos has its input an MPEG-encoded stream and the output is an uncompressed RGB video stream. Figure 13.9 shows the filter graph of the panoramic video viewer for each user. Multiple data streams associated with a single panoramic video are retrieved from local storage devices or from the video server. Each data stream is then decoded using the Elecard MPEG-2 Multiplexer and Video Decoder filter [202]. The decoded video frames are copied to the texture buffer of the Panoramic Video Renderer filter for rendering. For fast rendering speed, Direct3D is used to render and display the output images in the Panoramic Video Renderer filter. More precisely, the decoded panoramic image is projected onto a geometry model, which can be cylindrical, spherical or cubical. Subsequent rendering of the scene at different viewing angles is handled by Direct3D APIs. The viewer allows the user to pan, zoom and navigate interactively in the video by choosing his/her viewing angles.

For virtual walkthrough applications, the modified MPEG-2 video decoder retrieves the panoramic images from the compressed bitstream. The rendering and display are also implemented using Direct3D APIs. The user interface for the virtual walkthrough application has two windows: the viewport and the plan map of the scene. The viewport renders the virtual camera view at the current location. The plan map indicates the current position of the virtual camera and the current viewing direction. The user can freely navigate in the static environment map or change its location along the path by clicking at the desired destination on the plan map.

Table 13.1. Compression performance of the panoramic video sequence Cafeteria. (25 fps, resolution: $352 \times 768 \times 6$)

Bit-rate (Mb/s)	Compression Ratio	Mean PSNR (dB) Y	U	V
5.727	162	42.16	45.69	44.74
8.596	108	45.40	47.10	46.60

Fig. 13.11. Example screenshots of the synthetic environment during the virtual walkthrough.

Table 13.2. Compression performance of the synthetic panoramic video sequence Village. (resolution: $256 \times 256 \times 6$)

Compression	Mean PSNR (dB)		
Ratio	Y	U	V
30.49	36.13	39.54	41.40
35.02	35.26	39.10	41.03

Experimental results

The compression results of the *Cafeteria* panoramic video sequence described in Section 13.2.1 is given to illustrate the performance of the algorithms for dynamic scenes. Although the *Cafeteria* sequence was captured from a static scene, it is used for simplicity to illustrate the algorithm in the dynamic situation. The six tiles of the panoramic video were encoded using the MPEG-2 video coding standard. Each stream has a Group of Picture (GOP) consisting of seven image frames with two *B*-pictures between successive *I*- or *P*- pictures as illustrated in Figure 13.6. Table 13.1 shows the compression performance of the panoramic video sequence using the modified MPEG-2 algorithm at different bit-rates (target bit-rate of 1 and 1.5 Mb/s per tile). Figures 13.3 and 13.10 show respectively a typical panorama and the decompressed tiles of the panorama. The results show good quality reconstruction with a compression ratio of 108.

When the compressed data is streamed from a remote PC through a 100 Base-T LAN, the rendering speed of the viewer is about 7 fps (neglecting network congestion) using a Pentium 4 1.8 GHz PC with 256 MB memory. For the virtual walkthrough (static scene) experiment, the synthetic panoramic video sequence *Village* was used. For simplicity, it was projected onto a cubic geometric model. Each environment map therefore consists of six images, one for each face of the cube. The image sequence of each face was compressed as a video stream. The appropriate image frames, according to the current viewing angle, were decoded during rendering.

Table 13.2 shows the compression performance of the synthetic panoramic video sequence. Example screenshots of the synthetic environment during the virtual walkthrough experiment are shown in Figure 13.11. The perceptual quality is quite good with a compression ratio of 30. The low compression ratio of the synthetic scene as compared with the real scene is due to its lower resolution, complicated textures, and sharp edges, which make coding more difficult.

For real-time rendering, 20 fps from raw data and 15 fps from compressed bitstream can be achieved using a Pentium 4 1.8 GHz PC with 256 MB memory. It is expected the frame rate can be increased after further optimization/enhancement of the C++ source program. The overall results demonstrate that panoramic videos are an efficient means for providing impressive 3D visual experience to the users. Next, we briefly outline the transmission aspect of panoramic videos over cable networks, LANs and the Internet.

13.2.3 Transmission of panoramic videos

In order to deliver the interactive virtual walkthrough experience offered by panoramic videos, the compressed data stream can be broadcast or transmitted using video-on-demand (VOD) systems over, for example, the Internet, LANs or cable networks. For broadcasting applications, say, over cable networks, the whole panoramic video can be transmitted through a few cable TV channels with each channel carrying one or more tiles of the video streams. The set-top box can be configured according to the user input so that appropriate tiles in the panoramic video will be decoded. Because the panoramic videos are divided into tiles, only a limited number of tiles, two in the system proposed in [210], have to be decoded. Additional hardware is required to render novel views from the decoded video streams. For broadcasting over LANs, the decoding and rendering are most likely performed by a workstation or PC. With present-day technology, real-time rendering and decoding of panoramic videos do not present significant problems. In applications where the channel has limited and/or dynamic bandwidth such as communications over the Internet, the tiles can be transmitted on an "on-demand" basis, where only the required video streams are transmitted. Further reduction of bandwidth for transmission can be achieved by creating a scalable bitstream using, for example, multiresolution techniques.

In [210], a video on demand (VOD) system for delivering panoramic videos over LAN using an "advanced delivery protocol" (ADP) is described. Interested readers are referred to [210] for further information.

13.3 Dynamic light fields and plenoptic videos

13.3.1 Introduction

As mentioned in Chapter 12, an important problem of dynamic image-based representations is the logistical difficulties in their capturing and transmission, which involve huge amounts of data. On the other hand, higher dimensional plenoptic function can provide more viewing freedom and multiple viewpoints at each time instant. Therefore, they can also be viewed as versatile generalizations of traditional images and videos, which might be further developed into new interactive or immersive television systems. Because of these potential advantages, there is a growing interest in the video coding, computer vision and graphics communities in developing efficient capturing, compression and rendering techniques and systems (also known as intermediate view synthesis in the coding community) for dynamic image-based representations.

In [36, 37], a system for real-time capturing, compression and rendering of a simplified dynamic light field (SDLF) for dynamic scenes was developed. Because of their close relationship with traditional videos, it is also referred to as the plenoptic videos. The system is based on parallel processing using inexpensive equipment so that the dynamic IBR can be captured and processed mostly in real-time, which is one of the major obstacles in dynamic IBR research. An MPEG-2-like compression algorithm employing both DCP and temporal compensation was also proposed for the efficient storage and transmission of plenoptic videos [35]. In a SDLF, the viewpoints of the user are constrained along line segments instead of a 2D plane in [160]. This greatly reduces the complexity of the dynamic IBR system. However, unlike panoramic videos, users can still observe significant parallax and lighting changes along the horizontal direction. Experimental results show that DCP improves the coding efficiency. Possible applications of the system are "interactive 3D electronic catalog or brochures," "short plenoptic video advertisement clips," and 3D videophone.

13.3.2 The plenoptic video

In this section, we shall briefly describe the construction, compression, and rendering of the plenoptic video system in [36, 37]. We then briefly summarize and compare a number of related approaches in capturing dynamic image-based representations at Section 13.3.5. Among the static image-based representations reported so far (i.e., panoramas, Concentric Mosaics, light fields and lumigraphs), panoramas, light fields and lumigraphs are simpler to be generalized to dynamic scenes. Although the Concentric Mosaic is an excellent representation for static scenes, capturing dynamic Concentric Mosaics can be very complicated. It is because the capturing of Concentric Mosaics usually requires taking more than one thousand pictures uniformly by an outward facing camera, which moves along a circle with a certain radius. One possible solution is to capture only a portion of the videos on the circumference of this circle so as to limit somewhat the users' viewpoint inside the circle. Contrarily, as we have discussed in Section 13.2, the generalization of panorama to dynamic scenes, i.e., the panoramic video, is considerably simpler. The basic problems of capturing, data compression and transmission of panoramic videos have been addressed quite satisfactorily in [138, 211] and summarized in Section 13.2. In order to provide users with the experience

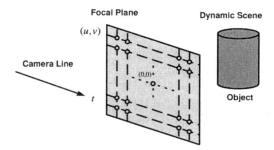

Fig. 13.12. 4D simplified dynamic light field (the plenoptic video): viewpoints constrained along a line (more generally on line segments) in a dynamic environment.

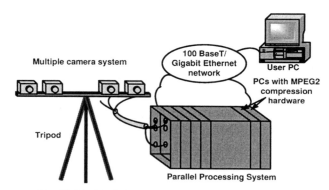

Fig. 13.13. Block diagram of the plenoptic video system.

of parallax and lighting changes, a simpler dynamic generalization of light fields was chosen in [37].

From [33] and Part II, the sampling rate of static light fields depends on the depth of the scene. In order to reduce the effect of aliasing, the number of cameras in a 2D arrangement can be very large, say 16×64. This creates hundreds of videos, which have to be compressed and stored in real-time. The calibration of such a large camera array is also problematic and very time consuming. To avoid this large dimensionality and the excessive hardware cost, the study in [37] was limited to light fields with viewpoints being constrained along a line (or line segments), as shown in Figure 13.12. This simplified dynamic light field, which is called the plenoptic video, has a dimensionality of four. Apart from the simplicity of the overall system, there are several reasons for such a choice. First of all, the user can still observe significant parallax and lighting changes along the horizontal direction. Secondly, the given number of cameras can be used to maximize the sampling rate along the horizontal direction and thus reduce the risk of insufficient sampling in a 2D configuration with the same number of cameras and horizontal panning range.

Fig. 13.14. Physical construction of the plenoptic video system [37].

Figures 13.13 and 13.14 show the block diagram and physical construction of the plenoptic video capturing and processing system. A set of synchronized video cameras is used to capture light field images at each time instant to form sequences of videos. The video signals are then fed to real-time video compression boards in the parallel processing system, which consists of a number of PCs connected together through a high-speed network such as the 100 BaseT or Gigabit Ethernet. With the advent of video compression hardware, inexpensive real-time MPEG-2 compression boards are now readily available. The compressed videos are stored directly to the hard disks of the PCs. Again thanks to the advent of PC technology, high-speed and inexpensive hard disks with 120 Gbytes of storage or more are now in common use. As a result, the parallel arrangement is able to capture 4D dynamic light fields for a fairly long period of time, say several hours.

To avoid unnecessary complication, the prototype system in [36] employs 8 (up to 10) cameras as a reasonable tradeoff between hardware complexity and performance. In contrast to the light field camera in [317], the system uses closely spaced CCD cameras to reduce problems due to insufficient sampling and avoid large variations of CMOS cameras, which usually complicate the camera calibration. This system is relatively easy to construct, as it requires only off-the-shelf components and readily available equipment. During construction, the camera lenses are carefully installed to the hardware stand and similar focuses and tilting angles are maintained. The cameras are then calibrated using the method in [346]. This method is originally proposed for calibrating a single camera and the relative position of the camera and the viewing angle with respect to a reference grid position that can be estimated. More precisely, five images (the grid images) of a certain grid pattern, which consists of squares evenly spaced at a regular grid, are taken by the camera at five different positions. The corners of the squares in each grid image are then determined in order to recover the intrinsic and extrinsic parameters of the cameras. This information allows us to correct the geometric distortion of the camera lens, determine the relative positions and viewing angles of the cameras with respect to the reference grid. After a reference camera plane is chosen, the images captured from all the other cameras

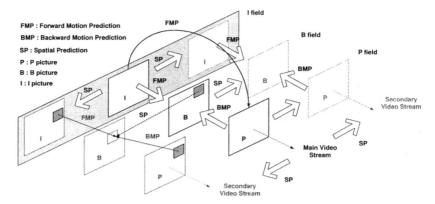

Fig. 13.15. Compression of 4D SDLF.

can be wrapped to the same coordinate of the reference camera. The rectified videos of the cameras are more amenable to rendering.

Another valuable feature of the system is its distributed nature, which allows capturing, compression, processing, and rendering of the plenoptic video to be performed efficiently. It is believed that parallel processing is essential to cope with the demanding storage and computational requirements of plenoptic videos and other dynamic image-based representations. The real-time rendering of the plenoptic video and rendering results will be described in more details in Section 13.3.4. Next we shall discuss the compression of the plenoptic videos.

13.3.3 Compression of plenoptic videos

The compression of plenoptic videos is closely linked to conventional video compression. However, as video streams in a plenoptic video are taken at nearby positions in a 1D array, they appear to be shifted relative to each other, because of the disparity of image pixels. In order to explore this correlation in the plenoptic video, the video streams are divided into groups and are compressed together using the temporal and disparity compensation. The MPEG-like algorithm proposed in [35] for coding the video streams in the plenoptic video has the advantages of good performance and relatively low implementation complexity. It employs both temporal and spatial prediction to better explore the redundancy in the video streams. This can be viewed as the generalization of the DCP techniques for coding of static light fields [164, 179, 298, 341] and stereo image coding [174, 214] to the dynamic situation.

The MPEG-like compression algorithm in [35] is shown in Figure 13.15. For simplicity, only three videos are shown, and it is called a group of field (GOF). To provide *random access to an individual picture*, a modified MPEG-2 video compression algorithm is employed to encode the image frames.

There are two types of video streams in the dynamic light field: *main* and *secondary* video streams. Main video streams are encoded using the MPEG-2 algorithm, which can be decoded without reference to other video streams. The image frames in a main stream are divided into *I*-, *P*-, and *B*-pictures, where *I*-pictures are coded using intra-frame DCT-based transform coding, while *P*-pictures are coded by hybrid motion compensated/transform coding using previous *I*- or *P*-pictures as references. *B*-pictures are coded by a similar method except that forward and backward motion compensation, which are indicated by the block arrow in Figure 13.15, are performed by using nearby *I*- or *P*-pictures as references. The images captured at the same time instant as the *I*-pictures in a main stream constitute an *I*-field. Similarly, the *P*- and *B*-fields are defined as the images containing respectively the *P*- and *B*-pictures of the main video stream. Pictures from the secondary stream in the *I*-field are encoded using *spatial prediction* (SP or DCP) from the reference *I*-picture in the *I*-field. Pictures from the secondary stream in a *P*-field are predicted using spatial prediction from adjacent *P*-pictures in the main stream, and the forward motion compensation from the reference *I*- or *P*- fields in the same secondary stream. Pictures from the secondary stream in a *B*-field are predicted using spatial prediction from adjacent *B*-pictures in the main stream, and the forward/backward motion compensation from nearby reference *I*- and/or *P*- fields in the same secondary stream.

For simplicity, only one main stream is included in each GOF. More sophisticated disparity compensation schemes such as bi-directional prediction with multiple main streams can be incorporated in a single GOF or successive GOBs. The scheme can also be generalized to 2D GOFs in the compression of 5D dynamic light fields, with main streams distributed on certain points in the 2D array, instead of a 1D array considered here. In order to maintain a more uniform reconstruction quality among the plenoptic videos, a higher bit rate is allocated to the main streams than the secondary streams because the *I*-pictures in the main streams usually require considerably more bits than *P*- and *B*-pictures. Furthermore, the rate control algorithm of the MPEG-2 Test Model 5 is used to prevent buffer overflow and underflow problems. To address the random access problem, *pointers* are embedded into the compressed data stream as in [268, 211]. During rendering, the required macroblocks are selectively decoded from the compressed data streams. This unavoidably adds to the overheads in the compressed data streams

Figure 13.16 shows several snapshots of two plenoptic videos captured by the system in [37] (rectified): "Glass Music Box" and "Crystal Dragon". They are extracted from a plenoptic video of about half an hour long. In the plenoptic video "Glass Music Box," a glass music box was placed at the center of the scene and it was rotating at a regular speed. A moving spotlight was used to change dynamically the lighting of the scene. It can be seen from the images that significant lighting changes, reflections, and parallax are captured. The "Crystal Dragon" sequence consists of a lead crystal in the shape of a dragon, which was placed on a wooden platform. Beside it is another crystal turtle, which was placed on a lighting platform that changed color periodically. A burning candle and a moving spotlight were also included to demonstrate the lighting changes and reflective properties of the scene. Each uncompressed video stream consumes about 30 Gbytes of storage. Figure 13.17 shows another synthetic plenoptic video, called the "Ball" sequence, which is generated using computer graphics techniques. It was rendered using the 3D Studio Max

Fig. 13.16. Four snapshots of the plenoptic videos "Glass Music Box" and "Crystal Dragon". Each row consists of the eight images taken from the cameras (form left to right) at a given time instant.

Fig. 13.17. Snapshots of the "Ball" sequence (only images from five virtual cameras are shown).

software and the data sets consist of 16×1 24-bit RGB videos with 320×240 pixels and 24 frames per second. Despite the relatively large depth variation, the use of a mean depth in rendering this plenoptic video does not introduce large rendering artifacts. Also, it was observed from the "Ball" sequence that the plenoptic video is not very sensitive to occlusion if the depth variation is not too large. For large depth variations, artifacts in the form of ghost images and image blurring will appear and more accurate geometrical information such as depth maps are required [33]. An effective approach is to employ the pop-up light field in Chapter 14 or the object-based approach described in Section 12.3.2, where the scene is segmented into layers with different depth values. The compression issues of the object-based plenoptic video will be discussed later in Section 13.4.

Before closing this subsection, let us evaluate the above compression algorithm using the "Glass Music Box," the "Crystal Dragon," and the synthetic "Ball" sequences. Coding results for different number of video streams in a group of fields

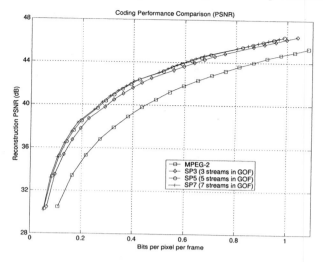

Fig. 13.18. Coding results of the synthetic plenoptic videos "Ball."

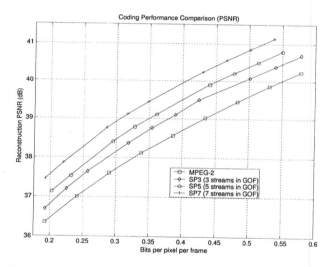

Fig. 13.19. Coding results of the plenoptic videos "Glass Music Box."

(GOF) were also given, and they are plotted in Figures 13.18, 13.19 and 13.20. For SP3, there are three video streams in the GOF as illustrated previously in Figure 13.15. For SP5 and SP7, there are five and seven video streams, respectively. As a comparison, all video streams of the synthetic and real plenoptic videos were also compressed by MPEG-2 algorithm independently. It can be seen that the performance of the algorithm using both temporal and disparity compensation has signif-

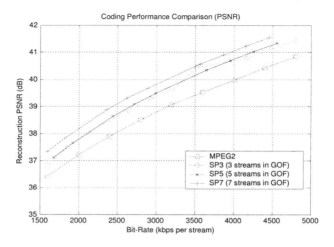

Fig. 13.20. Coding results of the plenoptic videos "Crystal Dragon."

(b)   (d)   (f)

Fig. 13.21. Typical reconstructed images. The "Ball" sequence in the main (a) and secondary (b) video streams (194 kbps per stream). The "Glass Music Box" sequence in the main (c) and secondary (d) video streams (583 kbps per stream); The "Crystal Dragon" sequence in the main (e) and secondary (f) video streams (624 kbps per stream).

icant improvement over the independent coding scheme. This shows that there is a significant amount of spatial redundancy among the video sequences. When the number of video streams in the GOF, and hence the number of secondary streams, is increased, the PSNR improves because less *I*-pictures are coded and better disparity prediction is obtained in the plenoptic video. However, the difference between SP5

Table 13.3. The number of macroblocks used for different types (the synthetic "Ball" sequence).

	Main Stream	Secondary Stream d = 1	d = 2	d = 3
97 kbps				
Intra MB	9.3%	0.1%	0.2%	0.3%
Temporary MB	90.7%	44.5%	49.0%	50.9%
Spatial MB	0.0%	55.4%	50.8%	48.8%
743 kbps				
Intra MB	9.0%	0.1%	0.2%	0.2%
Temporary MB	91.0%	62.1%	66.8%	67.4%
Spatial MB	0.0%	37.8%	33.0%	32.4%
1.78 Mbps				
Intra MB	9.0%	0.1%	0.2%	0.2%
Temporary MB	91.0%	64.7%	69.4%	70.4%
Spatial MB	0.0%	35.2%	30.4%	29.4%

and SP7 is small because disparity compensation will be less effective when video streams are far apart. Figure 13.21 shows several typical reconstructed images. They show good quality of reconstruction at: 194 kbps per stream for the synthetic "Ball" sequence (compression ratio = 204); 583 kbps per stream for "Glass Music Box" sequence (compression ratio = 341) and 624 kbps per stream for "Crystal Dragon" sequence (compression ratio = 319).

In order to evaluate the performance of spatial prediction, the number of macroblocks used in different prediction types for the synthetic "Ball" sequence were calculated and they are summarized in Table 13.3. At a bit rate of 1.78 Mbits/s per stream, secondary video streams which are next to the main video stream (distance $d = 1$) have 35.2% of their macroblocks predicted by disparity compensation prediction. When the distance (d) increases, there are fewer macroblocks predicted spatially. This drops to 29.4% when the distance is increased to 3. The reason is that the prediction will become less effective when the distance from the main video stream increases. This might be improved by using bi-directional disparity compensation prediction. Furthermore, it is noted that this percentage depends on the target bit rate. For example, when the bit rate decreases, more macroblocks (up to 50%) will employ spatial predication.

13.3.4 Rendering of plenoptic videos

There are several major considerations and challenges in the real-time rendering of plenoptic videos. Due to the difficulties in controlling the positions of the image sensors inside the cameras, the optical centers of the cameras do not usually lie on a straight line or even on the same plane. This problem is less serious in capturing static light fields where the relative positions of the camera can be accurately controlled. Fortunately, the relative positions of the cameras can still be recovered from

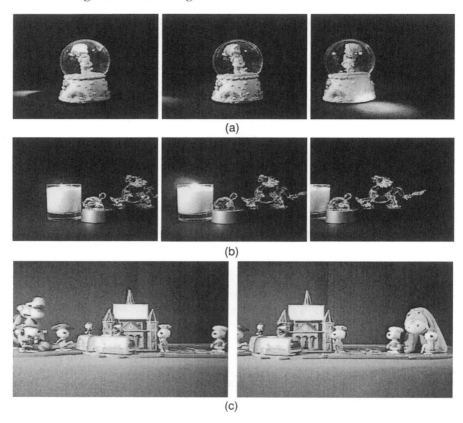

Fig. 13.22. Renderings from the real-time plenoptic video renderer. (a) and (b) show respectively three virtual views at two different time instants for the "Glass Music Box" and the "Crystal Dragon" sequences. (c) shows two virtual views of the "Train" sequence.

the camera calibration described in Section 13.3.2. Since the coordinates calculated do not lie on a straight line, unstructured lumigraph rendering as proposed in [22] or object-based approach have to be used. In [37], it was found that the geometric distortion and the rotation of the cameras could be satisfactorily compensated, partially because of the manual adjustment of the cameras prior to the capturing.

The second problem concerns with the artifacts encountered due to the incorrect depth estimation. For the "Glass Music Box" and "Crystal Dragon" sequences, the depth variation is relatively small and according to the plenoptic sampling analysis [33], the rendering artifacts will be small as long as the focal plane is chosen as the mean depth of the scene. For more complicated scenes, more geometry information would be required.

The final problem is the real-time rendering of the plenoptic video. If the plenoptic video is decoded into raw images and stored in a hard disk, real-time rendering can readily be achieved. However, the memory requirement is very large and the playback time is limited. If the plenoptic video is rendered from the compressed bit stream, then even with the use of selective decoding the computational requirement for the decoding and rendering will become very large. The basic idea of selective transmission/rendering is to decode in parallel the multiple streams of the videos in a network of computers, and transmit those pixels required to the rendering machine over the network, possibly with simple compression. This offloads the rendering machine at the expense of longer user response time. However, we believe that selective transmission is essential to the distribution of plenoptic videos in future applications. Using selective transmission, it is possible to stream continuously a plenoptic video with (256 × 256) resolution at a frame rate of 15 fps over the network. Due to network delay, there is a slight delay in the user response. The frame rate and the resolution can be increased if the raw data stream is compressed by a simple coding method such as vector quantization. For rendering from raw data in the hard disk, real-time rendering can be achieved with a resolution of (720 × 480) and a frame rate of 15 fps.

Figure 13.22 shows several virtual views rendered from the "Glass Music Box," the "Crystal Dragon" and the "Train" plenoptic videos . It can be seen that the lighting changes and reflective properties of the glass and lead crystal are well captured. The "Train" sequence demonstrates that scenes with more complicated details, occlusion, and moving objects (the toy train in the middle) can be rendered with reasonably good quality. It was found that slight artifacts, in the form of ghosting and blurring, are still present in some of the rendered images, because of the difficulty in determining exactly the camera positions and inaccurate depth values. It was also found that the artifacts are less noticeable if the objects are farther away from the camera planes because of the reduced resolution of the images as well as the reduced sensitivity of the image pixels to the errors due to camera calibration and depth values.

13.3.5 Other approaches

There were also previous attempts to construct light field video systems and a summary has been given in Chapter 3. Systems that are related to the plenoptic videos include the Stanford multi-camera array [317], the 3D rendering system of Naemura *et al.* [205], and the (8 × 8) light field camera of Yang *et al.* [335]. The Stanford Multi-Camera Array project [317] was probably the first attempt to develop a large-scale camera array and capturing hardware towards the difficult problem of dynamic IBR modelling. It employs low cost CMOS sensors and dedicated compression hardware. A preliminary six camera-array was reported in [317] and later extended to include more than one hundred cameras. Subsequently, B. Goldlücke *et al.* [87] used the video sequences captured by the (3 × 2) array in [317] to investigate the rendering of light field videos. The images with a resolution of 320 × 240 from the 6 cameras are warped and blended, according to a pre-computed disparity map, to synthesize the novel views at (640 × 480) resolution with approximately 14 fps (block size = 4). The system can handle larger depth variations but also potentially introduce artifacts due to inaccurate depth estimation. The objectives and design tradeoffs of the system in [37] and that in [317] are quite different. One advantage of the system in [37] is its good rendering quality, which is largely attributed to the higher resolution and better quality of the CCD sensors, smaller camera spacing, and camera calibration employed. On the other hand, the system in [317] is more concerned with large-scale modelling. In [218], the image pixels for rendering a given view are retrieved using hardware from an array of CMOS imaging sensors in order to avoid the high data rate for online rendering.

The system of Naemura *et al.* [205] consists of 16 closely spaced CCD cameras in a (4 × 4) 2D arrangement (can be reconfigured to a linear array similar to [37]). It does not incorporate real-time data compression as for the systems in [317, 37]. Instead, dedicated processors (Sony YS-Q430) are used to combine the video sequences from four cameras to form a video sequence divided into four screens. Therefore, the resolution is significantly reduced because of the bandwidth constraint. The final rendered view, using an Onyx2 workstation, has a resolution of only 180 × 120. On the other hand, the captured plenoptic videos in [37] have a resolution of 720 × 480 pixels at 30 fps. No camera calibration is performed in [205] and motion parallax is suppressed using linear translation operations. One distinct feature of this system is the use of a real-time depth map estimation board from Komatsu, FZ930 board (280 × 200 pixels, 8-bits depth map) at 30 fps, to divide the image into 3 layers for rendering (10 fps at 180 × 120 resolution). The system in [37] does not address scenes with large depth variation and by limiting the depth variation and using light field rendering with a single mean depth, fairly high quality real-time rendering of raw video data at a resolution of 720 × 480 with 15 fps can be achieved. For rendering from compressed data, the resolution is reduced to 256 × 256, due to limitation of processing power and transmission bandwidth over the 100 Base-T network without transcoding.

The (8 × 8) light field camera of Yang *et al.* [335] is mainly designed for interactive image-based rendering. Unlike the Stanford light field camera and the system in [37], all the videos from the video cameras are not recorded or stored due to difficulties in compressing the videos in real-time. Images from the cameras are divided into fragments and those fragments required to synthesize a given view are transmitted to a compositor for rendering. It is impossible to replay the videos as in [37], which resembles a traditional video system with continuous multiple viewpoints along a trajectory. The camera spacing is also very small to avoid aliasing. Camera calibration is done by first calibrating one of the cameras using Zhang's algorithm [346]. The rest of the cameras are calibrated using a structure from motion algorithm. Finally, a large nonlinear optimization is performed to cater for non-identical intrinsic parameters of the cameras. Manual color control adjustment in some of the sensors is necessary in [335]. The main features of the above systems are summarized in Table 13.4.

Table 13.4. Comparison of our system with the state-of-the-art systems.

	Plenoptic Video	[205]	[335]	[87]
Sensor	1/4 inch color CCD (CCX-Z11)	1/4 inch color CCD (XC-333)	1/4 inch color CCD (iBOT)	CMOS (OV7620)
Data Resolution	720×480 30fps	720×480 30fps (will be decimated due to no compression)	640×480 30fps	320×240 30fps
Rendering Resolution	From raw data 720×480 15fps, streaming 256×256 15fps.	180×120 10fps (3 depth layers)	320×240 18fps	640×480 13.5fps (block size 4)
Hardware Required	8×1 cameras, 2.5 cm apart. 9 Pentium 4 PCs, 8 Pinnacle PCTV boards, 100 BaseT LAN.	16 (4×4) cameras 3.1 cm apart. Dedicated processors (Sony YS-Q430), Onyx2 workstation (4 400MHz R12000) with a DIVO, real-time depth map estimation board, Komatsu, FZ930.	8×8 cameras, farther apart than [205]. 7 Pentium 4 PCs connected by firewire.	3×2 demo array. Distance between cameras not available but farther apart than others. Embedded microprocessor board, MPEG-2 video encoder, IEEE1394 interface to Ultra160 SCSI disk drives.
Rendering Method	Light field rendering with mean depth. Can be extended to include depth map.	A 3-layer depth map is used to blend the images for rendering.	Images divided into rectangles and rendered using different focal plane.	Images divided into meshes and are warped and blended using dense disparity map.
Camera Calibration	Zhang's algorithm [346] for all cameras using calibration patterns. White balancing for color correction.	Nil	Zhang's algorithm [346] for one camera and others with structure from motion algorithm. Might need manual color control adjustment.	Camera's color reproduction by calibration matrices.
Real Time Compression for Storage	MPEG-2 (1:240 compression ratio). Streamed to harddisk of individual PCs. Modified MPEG-2 algorithm for efficient storage and transmission.	For interactive image-based rendering. Very low resolution.	Nil. For interactive image-based rendering.	MPEG-2 compression (5 Mbytes /sec per video). Stream to host PC's SCSI hard drive.
Applications	Interactive 3D electronic catalog or brochure, short advertising clips, head and shoulder-type 3D videophone.	Interactive image-based rendering.	Interactive image-based rendering.	3D movie for large environment.

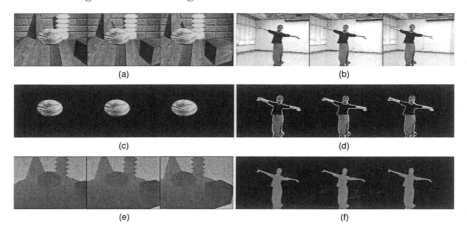

Fig. 13.23. Snapshots of (Top) (a) synthetic PV "synthesis" and (b) real-scene "Dance"; (Middle) the IBR objects (c) "Ball" and (d) "Dancer" extracted; (Bottom) the depth maps of the synthetic PVs (e) "Ball" and (f) the IBR object "Dancer."

13.4 Object-based compression of plenoptic videos

As mentioned earlier in Section 12.3.2, the object-based approach to light fields offers many desirable features such as better rendering quality and object-based functionalities. In this section, the problem of object-based compression of the plenoptic videos will be described. We shall base our discussion on the work in [327].

To extract these IBR objects, a semi-automatic segmentation tool such as Lazy Snapping in [165] can be employed to provide an initial segmentation of the objects in a few light field images. Object tracking algorithms such as the level-set methods described in Section 12.3.4 can then be used to segment the objects in adjacent light field images and subsequent time instants. Figure 13.23 shows several snapshots from two plenoptic videos and the IBR objects segmented from the scenes. Figure 13.23(a) shows a synthetic sequence called *Synthesis*, while Figure 13.23(b) is a real-scene PV called "Dance." The "Ball" and the "Dancer" in the scenes are segmented to form two IBR objects as shown in Figures 13.23(c) and (d).

Because of the similarity of the object-based plenoptic videos to traditional object-based coding in the MPEG-4 standard, it is natural to develop a MPEG-4 like object-based algorithm for compressing the video texture associated with the alpha maps, shape, and depth maps of the IBR objects [327]. The major difference between the two coding schemes is that: the IBR objects in the light fields and the plenoptic videos, and in general IBR compression, have to incorporate other important information such as additional geometry information in the form of depth maps and alpha maps, etc, to facilitate their rendering. As a result, under the object-based framework, multiple video streams in the plenoptic videos can be encoded into user-defined IBR objects, and flexibly reconstructed at the decoder for display and rendering at either

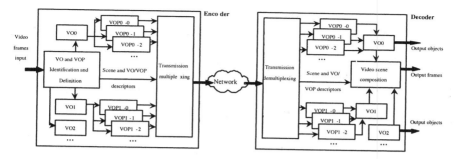

Fig. 13.24. Generic codec structure of the object-based compression system for plenoptic videos.

the object level or frame level. We now describe the compression algorithm proposed in [327].

13.4.1 System overview

Once the IBR objects have been identified, defined, and then extracted from the plenoptic videos (e.g., the objects "Ball" from the PVs *Synthesis* in Figure 13.23), they can be compressed individually to provide functionalities such as scalability of contents, error resilience, and interactivity with individual IBR objects. For example, different IBR objects might be given different numbers of bits (and different amounts of channel coding) and hence different reconstruction qualities (error resilience). They might also be transmitted at different frame rates to achieve object scalability. Figure 13.24 shows the generic codec structure of the object-based coding system in [327], which shares many useful concepts with the MPEG-4 video object coding. A video object (VO) includes the video object planes (VOPs) distributed in all the streams involved in the plenoptic videos, each containing its corresponding binary shape mask, grayscale shape map (alpha map) and depth map. Each VOP is then encoded based on its shape and motion. The scene and VO/VOP descriptors for the plenoptic videos are also encoded and multiplexed together with the VOPs, which are used to compose the video scenes at the decoder. Via the channels of the networks, the decoder can demultiplex and decode the VOPs for display or rendering. Of course, the reconstructed VOPs can also be further composed into a frame for presentation and other operations.

Figure 13.25 shows the encoder diagram of a VOP in an IBR object. It consists of four major components: texture coding, binary shape coding, grayscale shape coding and depth map coding. Texture coding is performed using Discrete Cosine Transform (DCT) based on motion prediction and compensation. The binary shape mask of the VOP is encoded using context-based arithmetic encoding (CAE) algorithm [18]. Grayscale shape information (alpha map), defined by an eight-bits number, is useful in matting VOs during VO composition and rendering at the decoder. Following MPEG-4, grayscale shape information (alpha map) is coded through alpha channels

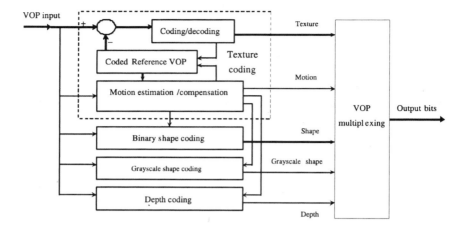

Fig. 13.25. Block diagram for encoding a VOP.

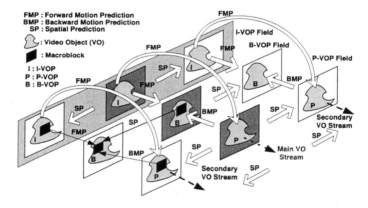

Fig. 13.26. The texture coding of an IBR object in the plenoptic videos.

in the same way as the luminance signal of texture. Each depth map, as a type of geometrical information, is encoded independently as a so-called "depth channel" in the object-based coding system. After these four parts are encoded, they are then multiplexed together as an entire encoded VOP. In the following, the details of the texture coding, binary shape coding and depth coding of VOPs will be described.

13.4.2 Texture coding

Because of disparity, adjacent light field images appear to be shifted relative to each other. Therefore, it is advantageous to employ both temporal or spatial predictions (also referred to as disparity compensated prediction (DCP)) to improve the coding efficiency as in traditional stereoscopic image coding [174] and coding of other image-based representations.

Figure 13.26 shows the block diagram for coding the texture of an IBR object in the object-based compression method in [327]. This is a generalization of the frame-based method in Section 13.3.3 and [35, 36] to the object-based framework. Likewise, it employs predictions in both the temporal and spatial directions. For simplicity only three video object (VO) streams are shown, and it is referred to as a group of video object field (GOVOF). In each VO stream, we have a view of the IBR object, which we refer to as the video object plane (VOP) as mentioned previously. There are two types of VO streams associated with each dynamic IBR object: main video object stream and secondary video object stream. Each main VO stream is encoded similar to the MPEG-4 algorithm, which can be decoded without reference to other VO streams. For better performance, bi-directional prediction for the B-VOPs are employed. To provide random access to individual VOP, the basic structure of a Group of VOP (GVOP) in MPEG-4 is employed in the main VO stream. A GVOP contains an I-VOP and possibly P-VOPs and/or B-VOPs between this I-VOP and the following I-VOP. I-VOPs are coded using intra-frame coding to provide a random access point without reference to any other VOPs, while P-VOPs are coded by motion predictive coding using previous I- or P-VOPs as references. B-VOPs are coded by a similar method except that forward and backward motion compensations are performed by using nearby I- or P-VOPs as references, which are indicated by the block arrow in Figure 13.26. The VOPs captured at the same time instant as the I-VOP in a main stream constitute an I-VOP field. Similarly, the P- and B-VOP fields are defined as the VOP field containing respectively the P- and B-VOPs of the main VO stream. A VOP from the secondary stream in the I-VOP field are encoded using disparity-compensated prediction (DCP) or "spatial prediction" from the reference I-VOP in the I-VOP field. As mentioned earlier, the disparity-compensated prediction has been used in the coding of static light fields. Therefore, the coding algorithm considered here can be viewed as their generalization to the dynamic IBR object context. Similarly, apart from using temporal prediction in the same stream, the secondary P/B-VOPs also employ spatial prediction from their adjacent P/B-VOPs in the main stream for better performance.

It can be seen that employing spatial prediction for coding the secondary VO streams can achieve a better prediction in comparison to the main stream, which only employs temporal prediction, especially for VOs having fast motions. However, introducing spatial prediction also increases the overhead used in selecting the prediction modes, since one more prediction mode will result in one more entry in the codeword table for the entropy coding of MB prediction modes. According to the occurrence probabilities of MB prediction modes, three codeword tables respectively for secondary I-VOPs, secondary P-VOPs, and secondary B-VOPs are constructed in [327], while keeping unchanged the codeword tables for the VOPs of main stream. Experimental results show that the extra overheads of the new prediction modes is paid off by better prediction offered by spatial prediction. Finally, the blocks which lie within the object are coded similar to traditional video coders, while blocks at the boundary of an object can either be coded using padding or shape-adaptive DCT.

13.4.3 Shape coding

Shape information is an important component in object-based coding. Following MPEG-4, there are two types of shape information: binary shape information and grayscale shape information. The former only provides the binary shape mask collocated with the luminance picture of the VOP, and it is used to indicate whether the pixels belong to that VOP or not. The latter one, also called the alpha map, provides pixels' transparency levels for a VOP, which is useful to the matting of VOs during composition and rendering after being decoded. Binary shape information generally can be coded using Context-based Arithmetic Encoding (CAE) algorithm [18]. As discussed in a previous subsection, grayscale shape information is coded by using DCT, similar to the coding of luminance signal, via the alpha channel. Hence, shape coding mainly refers to coding binary shape information. There are two coding modes in the CAE algorithm : Intra-CAE and Inter-CAE modes. The Intra-CAE mode codes shape information in intra mode, without using motion prediction, and therefore is mainly used for I-VOP in the main stream. In contrast, the inter-CAE mode makes use of motion prediction from a shape mask reference, and therefore it is used in other types of VOPs except I-VOP. In coding a B-VOP in the main stream, the inter-CAE mode will select a shape mask in the nearest preceding I-VOP/P-VOP or future I-VOP/P-VOP as its reference in order to perform shape motion prediction and compensation.

For the shape coding of a VOP in a secondary stream, it is possible to select the reference from either a VOP in this secondary stream or another in the main stream at the same time instant. In general, the shapes of the VOP in secondary streams are very similar to the VOP of the main stream at the same GOVOF, because they are captured by two cameras at the same time instance. As a result, selecting the VOP of the main stream as reference usually performs better than selecting the VOP in the same secondary stream. However, if the object is static, or moving very slowly, the shape motion prediction performed in the same secondary stream (i.e., intra stream mode) can achieve a better result than that performed between the secondary stream and the main stream (i.e., inter stream mode). To achieve a better shape coding result, both modes are incorporated. They are selected by performing the shape coding for each VOP in both modes, and the better one will be chosen. This method is referred to as the hybrid mode, and its improvement will be illustrated in Section 13.4.5.

13.4.4 Depth coding

In MPEG-4, alpha channels are provided to encode a set of grayscale shape information in the same way as the luminance component of textures. The depth map of a VO resembles closely the alpha information of a VO. Hence, it can be coded in a similar way as the alpha map, except for some pre-processing to be described below. Similar to "alpha channel," the data in the final encoded bitstream for storing the encoded depth map are called the "depth channel". We now describe the pre-processing of the depth map for better coding performance. Firstly, since the dynamic range of the depth values can be quite large, it is advantageous to scale it appropriately before

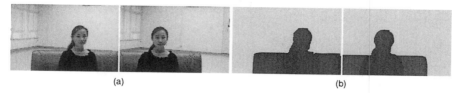

Fig. 13.27. (a) Snapshots of the sequence "Poem" captured by the system in [327]. (b) 8-bit depth maps of a snapshot of this sequence.

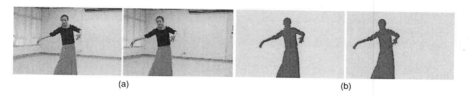

Fig. 13.28. (a) Snapshots of the sequence "Dance" captured by the system in [327]. (b) 8-bit depth maps of a snapshot of this sequence.

coding. Secondly, for a large object, its depth values might vary significantly, and the depth pixels with small values are commonly more important since they result in large disparity of image pixels in rendering the VO. To avoid introducing too much distortion in encoding depth pixels with small values after scaling, companding [129] is also applied to the depth map. A usual companding approach for coding depth values is to calculate the reciprocal of a depth pixel value Z, where the companded value Z' is given by $Z' = 1/Z$. Taking into account the scaling and companding operations mentioned above, the final value of a depth pixel Z_f, before feeding to the encoder, is given by:

$$Z_f = \frac{Z'}{Z'_{max}} \cdot S_{max} = \frac{1/Z}{1/Z_{min}} \cdot S_{max} = \frac{Z_{min}}{Z} \cdot S_{max}$$

where Z'_{max} is the maximum value of the companded depth maps, which also corresponds to Z_{min}, the minimum depth values of the VOPs, and S_{max} is the maximum scaling value. If 8 bits is used to represent a pixel for encoding, then S_{max} would be 255. Similarly, for 12 bits, S_{max} would be 4095.

After companding and scaling of the original depth values, the resulting depth map is then encoded using temporal/spatial prediction, similar to its corresponding texture and alpha map.

13.4.5 Compression results and performance

Figures 13.27(a) and 13.28(a) show the snapshots of two sequences "Poem" and "Dance" captured by the system in [79]. The resolution of these videos is 720 × 576. The "Poem" sequence represents a typical head and shoulder scene frequently

Fig. 13.29. Rendering results of the sequence "Poem" obtained by the algorithm in [327].

Fig. 13.30. Rendering results of the sequence "Dance" obtained by the algorithm in [327].

encountered in 3D videophone, and the "Dance" sequence is an example of life performance and other sport events. The corresponding depth maps are shown in Figures 13.27(b) and 13.28(b). Figures 13.29 and 13.30 show the rendering results obtained by the system. The rendering time for one frame of the "Poem" and "Dance" sequences in a Pentium 4 2.46 Hz computer are 27.5 ms and 128.3 ms respectively. In other words, the rendering of object-based plenoptic videos is nearly real-time.

The performance of the object-based coding scheme for the plenoptic video (PV) is illustrated using a synthetic and a real-scene PVs. The synthetic PV "Ball" is produced by the 3D Studio Max software with a resolution of 320×240 pixels and 24-bit RGB components per pixel. The real-scene PV "Dance" has a resolution of 720×576 pixels in 24-bit RGB format. It was captured by the multiple video cameras system described in Section 12.3.5, which consists of two linear arrays of cameras each hosting 6 JVC DR-DVP9AH video cameras. The corresponding depth maps are generated with 16 bits per pixel. Figure 13.23 also shows a few snapshots of two PVs and two IBR objects extracted from the "Ball" and the "Dance" sequences. The "Ball" and "Dancer" have respectively 40 frames/VOPs and 50 frames/VOPs in each stream. Due to space limitation, snapshots for only 3 streams are shown in Figure 13.23, despite that the "Ball" and "Dance" contain 9 and 6 streams, respectively.

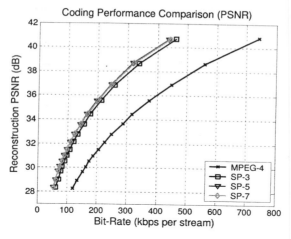

Fig. 13.31. Object-based coding result for the IBR object "Ball."

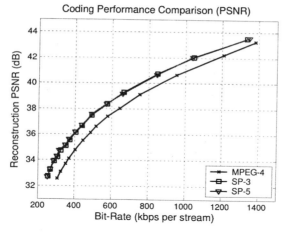

Fig. 13.32. Object-based coding result for the IBR object "Dancer."

Figures 13.31 and 13.32 show the combined coding results with respect to the PSNR of the texture and shape coding for IBR objects "Ball" and "Dancer" at different bit rates by using the VM rate control algorithm [124]. The frame rates used for the PVs are 24 frames per second. For illustration purposes, a Group of VOPs (GVOP) structure consisting of 12 VOPs (1 *I*-VOP, 3 *P*-VOPs and 8 *B*-VOPs) is employed. The curves denoted by "MPEG-4" represent the results using MPEG-4-like algorithm without spatial prediction, while those denoted by "SP-3," "SP-5," and "SP-7" represent the coding results using the coding scheme with 3, 5 and 7 VO streams within a GOVOF, respectively. It can be seen from Figure 13.31 that, for the synthetic IBR object "Ball," there is a considerable improvement in PSNR perfor-

Fig. 13.33. Typical rendering results for the IBR object "Ball."

Fig. 13.34. Typical rendering results for the IBR object "Dancer."

mance (4 dB) of the object-based coding scheme over the direct application of the MPEG-4 algorithm to individual VO stream. The coding performances of SP-5 and SP-7 are slightly better than that of SP-3, while the former two are very close to each other. This is to be expected because when the disparity between two video streams increases, spatial prediction becomes less effective. The performance improvement for the real-scene IBR object "Dancer," as shown in Figure 13.32, is less significant compared with the synthetic sequence. This is mainly due to the slight position errors introduced by imperfect camera calibration, which destroys somewhat the correlation between the video streams. Therefore, the results for SP-3 and SP-5 are very close to each other.

Table 13.5 compares the shape coding results produced in different types of prediction modes for different VO streams extracted from the synthetic PVs *Synthesis* (Figure 13.23). It is measured by the average number of bits used per VOP. Stream 2 is the main stream, and others are the secondary streams. The object "Ball" has a lot of motion, whereas the object "Pyramid" is static and the object "Green Hose" moves very slowly. From Table 13.5, we can see that stream 1 and stream 3 have better shape coding results than stream 0 and stream 4. This is because the disparity of the formers with respect to the main stream is much smaller than those of the latters. It can be seen that the hybrid mode achieves the best performance than using intra or inter stream mode alone. Since the variations in the depth map within the IBR object is much less than the texture information, the depth map can be coded with a higher compression ratio than the latter. The rendering examples displayed in Figure

13.33 are rendered using the reconstructed depth maps, where the average compression ratio of the depth map for the IBR object "Ball" is about 500 at a PSNR of 40 dB. Finally, to further demonstrate the object-based functionality of the codec, the renderings from the real-scene PV "Dance" at both the frame and the object levels are also shown in Figure 13.34. In closing, we note that the performance of the system can be further improved if more tools of MPEG-4 such as four motion vectors for a MB, direct prediction mode and so forth are incorporated in coding the secondary VO streams. Moreover, it would be valuable to incorporate other advanced coding tools in the recently introduced H.264 standard into the compression scheme for better coding performance. These will be a fruitful area of future work.

Table 13.5. Comparison of binary shape coding results using different shape prediction modes [327].

VO	Stream name	Shape prediction mode (bits/VOP)		
		Intra mode	Inter mode	Hybrid mode
Ball	Stream 1/3	409	287	287
	Stream 0/4	407	307	307
Hose	Stream 1/3	388	351	344
	Stream 0/4	405	368	359
Pyramid	Stream 1/3	143	356	143
	Stream 0/4	148	401	148

13.5 Future directions and challenges

Tables 13.6 and 13.7 summarize the various IBR compression methods described earlier. Despite the significant progress achieved in IBR compression over the last few years, many research problems still remain. We envision that the data compression and transmission of the various image-based representations described in this paper and related representations (such as the compression of LDIs [68]) will continue to be important issues in IBR research. For example, the integration of model-based coding with traditional video coding approaches for light field compression [85, 180, 181] is an interesting area of research.

Methods for capturing, compression, and transmission of dynamic IBR functions have not been well explored yet. The panoramic video, as discussed earlier, is a 3D dynamic image-based representation that is relatively simple to manipulate. As a result, this representation will be easier to use in a commercial setting. Dynamic generalizations of the light field and Lumigraph, which we called the plenoptic video, will likely involve scores of synchronized videos for them to be effective and compelling. It would be very challenging to efficiently compress and transmit them. The prototype systems described in this chapter demonstrate the feasibility and the potential applications of IBR in immersive TV and other related applications.

Table 13.6. Summary of IBR compression techniques for static scenes. Note: DCP = disparity compensation prediction, VQ = vector quantization, MRFP = multiple reference frame prediction, MB = macroblock, C.R. Compression ratio.

STATIC SCENES (Concentric Mosaics: Random access at line level)				
References	**Method**	**Random access**	**C. R.**	**Remarks**
[267]	VQ	Simple	Low	Simple, fast rendering
[150, 230]	Haar wavelet decomp. and thresholding	Tree of wavelet coeffs. to speedup data access	Moderate	Scalable bit stream
[268]	Modified MPEG-2	Pointer to line of MB	High	Real-time rendering
RBC [164, 340]	Modified MPEG-2	Simple (pointer)	High	Real-time cache to enhance speed
[175, 328, 329]	Wavelet	Complicated	High	Less rendering speed. Real-time with cache enhancement [328]. Good compression efficiency using smart rebinning [329].

STATIC SCENES (Light Field: Random access at pixel level)				
References	**Method**	**Random access**	**C. R.**	**Remarks**
[160]	VQ	Simple	Low	Simple, fast rendering
HDCP [298]	DCP, VQ	Pointer to regions	High	Simple, fast rendering
V-coder [177, 179]	DCP, DCT-based coding	Complicated	High	
D-coder [178, 179]	HDCP using disparity map, DCT-based coding	Complicated	Slightly inferior to V-coder	Possible to interpolate intermediate missing picture
RBC [164, 341]	Modified MPEG-2, MRFP	2-level index table	High	Real-time rendering. Also for Lumigraph compression. Significant overheads of indexing at high compression ratio.
Model-aided coder (MAC) [180]	Approx. 3D geometry model and DCP	Complicated	Better than MBC	
Model-based coder (MBC) [181]	Use scene geometry to convert images to texture maps, which are coded using a modified SPIHT algorithm [250]	Random access to arbitrary texture segments	High	Supports progressive decoding. Graphics hardware can be used to accelerate rendering.

Table 13.7. Summary of IBR compression techniques for dynamic scenes. Note: DCP = disparity compensation prediction, MB = macroblock, C.R. Compression ratio.

DYNAMIC SCENES (Panoramic Video: Random access at tile level)				
Reference	Method	Random access	C. R.	Remarks
[211]	Modified MPEG-2	B-pictures in MPEG-2, pointers to tiles	High	Panoramas divided into tiles for selective decoding and reception
DYNAMIC SCENES (Simplified Dynamic Light Field: Random access at line level)				
Reference	Method	Random access	C. R.	Remarks
[35]	Modified MPEG-2, DCP	Pointers to line of MB	High	
[327]	Modified MPEG-4	Pointers to line of MB	High	Object-based approach

We predict that future virtual reality and gaming systems will rely heavily on image-based representations to render photo-realistic real-world scenes. Realistic-looking synthetic scenes that are expensive to render may be prerendered instead and stored as image-based representations in such systems as well. However, before such systems become a reality, the high level of interactivity associated with 3D gaming will have to be enabled. This is a challenging and interesting topic that will need to be adequately addressed.

In addition, the amount of digital data associated with future image-based representations will become so large that selective decoding, reception, and streaming techniques for transmission will continue to play a major role in their processing. This again calls for sophisticated random access methodology to retrieve these components with a wide range of characteristics.

Part IV

Systems and Applications

In the last part of the book, we detail four systems that acquire and render scenes using different types of representations. The representations featured in this part are geometryless (Chapters 14 and 17), with geometric proxies (Chapter 15), and layers (Chapter 16).

Chapter 14 (Rendering by Manifold Hopping) shows how manifold mosaics or multiperspective panoramas can be made more compact. This is accomplished by using the simple observation that humans perceive motion as continuous if the change in scene is small enough. By capitalizing on this observation, the minimal number of sampled manifolds can be derived; the scene is then rendered by "hopping" across manifolds. No explicit geometry is used.

Constructing an image-based representation of large environments would, in principle, require a massive amount of image data to be captured. However, we can usually make the reasonable assumption that some parts of the environment are more interesting than others. Areas of higher interest would then be captured using more images and afforded higher degrees of freedom in (local) navigation. Chapter 15 (Large Environment Rendering using Plenoptic Primitives) describes an authoring and rendering system that generates a combination of panoramic videos and Concentric Mosaics (CMs). CMs are placed in areas of more significant interest, and they are connected by panoramic videos.

Chapter 16 (Pop-Up Light Field: An Interactive Image-Based Modeling and Rendering System) shows how a layered-based representation can be extracted from a series of images. The key is to allow the user indicate areas where artifacts occur. Alpha matting at the layer boundaries and highlighted areas are then estimated to ensure coherence across the images and artifact-free rendering. Hardware-accelerated rendering ensures that the editing process is truly interactive.

One characteristic of a light field is that it is difficult to edit and manipulate. This limits the appeal of light fields. Chapter 17 (Feature-Based Light Field Morphing) describes an interactive system that allows users to morph one light field into another using a series of simple feature associations. This technique preserves the capability of light fields to render complicated scenes (e.g., non-Lambertian or furry objects) during the morphing process.

Additional Notes on Chapters

Chapter 14 (Rendering by Manifold Hopping) has appeared in *International Journal of Computer Vision (IJCV)*, volume 50, number 2, pages 185–201, November 2002. The article was co-authored by Heung-Yeung Shum, Lifeng Wang, Jin-Xiang Chai, and Xin Tong.

Most of Chapter 15 (Large Environment Rendering using Plenoptic Primitives) has appeared in *IEEE Transactions on Circuits and Systems for Video Technology*, volume 13, number 11, November 2003, pages 1064–1073. The co-authors of this article are Sing Bing Kang, Mingsheng Wu, Yin Li, and Heung-Yeung Shum.

Heung-Yeung Shum, Jian Sun, Shuntaro Yamazaki, Yin Li, and Chi-Keung Tang originally co-wrote the article "Pop-Up Light Field: An Interactive Image-Based Modeling and Rendering System," which appeared in *ACM Transaction on Graphics*, volume 23, issue 2, April 2004, pages 143–162. Chapter 16 is an adaptation of this article.

Finally, Chapter 17 (Feature-Based Light Field Morphing) first appeared as an article in ACM SIGGRAPH, July 2002, pages 457–464. The co-authors are Zhunping Zhang, Lifeng Wang, Baining Guo, and Heung-Yeung Shum.

14

Rendering by Manifold Hopping

In Chapter 2, we surveyed image-based representations that do not require explicit scene reconstruction or geometry. Recall that representations such as the light field [160], Lumigraph [91], and Concentric Mosaics (CMs) [267] densely sample rays in the space based on the plenoptic function [2] with reduced dimensionalities. They allow photorealistic visualization, but at the cost of a large database requirement.

An effective way to reduce the amount of data needed for IBR is to constrain the motion or the viewpoints of the rendering camera. For example, the movie-map system [170] and the QuickTime VR system [41] allow a user to explore a large environment only at pre-specified locations. Even though a continuous change in viewing directions at each node is allowed, these systems can only jump between two nodes that are far apart, thus causing visual discontinuity and discomfort to the user. However, perceived continuous camera movement is very important for a user to smoothly navigate in a virtual environment. Recently, several panoramic video systems have been built to provide a dynamic and immersive "video" experience by employing a large number of panoramic images.

This chapter shows that it is possible to achieve significant data reduction, not through more sophisticated compression, but rather by *strategically subsampling*. This is shown in the context of CMs, which is made up of densely sampled cylindrical manifold mosaics. By strategically sampling a small number of cylindrical manifold mosaics, it is still possible to produce *perceptually continuous navigation*; the resulting rendering technique is called *manifold hopping*. The term manifold hopping is used to indicate that while motion is continuous within a manifold, motion between manifolds is discrete, as shown in Figure 14.1.

Manifold hopping has two modes of navigation: moving continuously along any manifold, and discretely between manifolds. An important feature of manifold hopping is that significant data reduction can be achieved without sacrificing output visual fidelity, by carefully adjusting the hopping intervals. A novel view along the manifold is rendered by locally warping a single manifold mosaic using a constant depth assumption, without the need for accurate depth or feature correspondence. The rendering errors caused by manifold hopping can be analyzed in the signed

Hough ray space. Experiments with real data demonstrate that the user can navigate smoothly in a virtual environment with as little as $88k$ data compressed from 11 cylindrical manifold mosaics.

Manifold hopping significantly reduces the amount of input data without sacrificing output visual quality, by employing only a small number of strategically sampled manifold mosaics. This technique is based on the observation that, for human visual systems to perceive continuous motion, it is not essential to render novel views at infinitesimal steps. Moreover, manifold hopping does not require accurate depth information or correspondence between images. At any point on a given manifold, a novel view is generated by locally warping the manifold mosaic with a constant depth assumption, rather than interpolating from two or more mosaics. Although warping errors are inevitable because the true geometry is unknown, local warping does not introduce structural features such as double images which can be visually disturbing.

14.1 Preliminaries

In this section, we describe view interpolation using manifold mosaics, warping manifold mosaics, and manifold hopping. Throughout this section, CMs are used as examples of manifold mosaics to illustrate these concepts. (The notion of manifold mosaic has been covered in Chapter 2; the reader is encouraged to review the chapter before proceeding.)

14.1.1 Warping manifold mosaics

High quality view interpolation is possible when the sampling rate is higher than Nyquist frequency for plenoptic function reconstruction [33] (see also Chapter 5). However, if the sampling interval between successive camera locations is too large, view interpolation results in aliasing artifacts. More specifically, double images are observed in the rendered image. Such artifacts can be reduced by the use of geometric information (e.g., [91, 33]) or by pre-filtering the light fields [160, 33], thus reducing output resolution.

A different approach is to locally warp manifold mosaics, which is similar to 3D warping of a perspective image. An example of locally warping CMs using an assumed constant depth is illustrated in Figure 2.8(b). Any rendering ray that is not directly available from a CM (i.e., not tangent to a concentric circle) can be retrieved by first projecting it to the constant depth surface, and then re-projecting it to the CM. Therefore, a novel view image can be warped using the local rays captured on a single Concentric Mosaic, rather than interpolated by collecting rays from two or more CMs.

According to Zorin and Barr [353], for humans to perceive a picture correctly, it is essential that the retinal projection of a two-dimensional image of an object should not contain any structural features that are not present in the object itself. In contrast, human beings have a much larger degree of tolerance for other important

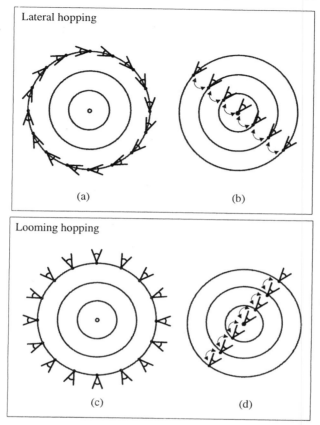

Fig. 14.1. Manifold hopping using Concentric Mosaics (CMs): a plan view. Manifold hopping has two modes of navigation: (a)(c) move continuously along any manifold, and (b)(d) discretely across manifolds. The arrows in (b)(d) indicate that the user can only hop to the viewpoints on the circle, but not stop anywhere in between. Two classes of manifold hopping are shown here: lateral hopping whose discrete mode of navigation is perpendicular to the viewing direction, and looming hopping whose discrete mode of navigation is along the viewing direction. Lateral hopping uses tangent CMs (Figure 2.6(a)), while looming hopping employs normal CMs (Figure 2.6(b)). Note that the rendering views are restricted on the circles. Rendering images have limited field of views.

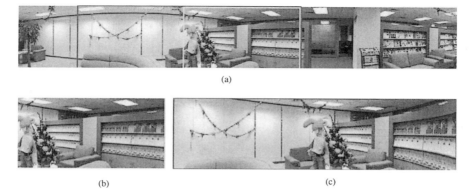

(a)

(b)   (c)

Fig. 14.2. Local warping with an assumed constant depth: (a) part of a Concentric Mosaic; (b) a rendered view with FOV = 45; and (c) another rendered view with FOV = 90. The distortion error towards the right edge of (c) can be clearly seen as straight lines become curved. The image is rendered column by column with local warping. Note that in (c), the vertical field of view is significantly reduced as some of the ceiling lights become invisible.

requirements for the perception of pictures, including the direct view condition, foreshortening of objects, relative size of objects, and verticality. Double images, which are common artifacts from view interpolation with poor correspondence, unfortunately result in mistakenly perceived structural features in the observed objects, e.g., more noticeable edges. This leads to uncomfortable visual perception.

On the other hand, locally warping a multiperspective image preserves structural features. Even though the local warping method, similar to classical 3D warping methods, introduces geometric distortions because of imprecise geometric information, the amount of geometry distortion can be tolerated by human visual perception as long as the field of view (FOV) of the rendering image is small. Consistent re-projection errors appear less disturbing than perceived structural changes to human beings. A single multiperspective panorama, for instance, has been widely used in cel animation to generate novel view images by projecting the multiperspective panorama to perspective images [324].

An example of locally warping a CM is shown in Figure 14.2, with images of different FOV's. The projection error in the rendered image caused by warping the CM with (incorrect) constant depth assumption increases as the field of view becomes larger. Note the distortion toward the right edge in Figure 14.2(b). The geometric distortions introduced by local warping methods because of imprecise geometric information are, however, tolerated by human visual perception when the field of view (FOV) of the rendering image is small (e.g., Figure 14.2(a)).

14.1.2 Hopping classification and issues

We now introduce the idea of manifold hopping using a small number of CMs to observe an environment from the inside looking out. Manifold hopping has two modes of navigation: moving continuously along any of the concentric circles as shown in Figure 14.1(a) and (c), but discretely along the radial direction as in Figure 14.1(b) and (d).

Manifold hopping works because moving continuously along any concentric circle uses local warping, which preserves structural features. In addition, moving discretely along the radial direction can be made perceptually smooth if the interval can be made reasonably small. A key observation is that there exists a critical hopping interval for users to perceive a smooth navigation. In other words, manifold hopping is able to provide users with *perceived continuous camera movement*, without continuously rendering viewpoints at infinitesimal steps. As a result, manifold hopping significantly reduces the input data size without accurate depth information or correspondence.

Figure 14.1(a) also shows that, at any point on a circle, the rendering view is constrained to be on the circle and the viewing direction along the tangent line to minimize the rendering errors caused by local warping. Note that no parallax is observed from these views generated on the same circle using the same CM. Parallax and lighting changes are captured in manifold hopping because of the viewpoint variations across different concentric circles, as shown in Figure 14.1(b).

There are two types of manifold hopping with CMs: lateral hopping, whose discrete mode of navigation (Figure 14.1(b)) is perpendicular to the viewing direction; and looming hopping, whose discrete mode of navigation (Figure 14.1(d)) is along the viewing direction. Note that for each type of hopping, there are two modes of navigation, namely the continuous mode along the manifold and discrete mode between manifolds. The type of hopping is named after the direction of discrete navigation.

Detailed analysis of manifold hopping is needed to address the following important questions.

- What is the largest field of view that still produces acceptable local warping error?
- How large can the hopping interval be so that continuous motion can be perceived?

Before addressing these questions, we first introduce the idea of the signed Hough ray space. The signed Hough ray space is important for the analysis of hopping.

14.2 The signed Hough ray space

The Hough transform is known to be a good representation for lines. However, it is not suitable for representing rays that are directional. The conventional Hough space can be augmented to a signed Hough ray space [31], or an oriented line representation [160], by using the following right-hand rule: a ray that is directed in a counter-clockwise fashion about the coordinate center is labeled positive, otherwise is labeled negative. A "positive" ray is represented by (r, θ), whereas its "negative" counterpart is $(-r, \theta)$ where r is always a positive number. Figure 14.3 shows four

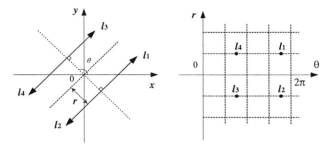

Fig. 14.3. Definition of the signed Hough ray space: each oriented ray in Cartesian space at the left is represented by a sampled point in the signed Hough space on the right.

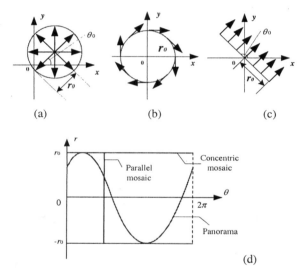

Fig. 14.4. Three typical viewing setups and their respective sampled curves in the signed Hough space: (a) a panorama at a fixed point; (b) a Concentric Mosaic (CM); (c) a parallel projection mosaic; and (d) their respective sampled curves in the signed Hough space. Two CMs (straight lines at r_0 and $-r_0$) are shown in (d) to represent rays captured at opposite directions along the circle. Note that a perspective image is only part of a panorama, thus represented by a segment of a sinusoidal curve in the signed Hough space.

different rays in a 2D space and their corresponding points in the signed Hough space.

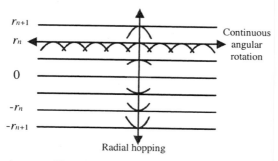

Fig. 14.5. Hopping between CMs along a radial direction in the signed Hough space. Continuous rotation is achieved along any of the concentric circles, but hopping is necessary across any radial direction. In radial hopping, the curve segment varies from r_n to r_{n+1} because the corresponding sine curve is different, as shown in Figure 6.

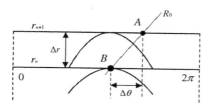

Fig. 14.6. Analysis of hopping size: horizontal parallax change due to viewpoint change.

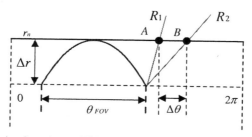

Fig. 14.7. Analysis of maximum FOV: warping error due to the incorrect depth value.

Figure 14.4 shows three typical viewing setups and their representations in the signed Hough space. For example, a panoramic image (i.e., rays collected at a fixed viewpoint in Cartesian space) is represented as a sampled sinusoidal curve in the parameter space, located at (r_0, θ_0) as shown in Figure 14.4(a). A CM shown in Figure 14.4(b) is mapped to a horizontal line, whereas parallel projection rays (Figure 14.4(c)) are mapped to a vertical line in the signed Hough space. Thus, captured perspective images can be easily transformed into samples in the parameter space.

Rendering a novel view in the scene is equivalent to extracting a partial or complete sinusoidal curve from the signed Hough space.

14.3 Analysis of lateral hopping

When the hopping direction is perpendicular to the viewing direction, as shown in Figure 14.1, it is called lateral hopping. In the signed Hough space, such a hopping is illustrated in Figure 14.5 where a segment of a sinusoidal curve is approximated by a line segment. Equivalently, at each rendering viewpoint, a perspective image is approximated by part of a CM.

Obviously, the smaller the hopping interval, the smaller the rendering error. On the other hand, the larger the hopping interval, the less data needed for wandering around an environment. It is possible to have a fairly large hopping interval for manifold hopping and still allow the rendering to be perceptually acceptable.

14.3.1 Local warping

When moving on a CM, the horizontal field of view should be constrained within a certain range so that the distortion error introduced in local warping from a multiperspective image to a perspective image will not cause much visual discomfort to the user.

The distortion threshold η_d is defined as the maximum allowable distance between point A and point B in Figure 14.7. These two points are projections of the rightmost pixel that are locally warped with distance R_1 (assumed distance) and R_2 (corrected distance), respectively. A radial hopping camera must satisfy the following:

$$\Delta\theta = \theta_B - \theta_A \leq \eta_d, \tag{14.1}$$

where $\theta_B = \sin^{-1}\frac{r_n}{R_2} - \sin^{-1}\frac{r_n - \Delta r}{R_2}$ is the angular difference when the object at R_2 distance viewed along circles of r_n and $r_n - \Delta r$. θ_A is defined similarly for object located at R_1. Thus,

$$\sin^{-1}\frac{r_n}{R_2} - \sin^{-1}\frac{r_n - \Delta r}{R_2} - \sin^{-1}\frac{r_n}{R_1} + \sin^{-1}\frac{r_n - \Delta r}{R_1} \leq \eta_d \tag{14.2}$$

If parallel interpolation is applied to local warping by assuming the depth R_1 at infinity, we can simplify the above constraint to

$$\sin^{-1}\frac{r_n}{R_2} - \sin^{-1}\frac{r_n - \Delta r}{R_2} \leq \eta_d \tag{14.3}$$

$$1 - \cos(\frac{Fov}{2}) = \frac{\Delta r}{r_n} \tag{14.4}$$

From the above two equations, we can derive the maximum FOV under parallel interpolation as

$$\cos(\frac{FOV}{2}) \geq \cos \eta_d - \sqrt{(\frac{R_2}{r_n})^2 - 1} \sin \eta_d \qquad (14.5)$$

The above equation shows that, under parallel interpolation, the maximum FOV for a hopping camera depends on the radius of the Concentric Mosaic, the scene depth, and the distortion error threshold. The field of view can be significantly increased when the object moves farther away. A smaller radius enables a larger FOV. For example, a panorama with a very large FOV can be rendered as the radius goes to zero. In addition, warping with constant depth (rather than infinite depth) can further increase the maximum FOV.

Consider a scene that is located along a circle whose radius is four times that of the outermost CM. If the distortion threshold is assumed to be $1°$ (that is a flow of 5 pixels for a mosaic with width 1800), the maximum allowable FOV is $42.42°$.

Fortunately human visual perception does not require a very large field of view for a hopping camera when wandering in a virtual environment. It has also been shown that $36°$ is close to perceptually optimal for most people [353]. It is well known that small FOV perspective images are generated from a large multiperspective panorama for the purpose of animation [324].

14.3.2 Hopping interval

The efficiency and effectiveness of hopping depend on the size of sample intervals along both the radial and angular directions. The angular direction is sampled uniformly and densely to ensure a continuous rotation. The maximum hopping interval Δr allowed for smooth visual perception is determined by the threshold of the horizontal pixel flow D_0 (in angular measurement) between two neighboring frames. The analysis of vertical parallax is ignored due to the nearly horizontal epipolar geometry between neighboring CMs.

Suppose that a point at a distance R_0 is seen in two CMs r_n and r_{n+1}, respectively. As shown in Figure 14.6, the horizontal parallax $\Delta\theta$ between two observed pixels A and B at the two CMs satisfies

$$\Delta\theta = \sin^{-1}(\frac{r_n + \Delta r}{R_0}) - \sin^{-1}(\frac{r_n}{R_0}) \leq D_0 \qquad (14.6)$$

which leads to the maximum hopping size

$$\Delta r = \sqrt{R_0^2 - r_n^2} \sin D_0 + r_n \cos D_0 - r_n \qquad (14.7)$$

$$= R_0 \sin(D_0 + \sin^{-1} \frac{r_n}{R_0}) - r_n. \qquad (14.8)$$

The above equation reveals that the sample interval along the radial direction depends on the depth (R_0), the smooth perception threshold (D_0), and the radius (r_n) of the CM. Specifically:

- Sampling along the radial direction is nonlinear. The smaller the radius, the larger the hopping intervals should be .

- The hopping interval can be increased with object distance. When objects are located at infinity, all CMs degenerate to the same panorama.
- A larger threshold D_0 allows for a larger hopping interval along the radial direction. As $\Delta r \to 0$, the hopping interval $D_0 \to 0$. This is equivalent to rendering with CMs [267]. On the other hand, if it is not required to observe parallax, a single manifold mosaic is enough for a user to look at any viewing direction.

The choice of threshold D_0 is closely related to the human visual system. It is well known that, for a human to observe smoothly moving pictures, the frame rate is 24 fps. Suppose that the average speed of rotation for a person to observe an environment is below 48°/second, then D_0 should be 2°. In other words, a person can tolerate 2° of average pixel flow for two neighboring frames and still observe smooth and continuous motion.

Consider a particular scene in which the radius of the outermost CM is 1 unit and the objects are located at a distance of 4 units. If D_0 is 1.5°, we have $\Delta r = 0.1$. Therefore, only 21 Concentric Mosaics are required (two for each concentric circle and one for the center). This is a significant reduction from 320 rebinned CMs needed in rendering with CMs [267].

14.4 Analysis of looming hopping using extended signed Hough space

In the previous section, we have analyzed manifold hopping where the hopping direction is perpendicular to the viewing direction. If the hopping direction is along the viewing direction, i.e., if the user moves forward and backward, the conventional CMs assembled by rays along the tangent lines of the concentric circle cannot be used. Instead, hopping with a looming motion can be achieved if we construct *normal* CMs that are formed by slit images with unit pixel width along the normal directions of the concentric circle, as shown in Figure 2.6(b). A novel view at any point on a circle can be rendered by locally warping rays from the normal CM near the viewpoint, as shown in Figure 14.1(c).

The signed ray space is no longer adequate for analyzing looming hopping. For a looming motion, points along the same ray need to be represented differently based on direction. The extended signed Hough space is used to account for direction. It is defined by a 3-tuple (r, θ, d), where d is the distance from the origin to the location where the ray is captured. Two points (P and P') along the same ray have identical (r, θ) but different values of d, as shown in Figure 14.8. Also, d will take the same sign as r to differentiate a "positive" ray from a "negative" one, similar to the signed Hough space. Although rays captured at P and P' are the same in the plan view of Figure 2.6(b), slit images captured at these two points are different.

Figure 14.8 also shows three different mosaics represented in the extended Hough space.

- A panorama: $r = d\sin(\theta - \phi)$;
- A tangent CM: $r = d = r_n$;

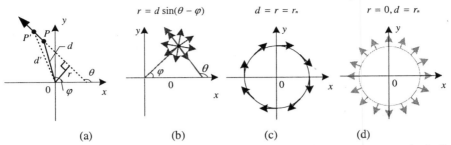

$$r = d\sin(\theta - \varphi) \qquad d = r = r_n \qquad r = 0, d = r_n$$

(a)   (b)   (c)   (d)

Fig. 14.8. (a) The extended signed Hough ray space is defined by three parameters (r, θ, d). Different points on the same ray have different d values. (b)(c)(d) A panorama, a tangent CM, and a normal CM are represented in the extended signed Hough space.

- A normal CM: $r = 0$ and $d = r_n$.

Note that ϕ is the constant angle for the viewpoint, and r_n is the diameter of one of the concentric circles. It becomes evident now why the location of the ray, which was ignored in lateral hopping (in the signed Hough space), should be considered in looming hopping because r is always zero under looming. Therefore, the (r, θ, d) representation is necessary and sufficient to index rays in 2D (plan view in Figure 2.6(b)) to capture the looming effect as the user moves forward and backward.

Figure 14.9 illustrates looming hopping in the extended signed Hough space. Similar to lateral hopping in the signed Hough space (Figure 14.5), rendering a novel view in looming hopping is also equivalent to approximating a partial sinusoidal curve by a line segment of a normal CM. Unlike lateral hopping, however, each sinusoidal curve is constructed at a different d. For clarity, we skip the looming hopping interval analysis in the extended signed Hough space, which is similar to the analysis in the signed Hough space in the previous section.

Lateral hopping is also illustrated in Figure 14.9. In the (r, θ, d) space, the plane for lateral hopping is $r = d$, but $r = 0$ for looming hopping. The sinusoidal curve segment is approximated around the maximum r in lateral hopping, and around $r = 0$ for looming hopping. If we project the lateral hopping plane in (r, θ, d) space onto the $d = 0$ plane, we obtain the (r, θ) counterpart for lateral hopping. There is therefore a duality between lateral hopping (r, θ) and looming hopping (d, θ).

14.5 Outside looking in

CMs are suitable for wandering around in an environment when a user is looking outwards. When the user is looking at an object, it is desirable to observe the object from outside at different viewing angles. In addition, the user may wish to view the object up close; the simplistic solution of simply zooming in will produce double images due to the effectively reduced resolution.

For simplicity of analysis and data capture, camera motion is assumed to be on a plane as an object rotates in front of a camera. A sequence of perspective images

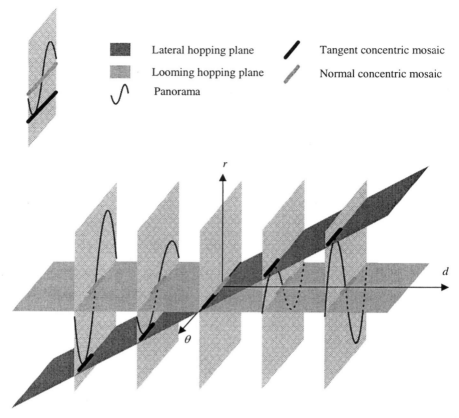

Fig. 14.9. Looming hopping with normal CMs, and lateral hopping with tangent CMs in the extended signed Hough space. Rendering a novel perspective view is equivalent to approximating a sinusoidal curve segment by a straight line segment representing part of a CM. In looming hopping, green segments are used to approximate the sine curve at different d values on the brown $r = 0$ plane. In lateral hopping, black segments are used to approximate the sine curve at different r (and d) values on the blue $r = d$ plane.

are then taken along the camera path (i.e., a circle). The rendering camera is also constrained to have continuous motion along the radial direction (moving towards and away from the object along a line) as shown in Figure 14.10(a), and discrete hopping motion in the angular direction as shown in Figure 14.10(b).

14.5.1 Hopping between perspective images

This rendering camera motion can be achieved by simply using the perspective images captured in the original sequence. Assuming a constant depth for the object, we can reproject perspective images to any novel views along the radial direction. However, only zooming effect, not the parallax effect, can be observed when the camera moves along the radial direction. When the camera moves away from the object, it is not possible to observe any additional part of the object around the boundary other than what is in the original image.

14.5.1.1 Angular hopping interval

Many previous systems have used multiple images to observe a single object from outside. It is, however, important to study how large the hopping interval should be to ensure a perceived smooth transition between the images.

As shown in Figure 14.11, two neighboring cameras A and B are located along the circle (with radius R) of camera path. The object is assumed to be at the circle (with radius r) of constant depth. $OA = OB = R$, and $OA_1 = OB_1 = r$. The camera spacing is $\alpha = AOB = A_1OB_1$. Let $\beta = A_1AO$, $2\beta = A_1AA_2$, and $\sin \beta = r/R$. The angular flow between two images can be approximated as

$$\Delta\theta \approx 2\beta \frac{\widehat{AB}}{\widehat{A_1A_2}} = 2\beta \frac{\alpha}{\pi - 2\beta}. \qquad (14.9)$$

Therefore, given the pixel flow threshold D_0, we obtain the camera spacing as

$$\alpha = (\frac{\pi}{2\beta} - 1)D_0. \qquad (14.10)$$

For example, if D_0 is $1°$, and $R = 3r$, then α is computed as $4°$. In other words, we need to capture 90 images along the circle.

14.5.2 Hopping between parallel projection mosaics

Another way to achieve continuous radial motion is to use parallel projection mosaics [32]. Parallel mosaics are formed by collecting all parallel rays in the same direction. This is called angular hopping with parallel mosaics.

Because parallel projection cameras are not commonly available, parallel mosaics are rebinned by taking the parallel rays from a dense sequence of perspective images taken along a circle outside the object. Figure 14.13 shows a projective image from the original sequence and the rebinned parallel mosaic. Note that the rebinned

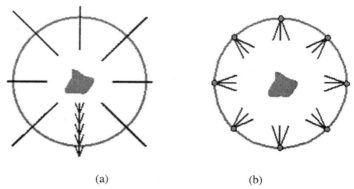

(a)   (b)

Fig. 14.10. Hopping from outside: (a) translating continuously in the radial direction (toward the object); (b) hopping discretely in the angular direction (around the object).

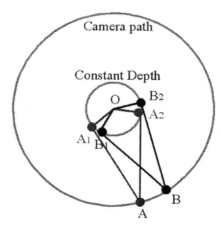

Fig. 14.11. Hopping interval between two perspective images (viewed from outside the object) is related to the field of view of the object ($A_1 A A_2$). The object is assumed to be bounded within the constant depth circle.

mosaic is called 1D parallel mosaic because the vertical direction is still perspective, only the horizontal direction is under parallel projection.

Assuming a constant depth for the object, we can reproject parallel mosaics to any novel view along the radial direction, as shown in Figure 14.12. Warping 1D parallel mosaic in Figure 14.13(b) using constant depth is shown in Figure 14.13(c). Even though warping errors are created, such as those around the boundary of the object, they are small enough to cause little visual distortion.

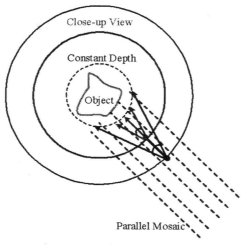

Fig. 14.12. Reprojecting a parallel mosaic (shown as parallel dotted green lines) to different perspective images along the radial direction using constant depth assumption. The image shown by blue arrows is viewed at a normal distance away, while the image with red arrows is a close-up view.

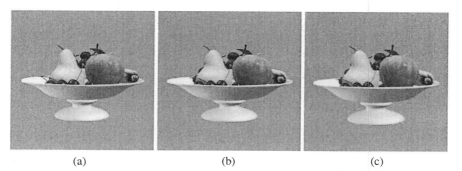

(a)   (b)   (c)

Fig. 14.13. Warping parallel projection images: (a) a perspective image; (b) a 1D parallel projection mosaic; (c) 1D mosaic of (b) warped with constant depth. (c) and (a) are mostly similar except around edges (e.g., on the left edge of the pear). An advantage of using parallel mosaics is to have higher resolution especially for close-up views.

14.5.2.1 Close-up views

Rendering a novel view with angular hopping using parallel mosaics can again be explained in the signed Hough space. Continuous motion along the angular direction is obtained by approximating a cosine segment using a line segment. When the viewpoint is far away, the parallel mosaic approximates the perspective view very well. The reprojection or warping error increases as the viewpoint approaches the object. In addition, the image size of the parallel mosaic determines how closely the rendering camera can get to the object. Hopping using parallel mosaics and hopping using perspective images have similar warping errors, especially if constant depth is assumed.

However, rebinned parallel mosaics can have a much higher resolution than the original image if a very dense sequence is captured. For example, a 1D parallel mosaic of 640×240 can be obtained from 640 original images with size 320×240. Close-up views rendered from rebinned parallel mosaics have better quality than simply zooming-in the original images.

14.6 Experiments

In this section, we show rendering results using images of both synthetic and real scenes.

14.6.1 Synthetic environments

A synthetic environment is represented with 41 CMs (with size 2400×288) on 11 concentric circles. There are 21 tangent CMs and 21 normal CMs. Note that the center mosaic degenerates to a single perspective panorama, as shown in Figure 14.15(a). At the outermost circle, the tangent CM is shown in Figure 14.15(b), while the normal CM is shown in Figure 14.15(c). By hopping between these mosaics, five images are rendered from the left, right, center, front and back viewpoints shown in Figure 14.15(d). Parallax effects (both lateral and looming) are clearly visible from the rendered images. Also, hopping between these mosaics provides a smooth navigation experience. However, one can only switch lateral motion and looming motion at the center. For conventional rendering with CMs, 720 such mosaics would have been used instead. Therefore, manifold hopping requires significantly less data for a similar viewing experience.

A much larger environment can be constructed by combining more mosaics captured at different locations. By carefully adjusting constant depths used for different sets, the user can hop smoothly from one circle to another, and inside a circle.

14.6.2 Real environments

A Sony Mini DV digital video camera was used to capture CMs of a real environment. The camera rotates along a circle. The video is digitized at the resolution of

720×576. A total of 5726 frames are captured for a full circle. The raw data for the video sequence amounts to a total of 7 Gigabytes. Instead of using 720 rebinned CMs of size 5726×576, only a small subset (typically 21) of resampled CMs were selected.

Three rebinned CMs are shown in Figure 14.16(a). Two high resolution images (with display size 500×400) rendered from 21 CMs are shown in Figures 14.16(b) and (c). Horizontal parallax around the tree and lighting change reflected from the window can be clearly observed. Constant depth correction is used in all the experiments.

The original CMs can be resized to reduce the amount of data used in manifold hopping. As shown in Figures 14.16(d) and (e), two images with low resolution 180×144 are rendered from 11 resized smaller CMs. It is important to note that simply resizing the original 11 CMs does not generate the expected CMs. Instead, mosaics of such small size should be resampled from the original dense sequence.

A predictive coding compression algorithm with fast selective decoding and random access was used to compress the CMs. With this compression algorithm, the above 11 CMs were reduced to $88k$ with a compression ratio of 78. Two corresponding rendered images using the compressed data are shown in Figures 14.16(f) and (g).

14.6.3 Outside looking in

In another experiment, a rotating object was captured in front of a stationary camera. Ninety parallel mosaics with resolution 645×288 were rebinned from the input sequence of 5277 images of resolution 360×288. These parallel mosaics have an angular hopping interval of 4 degrees. A perspective image from the input sequence is shown in Figure 14.17(a). Using the Lagrange interpolation, the rebinned $1D$ parallel mosaics are rather smooth, as shown in Figure 14.17(b).

Figures 14.17(c) and (d) show two warped images from the $1D$ parallel mosaics. Experiments showed that the hopping angular interval of 4 degrees provided a very perceptually smooth virtual camera movement. Two close-up views along the same viewing direction of Figure 14.17(c) are also shown in Figures 14.17(e) and (f). Because parallel mosaics have a higher resolution than the original images, close-up views provide details that would not be possible by simply zooming-in on the original images.

A sequence of $2D$ parallel mosaics were then synthetically generated using an angular hopping interval of 4 degrees in both longitudinal and latitudinal directions. Hopping between this collection of parallel mosaics again provides perceived smooth camera movements in two dimensions.

14.7 Discussion

While reducing data significantly, manifold hopping limits the freedom of user movement. In hopping with CMs, for instance, a user can only rotate along one of the concentric circles. The user is not allowed to rotate at any given viewpoint except in the center. As shown in the synthetic experiments, the user can only change from lateral hopping to looming hopping at the center. If the number of CMs is sufficiently large, it is also possible to hop around any fixed point in the angular direction by warping different CMs. In the signed Hough space, it is equivalent to finding segments from different r lines that approximate a sinusoidal curve.

Manifold hopping is not restricted to hopping with CMs or with lateral or looming movements. There are many other choices for manifolds and hopping directions. For example, hopping between panoramas has been used in QuickTime VR [41] using "hotspots". When panoramas are closely spaced, hopping between them can also achieve a smooth transition. Figure 14.14 shows two examples of hopping between panoramas. Figure 14.14(b) shows the signed Hough representation of a line of panoramas as in Figure 14.14(a), and Figure 14.14(d) shows the signed Hough representation of a circle of panoramas as in Figure 14.14(c).

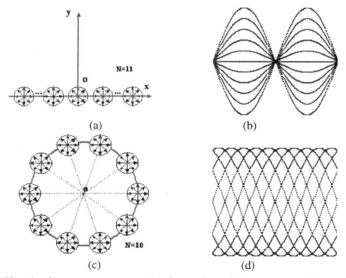

Fig. 14.14. Hopping between panoramas: (a) along a line of 11 panoramas; (b) ray distribution of (a) in signed Hough space; (c) along a circle of 10 panoramas; (d) ray distribution of (c).

There are two major differences between manifold hopping with CMs and hopping with panoramas. The first difference is in capturing. Panoramas can capture similar rays to CMs as the number of panoramas increases. However, the same result will require capturing panoramas many times at different locations, as opposed to rotating the camera only once for capturing CMs.

The second and perhaps more important difference is in sampling. Each manifold mosaic is multiperspective, while each panorama has only a single center of projection. Since different viewpoints can be selected as the desired path for the user, a multiperspective panorama could be more representative of a large environment than a single perspective panorama. If the multiperspective image is formed by rays taken along the desired path of the user, the warping error from a multiperspective image is, on average, smaller than that from a perspective image (e.g., a panorama).

CMs are suitable for the inside looking out. To observe objects from the outside looking in, parallel mosaics can be used for manifold hopping. For CMs, the manifold is a cylindrical surface. For parallel mosaics, the manifold is a plane originating from the object center. We have discussed manifold hopping in two dimensional space by constraining the rendering camera on a plane. The concept of manifold hopping can be generalized to higher dimensions. The analysis in higher dimensions is very similar to the two-dimensional cases. However, it is difficult to capture such manifold mosaics in practice.

14.8 Concluding remarks

In this chapter, we describe an IBR technique called *manifold hopping* that has the following properties:

- It does not require a large amount of image data, and yet the user can perceive continuous camera movement.
- It requires neither accurate depth nor correspondence, yet generates perceptually acceptable rendered images.

Manifold hopping renders a novel view by locally warping a single manifold mosaic, without the need for interpolating from several images. The key is that warping a single multiperspective image to a perspective image with a regular field of view causes insignificant distortion to the human eye, even with warping errors resulting from incorrect depth information. Furthermore, local warping does not introduce structural errors such as double images which are perceptually disturbing.

Most importantly, manifold hopping requires relatively little input data. Capturing manifold mosaics such as CMs is also easy. By sparsely sampling the CMs, the amount of data from the original CMs can be reduced by more than an order of magnitude. While manifold hopping provides only discrete camera motion in some directions, it provides reasonably smooth navigation by allowing the user to move in a circular region and to observe significant horizontal parallax (both lateral and looming) and lighting changes. The ease of capture and the very little data requirement make manifold hopping very attractive and useful for many virtual reality applications, in particular those on the Internet.

Table 14.1 compares how manifold hopping differs from previous IBR systems, in terms of their geometric requirements, number of images, rendering viewpoints and perceived camera movement. Manifold hopping stands out in that it ensures a perceived continuous camera movement even though rendering viewpoints are discrete. It builds on the observation that a fairly large amount of viewpoint change is allowed, while maintaining perceptually continuous camera movement to humans. This observation of "just-enough hopping" for reducing image samples is, in spirit,

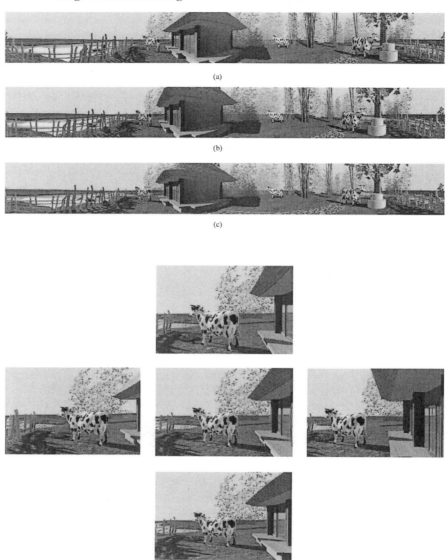

Fig. 14.15. Lateral and looming hopping between CMs of a synthetic environment: (a) a tangent CM; (b) the middle panorama; (c) a normal CM; each mosaic has the size of 2400 × 288. (d) five rendered views from manifold hopping at the left, center, right, forward and backward locations. Note that the horizontal parallax is clearly visible between the left and right views; the looming effect can be seen from the forward and backward views.

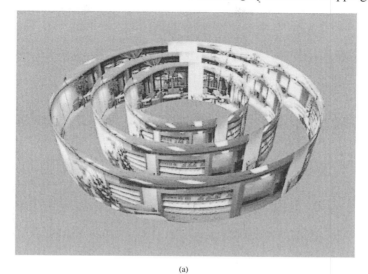

(a)

(b)   (c)

(d)   (e)   (f)   (g)

Fig. 14.16. Hopping between CMs: (a) three CMs projected onto cylinders; (b)(c) two rendered images at a high resolution 500×400; (d)(e) rendered images with a low resolution 180×144; (f)(g) low resolution rendered images using $88k$ compressed data.

(a)

(b)

(c)

(d)

(e)

(f)

Fig. 14.17. Hopping between parallel mosaics: (a) a perspective image from the original sequence; (b) a rebinned 1D parallel mosaic with higher resolution; (c)(d) two rendered images from different viewing directions; (e)(f) close-up views along the viewing direction of (c).

	Geometry	Images	RV	PM
Light fields [160, 91, 267]	no/approximate	very large ($100 \sim 10000+$)	C	C
3D Warping [189, 192, 264, 39]	accurate	small ($1 \sim 10+$)	C	C
View interpolation [40, 194, 260, 6]	accurate	small ($2 \sim 10+$)	C	D
Hopping [170, 41]	no	moderate ($10 \sim 100+$)	D	D
Manifold hopping	no/approximate	moderate ($10 \sim 100+$)	D	C

Table 14.1. A table of comparison for different IBR techniques (with representative citations): geometry requirements, number of images, rendering viewpoints (RV), and perceived camera movement (PM). Note: C = continuous, D = discrete.

similar to the "just-necessary effort" adopted by perceptually based techniques [241] on realistic image synthesis to reduce computational cost.

15

Large Environment Rendering using Plenoptic Primitives

One of the most difficult tasks in computer graphics is to enable virtual walkthroughs in very large and complicated environments that are photorealistic, seamless, and in real-time. Most current IBR techniques, while capable of photorealism and interactive speeds, have failed in practice to extend to visualizations of such environments. (One exception is the work of Aliaga and Carlbom [4], but their approach requires a dense set of images to be taken.) In this chapter, we describe an approach that defines a virtual walkthrough experience using *plenoptic primitives* (PPs). A PP can be any type of local visual experience: 360° static panorama, panoramic video (PV), Lumigraph/Light Field representation, or Concentric Mosaics (CM). By combining them judiciously, user experience can be authored with significantly reduced effort while maintaining high-quality user experience. In this chapter, results for synthetic and real environments using PVs and CMs are shown; they illustrate how the issue of achieving smooth transitions among PVs and CMs can be addressed by using position-dependent local geometries.

For purely IBR techniques, the data acquisition process is generally not trivial, since the camera parameters associated with each input image need to be known accurately. In addition, the size of the database is often very large, especially for the light field, Lumigraph, and Concentric Mosaics representations. Furthermore, none of these methods have been demonstrated on continuous navigation within a large and complicated environment, even though it is possible in principle. In practice, both the image acquisition process and resulting size of the augmented database will quickly become unmanageable. The other option of using a global 3D model is not very attractive from two perspectives: One, it is unlikely to produce photorealistic views, and two, generating a reasonably accurate global model of a scene on a wide scale is extremely difficult without good initial estimates from non-image-based sources. An example of this approach is MIT's City Scanning Project [296].

Aliaga and Carlbom [4] describe an approach for rendering a large environment. They use a remotely guided robot fitted with a catadioptric omnidirectional camera to acquire sequences of images along a criss-crossing path. The smallest enclosing camera path loop is used to synthesize a virtual view. Their approach differs from the PP-based system because they sample the environment almost uniformly and using a homogeneous representation throughout (i.e., treating all spaces as equally important or interesting). In the PP-based approach, the visual experience is customized by using spatially-dependent representations.

15.1 Customized visual experience

The PP-based approach relies on the assumption that different locations in a scene usually have different visual appeal, and that the mode of visual interaction should reflect this. Consequently, by customizing visual experience at each scene location, the user can reduce the effort in image acquisition and resulting size of the total database without degrading the quality of visual experience. The customization is in the form of prespecifying *plenoptic primitives* (PPs), in a spirit similar to the concept of "visual tunnel primitives" [137].

Note that while the size of the environment plays a part in how a person can be influenced in customizing the visual experience, the complexity of the environment is probably a more significant factor. For example, a 100 ft x 100 ft bare room can be easily represented by just a set of Concentric Mosaics (CMs) or a panoramic video (especially if the room looks uninteresting). However, the floor of a building with the same exact dimensions, but with many rooms and more detailed structures could require a significant number of panoramic videos and CMs.

15.2 Organization of chapter

We start by describing PPs in Section 15.3. Section 15.4 delineates the steps required to construct and render the environment. We then describe the authoring process in Section 15.5, the user interface in Section 15.6, and the rendering process in Section 15.7. Section 15.8 shows image snapshots of a virtual walkthrough in progress with descriptions of rendering performance. Practical issues regarding large-scale environment visualization and areas for future work are discussed in Section 15.9, followed by concluding remarks in Section 15.10.

15.3 Plenoptic primitives (PPs)

A PP can be a 360° static panorama, panoramic video (PV), Lumigraph/Light Field representation, or Concentric Mosaics (CMs). By prespecifying user experience as combinations of PPs, the effort required for data capture and authoring can be reduced while maintaining high-quality visual impact and walkthrough experience. In this chapter, PVs and CMs of synthetic and real environments are used to illustrate this idea; PVs and CMs are used because of their ability to allow full panoramic viewing (i.e., inside looking out and all around).

Each PP has its own range and degree of freedom of visualization. It is known that an effective way of reducing data is to constrain the motion or viewpoints of the virtual camera. One of the earliest example of this is the Movie Map system [170]. Here the user plays back a prerecorded sequence of images to create the illusion of navigating along a spatial path. Another well-known example is the QuickTime VR system [41], where users are afforded panoramic experiences at discrete locations. In the authoring process, locations which are deemed less interesting will be assigned PPs with less input data requirements (and consequently more impoverished range of virtual camera motion).

A number of adventure/mystery games, such as Myst® and Riven®, feature walkthroughs within realistic-looking environments. They also provide a mix of different kinds of experiences, such as panoramas and texture-mapped 3D scenes. There are two important differences between these applications and the PP-based system. For one, the its rendering mechanism is primarily IBR-based, and two, its rendering is visually smooth throughout the course of a virtual walkthrough.

15.3.1 Panorama and panoramic video

A panorama is a 2D image with a 360° field of view along at least one angular direction. A lot of research work has been done in generating panoramas (e.g., [41, 291, 330]), and many commercial hardware and stitching software products to create panoramas are also now available. A panoramic video (PV) refers to a sequence of panoramas created at different locations along a path in space. While the virtual camera location is constrained to be along the path of image acquisition, the PV is still effective in providing seamless walkthrough. A PV can also be used to capture dynamic scenes at a stationary location.

15.3.2 Lumigraph/Light Field representations

The light field [160] and Lumigraph [91] are 4-parameter subsets of the plenoptic function [2]. By using a two-slab representation and dense image sampling, they have demonstrated that interactive photorealistic visualization is possible. However, in addition to requiring a large database, their field of view is also limited by virtue of the two-slab representation. Extensions to the light field such as the Spherical Light Field [113] exist, but data capture for such extensions is usually not trivial.

15.3.3 Concentric Mosaics

The disadvantage of the small field of view of the light field and Lumigraph can be overcome by sampling along a horizontal circular path and rebinning light rays in a 3D representation called Concentric Mosaics (CMs) [267]. This representation allows panoramic visualization within a planar disk. While it does require some depth correction along the vertical axis for visually acceptable scaling, its ease of data capture and compelling local visual experience make the CM an attractive representation.

The comparisons between the different PPs are shown in Table 15.1. In this chapter, the PP-based concept of representing visual experience in a large environment is

Table 15.1. Comparison of different PPs. "Dims." refers to the number of dimensions for representation.

Plenoptic Primitive	Dims.	Navigation Mode
Panorama	2	Orientation at a point
Panoramic video	3	Orientation along a 3D trajectory
Concentric Mosaics	3	Panoramic within disk
Light Field/Lumigraph	4	Limited 6 DOF

illustrated using PVs and CMs. This is primarily because both are relatively easy to capture, and they both provide panoramic visibility from the inside looking out.

15.4 Constructing and rendering environments

There are a number of steps that must be taken before a virtual walkthrough in a large environment can take place. These steps (shown in Figure 15.1) are as follows:

- *Calibrate.* The first step is to calibrate the camcorders in order to extract their projection characteristics. Calibration techniques described in [136] and [346] were used to estimate the radial distortion and camera intrinsic parameters, respectively. The technique described in [136] involves just manually drawing contours approximately corresponding to projections of 3D straight lines; the contours are automatically refined while the radial distortion parameters are computed. The technique of [346] requires only a small number of snapshots of a flat calibration pattern at different orientation. These two techniques are very simple to use.
- *Capture.* For the synthetic environment case, 3D Studio Max® was used to generate the input images, while a special rig was used to acquire images for a real environment. In each case, the inputs are five image streams. More details are given in their respective subsections in Section 15.8.
- *Preprocess.* Before the authoring process can begin, the five image streams have to be corrected for distortion, recentered, synchronized, and composited frame by frame.
- *Author.* This step involves the specification of PVs and CMs, and how they are related to each other spatially. It is described in Section 15.5.
- *Render.* This online step is described in Section 15.7. Among others, it involves the important issue of ensuring smooth transitions between different PPs.

All steps except the render step are done offline.

15.5 The authoring process

The authoring process requires the specification of *visual nodes* and *visual paths* that involves single panoramas, PVs, and CMs. A visual node signifies transition points or

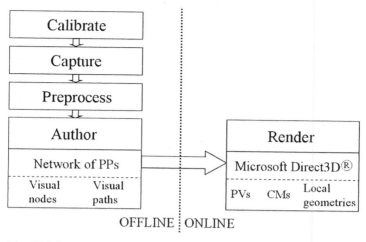

Fig. 15.1. Processes involved in constructing and rendering environments.

Feature	Specification
Plan map	Image, extent
Visual spots	Count, location, index to local geometry
Visual paths	Count, link to visual spots
Panoramic Videos	Count, filename format, links to panoramic image frames, number of rendering faces, orientation of rendering faces
Concentric Mosaics	Count, filename format, links to visual spots, camera parameters, frame size
Local geometries	Count, topology, vertex locations

Table 15.2. Feature placement information. It has six parts.

points of interest. It can be a CM center, center of a panorama, where a PV starts and ends, where a PV intersects another PV, and where local scene geometry is specified. In addition, it encodes relative camera orientations between connected PPs. If these orientations are different at the two ends of a PP, they are linearly interpolated in-between. A visual path is an entity that connects any two visual nodes. This permits the user to create a network of PPs. Figure 15.2 shows a plan map of an environment with visual nodes and paths. The feature placement information is specified in a descriptive text file, which has six parts (see Table 15.2).

The location-dependent geometries (last item in Table 15.2) are used for CMs, and for PV-CM and CM-CM transitions. It is assumed that the image data capture and online viewing occur at the same horizontal level.

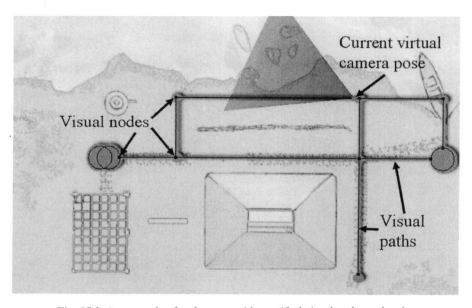

Fig. 15.2. An example of a plan map with specified visual nodes and paths.

Once the authoring process is complete, with the network of PPs fully specified, the system will then be ready for interactive rendering of the environment. Before we provide details on the rendering process, let us first briefly describe the user interface and how the user can navigate with it.

15.6 User interface

The user interface for a virtual walkthrough is composed of the viewport on which the current virtual camera view is displayed, and the plan map, which indicates the current virtual camera location and orientation relative to the environment. The interface is shown in Figures 15.5 and 15.11 for two different environments.

The user interface allows bidirectional control over the virtual camera position and motion between the viewport and map. Within the viewport and using the mouse, the user is able to pan and tilt the camera, zoom in or out, and move forward or backward. Within the map, the user can jump directly to a location by just clicking over the desired destination. The user can also move continuously by holding the mouse down and move along a PV, within a CM, or transition between them. In addition, the map can also be reoriented or zoomed in or out within the screen plane.

15.7 Rendering issues

In this section, we describe how various PPs are rendered and how smooth transitions between different types of PPs are achieved.

15.7.1 Rendering PVs

Each frame of a PV is cast as a cubic texture, with each texture face mapped to a face of a cube and rendered. The frames are accessed on demand directly from disk using the file cache of the OS. Panning and tilting is achieved by rotating the viewing camera, and moving is by translating the cube center to the current viewing position (i.e., loading the appropriate texture maps).

15.7.2 Rendering CMs

Full details of rendering CMs can be found in [267] (a summary is given in Chapter 2). The basic rendering algorithm is extended in the PP-based system by incorporating the notion of position-dependent local scene geometries (described in more detail in Section 15.7.3) similar in concept with [239]. The steps required for rendering CMs are as follows:

- For a given camera viewpoint, compute z-buffer for the local scene geometry and lock the z-buffer. The z-buffer has to be locked before it can be read. (GPU)
- Read the z-buffer and convert to floats. (CPU)
- For each column of the output view, render the slit image with the depth correction using the z-buffer depths. (CPU)

To optimize rendering speed, Microsoft Direct3D® is used to render the mesh without displaying it.

15.7.3 Ensuring smooth transitions

This is actually non-trivial and posed a significant challenge in the PP-based work. The key is to use local geometries. These local geometries can be placed at positions where the user deems to be necessary or appropriate. Examples of placements are shown in Figure 15.3. In the implementation, the number, placement, and complexity of the local geometries were manually chosen.

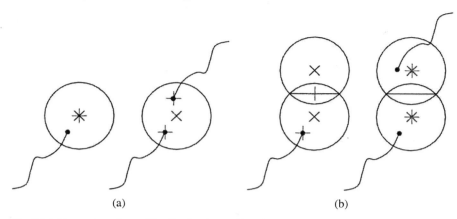

Fig. 15.3. Examples of possible distributions of local geometries for depth correction (many other possibilities exist). (a) For PV-CM, (b) For PV-CCM. A CM is represented by a circle while a PV is represented by a curve segment. Places of transition are in red. The center of a CM is marked × while the location of a local geometry is marked a green +.

PV-PV transition

This is the simplest transition. In the authoring process, frame indices are specified in transitions between PVs. Along a PV, moving forwards or backwards has the effect of stepping ahead or behind in time in the current panoramic image sequence. Moving from one PV to a different PV via a prespecified visual node is equivalent to reading a different set of image streams, possibly starting at an intermediate frame. In either case, the virtual camera orientation is maintained during the transition.

CM-CM transition (Concatenated CM, or CCM)

Smooth transitioning between CMs can be regarded as a special case of image morphing. Without depth information, visual discontinuity will very likely be observed when the user switches from one CM to another, or double images in the case of blending. In real scenes, accurate depth is usually not known or easily recovered; however, as we will show later, approximate depth is more often than not sufficient.

Constant depth has been used for rendering CMs, and while this is adequate for single CMs, visual artifacts can be significant with multiple CMs. As a result, location-dependent approximate scene geometry are used—in the form of vertically stacked *radial depth polygons* (RDPs), as shown in Figure 15.4. The vertices of each RDP are distributed uniformly along the angular direction. The 3D surface representing the scene is created by linking these stacked RDP vertices. In the implementation, each local scene geometry is represented by 13 stacked RDPs, with each RDP having 128 vertices, yielding a total of 1658 vertices.

The shape of the local scene geometry is initialized by applying stereo on the CM in a similar manner as described in [272], except that 1D matching is used in conjunction with tensor voting [195]. This produces reasonably good quality stereo

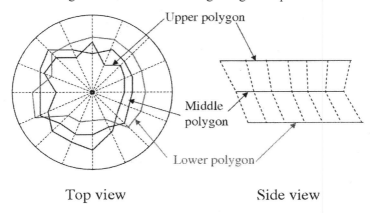

Top view   Side view

Fig. 15.4. Local scene geometry represented as stacked radial depth polygons (RDPs). Only three RDPs are shown here.

reconstruction (the measure in terms of RMS reprojection error), even in the presence of discontinuities.

PV-CM and PV-CCM transitions

Within a CM (or CCM, i.e., concatenated Concentric Mosaics), prior to transitioning to a PV, the image is rendered in the following manner. At the current virtual camera location, the environment is rendered using each local geometry separately first. The final view is produced by linearly combining these rendered views using weights that are based on proximity of the local geometry centers to the current virtual camera position. More specifically, the weights are inversely proportional to the distance of the virtual viewpoint to the local geometry centers. As a result, the rendering time increases linearly with the number of local geometries. In the implementation, the number of local geometries within each PV-CM, PV-CCM overlap were restricted to at most two, and for the two environment examples, this restriction was adequate.

Once the local geometry position has been decided, the time taken to generate a single local geometry takes between 20-50 minutes. This includes stereo computation (done only once), verifying transitions to be visually acceptable, and modifying the complexity of the local geometry if necessary.

15.8 Experimental results

The results for two environments are used to illustrate the PP-based concept of customizing visual experience. One is a relatively complicated synthetic environment. The other is a real environment inside a small museum. Both environments were rendered using a 866MHz PC and a Hercules 3D Prophet II GTS graphics card with 64MB DDR RAM. The rendering performance is shown in Table 15.3.

15.8.1 Synthetic environment

The input images used for the synthetic environment experiment were rendered using 3D Studio Max®. Each input image has the resolution of 512×512, and a total of about 5000 images were used (3000 for three CMs and 2000 for PVs). The rendering time for all of these images took about 30 hours using 8 PCs. Example screenshots of the synthetic environment during the course of a virtual walkthrough are shown in Figure 15.5 and 15.6. An example of the result of using local geometries for transitioning is shown in Figure 15.7. A close-up view of the transition frames in Figure 15.8 shows that using local geometries produced a more visually correct effect.

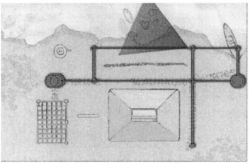

Fig. 15.5. Display of a synthetic environment. Top: Viewport (700×480), Bottom: Plan view of the virtual camera relative to the environment.

Fig. 15.6. Sample screenshots of a walkthrough for synthetic environment.

Fig. 15.7. Transitioning between PVs and CMs for synthetic environment. The virtual camera is moving forward. Top row: Transition using only constant depth. Bottom row: Transition using local geometry. The first two left images are from PVs while the last two images are from CMs.

Fig. 15.8. Close-up view of transition between PVs and CMs (second and third columns of Figure 15.7). Note that the virtual camera is moving forward. Top row: Using only constant depth. Notice the scaling is significantly more in the x-direction than in the y-direction. Bottom row: Using local geometry. Here the scaling looks more visually correct. Compare the shadow of the fence at the bottom of the images.

15.8.2 Real environment

The interior of a local museum is used as the real environment example. Example screenshots of the real environment during the course of a virtual walkthrough are shown in Figure 15.11 and 15.12. An example of the result of using local geometries for transitioning is shown in Figure 15.13. As shown in the close-up version in Figure 15.14, using local geometries produced a significantly less abrupt change in viewpoint during the transition from a panoramic video to Concentric Mosaics.

Acquiring the data

Fig. 15.9. Capturing device. It has five camcorders attached to a rotating arm, a motor, and an uninterruptable power suppy (UPS).

The capturing system shown in Figure 15.9 was used to acquire the input images for visualizing a real environment. It has five Elura digital camcorders fitted with wide-angle lenses attached to a platform. This platform is attached to the end of an arm capable of being rotated by a motor. The length of the arm is about 2.5 feet. All these are placed on a movable cart. An uninterruptable power supply (UPS) is used as a portable power supply for the motor.

Since it is not possible to genlock the camcorders, synchronization is performed manually by the operator positioning between two cameras while maintaining visibility and clapping hands. This is done for every adjacent pair at the start of recording. The cart was then manually pushed around during the recording of the PVs, and was kept stationary with the motor switched on for the recording for CMs. The operator was careful to get out of the visual range of the camcorders as much as possible; this was done by crouching behind the cart during recording. Each image has a resolution of 688 × 480, and a total of about 4800 frames per camcorder were used in this rendering experiment.

Data processing

Since the images were acquired using camcorders with wide-angle lenses, they were quite severely distorted, as can be seen in Figure 15.10. Calibration techniques described in [136] and [346] were used to extract the radial distortion and camera intrinsic parameters, respectively.

Once the images have been corrected and calibrated, the optimal relative transforms between the cameras were found using the image overlaps. There is some parallax due to the camcorders having different centers of projection, but this problem was not severe since the objects in the scene were distant. Once the camera relative transforms have been estimated, the multiple video streams are composited frame by frame. The parallax problem was reduced by applying a deghosting technique as described in [291].

Fig. 15.10. Frames from two of the five digital camcorders.

Rendering performance

Table 15.3 shows the rendering performance for the PP-based system (a 866MHz PC and a Hercules 3D Prophet II GTS graphics card with 64MB DDR RAM). Not surprisingly, virtual navigation within a CM or CCM is relatively slow. This is because of the processing necessary for depth compensation using the local geometries (there are two local geometries per CM or CCM). The rendering speed is inversely proportional to the total pologon count in the local geometries. One direct way of increasing this speed is to reduce the polygon count associated with the local geometry, but this will very likely degrade the PV-CM rendering transition. The number of local geometries and their complexity are currently selected *a priori*.

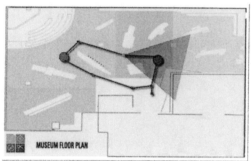

MUSEUM FLOOR PLAN

Fig. 15.11. Display of a real environment. Top: Viewport (700 × 480), Bottom: Plan view of the virtual camera relative to the environment.

Table 15.3. Average rendering performance for a viewport size of 700 × 480.

Type of navigation	Frame rate (fps)
PV move	8 (first time, retrieve from disk)
	20 (when frames are cached)
PV rotate	63
CM move	10
CM rotate	10

Fig. 15.12. Sample screenshots of a walkthrough for real environment.

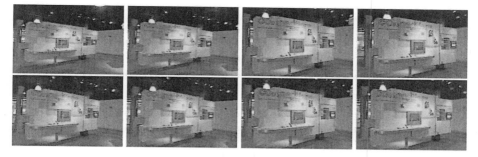

Fig. 15.13. Transitioning between PVs and CMs for real environment. The virtual camera is moving at an angle to the principal view direction. Above: Transition using only constant depth. Below: Transition using local geometry. The first two left images are from PVs while the last two images are from CMs. Notice the shape of the monitor near the middle of the image before and after the transition.

15.9 Discussion

There remains a number of issues to contend with. For example, the issue of capture is a difficult one, especially when full panoramas need to be captured in real-time. Because multiple camcorders were used to capture the environment, parallax was an issue. The solution to this is to use specialized hardware for image acquisition, such as that used by iMove, Inc.[1] for capturing their panoramic video. In their system, six CCDs are assembled close to one another to minimize parallax. Each CCD is arranged to look out of a different face of a cube. Their outputs are also synchronized and streamed directly to disk.

It is clear also that for optimal visual experience, it is necessary to put the human in the loop to decide the importance (or attach degree of visual appeal) to the

---

[1] http://www.imoveinc.com/

Fig. 15.14. Close-up views of the transition frames (second and third columns in Figure 15.13). Note that the virtual camera is moving at an angle to the principal view direction. Leftmost pair: Using only constant depth. Rightmost pair: Using local geometries. The appearance change is much less abrupt when local geometries are used.

environment. This puts the control of customizing right to the user or environment author. This is critical in the case of authoring walkthroughs for very large or complicated environments, such as the interior of an entire building or a city. Allowing the user in the loop will help reduce the size of the database required.

The use of PPs to represent the environment avoids the difficult problem of global reconstruction of 3D models. While local 3D reconstruction has been relatively successful, it has yet to be demonstrated on scenes on a very wide scale. An example of such a project is MIT's City Scanning Project [296], but it uses GPS readings and spatially sparse image capture. The output is a set of texture-mapped 3D models. The "plenoptic stitching" work of Aliaga and Carlbom [4] is a good approach to rendering large environments, but this approach was demonstrated for a small number of connected rooms at a time. The size of the database becomes an issue if the environment involves, say, an entire floor of a building.

Local geometries are used to enhance the quality of transition between separate PVs. However, the degree of accuracy of each local geometry and number of them to be used for acceptable transition is still an area of future work. While Chai *et al.* [33] have analyzed the rendering quality with respect to the sampling density and accuracy of geometry, their analysis does not take into consideration occlusion events, and they assume regular sampling. In the implementatino, the number of local geometries and the complexity of the local geometries were chosen by hand to produce a visually acceptable transition.

The performance of the rendering system can be improved by using a faster PC and graphics accelerator (obviously), or by using multithreading in a multi-processor

PC. There are a number of rendering processes (such as z-buffer conversion to float and depth-correction of vertical CM slits) that can benefit from multithreading.

15.10 Concluding remarks

This chapter illustrates the concept of customizing a visual experience that is realistic, interactive, and equally important, smooth. This concept relies on plenoptic primitives (PPs), which permit the author of the environment to minimize labor involved in acquiring input and processing requirements without sacrificing quality of experience.

Other PP types such as the two-slab light field and Lumigraph can be added. In addition, adding local geometries along PVs would allow the navigation to deviate a little from the sampling camera path. Stereo data can also be used to automatically estimate these local geometries.

16

Pop-Up Light Field: An Interactive Image-Based Modeling and Rendering System

In this chapter, we describe an image-based modeling and rendering system called *pop-up light field*. It models a sparse light field using a set of *coherent layers*. In this system, the user specifies how many coherent layers should be modeled or popped up according to the scene complexity. A coherent layer is defined as a collection of corresponding planar regions in the light field images. A coherent layer can be rendered free of aliasing, all by itself or against other background layers. To construct coherent layers, we introduce a Bayesian approach, *coherence matting*, to estimate alpha matting around segmented layer boundaries by incorporating a coherence prior to maintaining coherence across images.

The system to construct the pop-up light field has an intuitive and easy-to-use user interface (UI). The key to UI is the concept of human-in-the-loop where the user specifies where aliasing occurs in the rendered image. The user input is reflected in the input light field images where pop-up layers can be modified. The user feedback is instant through a hardware-accelerated real-time pop-up light field renderer. Experimental results demonstrate that the system is capable of rendering anti-aliased novel views from a sparse light field.

16.1 Motivation and approach

Here is an interesting question: Can we use a relatively sparse set of images of a complex scene and produce photorealistic virtual views free of aliasing? A straightforward approach would be to perform stereo reconstruction or to establish correspondence between all pixels of the input images. The geometric proxy is a depth map for each input image. Unfortunately, state-of-the-art automatic stereo algorithms are inadequate for producing sufficiently accurate depth information for realistic rendering. Typically, the areas around occlusion boundaries [147, 140] in the scene have the least desirable results, because it is very hard for stereo algorithms to handle occlusions without prior knowledge of the scene.

One can approach this problem by suggesting that it is not necessary to reconstruct accurate 3D information for each pixel in the input light field. A reasonable solution would then be to construct a pop-up light field by segmenting the input sparse light field into multiple coherent layers. A pop-up light field differs from other layered modeling and rendering approaches (e.g., [9, 156, 264]) in a number of ways. First, the number of layers needed in a pop-up light field is not pre-determined. Rather, it is decided interactively by the user. Second, the user specifies the layer boundaries in key frames. The layer boundary is then propagated to the remaining frames automatically. Third, the representation is simple. Each layer is represented by a planar surface without the need for per-pixel depth. Fourth and most importantly, the layers are coherent so that anti-aliased rendering using these coherent layers is achieved. Each coherent layer must have sufficiently small depth variation so that anti-aliased rendering of the coherent layer itself becomes possible. Moreover, to render each coherent layer with its background layers, not only accurate layer segmentation is required on every image, but segmentation across all images must be consistent as well.

To segment the layers, the user interface is key. A good user interface can enable the user to intuitively manipulate the structure of a pop-up light field so that virtual views with the desired level of fidelity can be produced. By having a "human-in-the-loop" for pop-up field construction, the user can specify where aliasing occurs in the rendered image. Then, corresponding layers are refined accordingly. More layers are popped up, refined and propagated across all images in the light field until the user is satisfied with the rendering quality (i.e., no aliasing is perceived).

16.2 Outline of chapter

The rest of this chapter is organized as follows. After reviewing related work in Section 16.3, we introduce in Section 16.4 the representation of pop-up light field, which consists of a set of coherent layers, and *coherence matting* that maintains the layer consistency across frames in the light field. Section 16.5 details the operations in the user interface. Section 16.6 describes the real-time hardware-accelerated pop-up light field rendering algorithm. Experimental results are presented in Section 16.7. Discussion and concluding remarks are provided in Sections 16.8 and 16.9, respectively.

16.3 Related work

Many image-based interactive modeling systems use only one image, which impose or assume certain geometric constraints on the scenes. For example, the "Tour in Pictures" system [110] models the scene by a simple spidery mesh. In the single view metrology work of Criminisi *et al.* [57], 2D projections of 3D parallel lines must be present in the input image, so that the user can click on them to compute vanishing points. An elaborate modeling system was proposed in [201] where depth values are assigned to pixels in a single picture. Interactive single view systems are difficult to generalize to a sparse light field, which still consists of many images.

Furthermore, while it is straightforward to perform interactive image segmentation on a single image, consistent propagation of image regions to different images is a challenging task.

The UI for designing a multiview/multiframe interactive modeling system has been a challenge. Most available movie editing tools are in fact manual systems, requiring frame-by-frame editing and consistency maintenance. The Façade system [61] is an interactive modeling system that makes use of 3D geometry derived from a sparse set of views (image-based) and a modeling program (geometry-based). Plenoptic editing [262] first recovers a 3D voxel model from a sparse light field, and then applies traditional 3D warping to the recovered model. Thus, this automatic system shares the same shortcomings with stereo reconstruction. Chapter 17 (based on [347]) describes a feature-based technique for morphing between two light fields. The key to this technique is an easy-to-use UI for feature specification.

Layers have been proved successful [9, 156, 264] in image-based rendering. Coherent layers [156] are constructed from 3D models for efficient rendering. A two-step rendering algorithm is used to render layered depth images (LDIs). While significant progress has been made in automatic reconstruction of layers [313, 9, 143] in the computer vision community, the layers are not accurate enough for artifact-free rendering. In a pop-up light field, the user interactively determines how many layers to be generated (or "popped up"), initializes the layers in key frames, and modifies the layers by inspecting the rendering quality.

Layered representations always demand accurate alpha matting along the layer boundary (e.g., the sprites in [264]). To estimate alpha matting from images, blue screen matting was proposed in [279]. Recently, a Bayesian approach for digital matting was proposed in [47], which provides an excellent survey on other patented matting algorithms [14, 197, 249]. Video matting based on the same Bayesian framework has been reported in [46]. However, video matting is not designed to have consistent alpha mattes across frames. More accurate alpha can be estimated when the multi-backgrounds are available [315, 191]. In image-based opacity hulls, multi-background matting is used to acquire alpha mattes to construct a visual hull with view dependent opacity.

16.4 Pop-up light field representation

If a light field is undersampled, conventional light field rendering [160] results in aliasing. The top row of Figure 16.1 shows the rendering of a sparse light field with 5×5 Tsukuba images. The top right image is rendered with the 5×5 sparse light field by setting a single focal plane in the scene [116]. Double images can be easily observed on the front objects.

The bottom row of Figure 16.1 shows that anti-aliased rendering can be achieved using four layers, each of which employs a simple planar surface as its geometric proxy. By splitting the scene into multiple layers, the depth variation in each layer becomes much smaller than that in the original sparse light field.

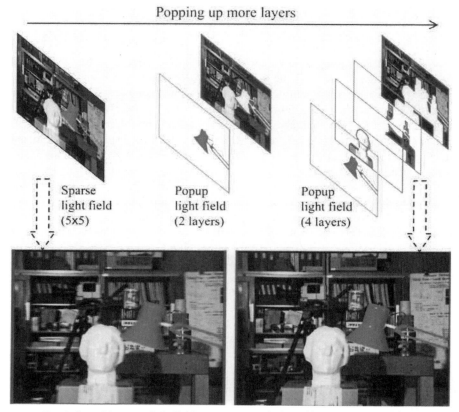

Fig. 16.1. An example of rendering with pop-up light fields. Rendering using the 5x5 Tsukuba light field data set is shown in the top left. Aliasing is clearly visible near front objects in the bottom left image because the input light field is sparse. The top row shows that the pop-up light field splits the scene gradually into four coherent layers, and achieves anti-aliased rendering as shown in the bottom right image.

See color plate section near center of book.

The pop-up light field is represented by a collection of *coherent layers*. A key observation in the pop-up light field representation is that the number of coherent layers that should be modeled or "popped up" depends on the complexity of the scene and how undersampled the input light field is. For a sparser light field, more layers need to be popped up for anti-aliased rendering.

16.4.1 Coherent layers

A coherent layer L_j is represented by a collection of corresponding layered image regions R_j^i in the light field images I^i. These regions are modeled by a simple geometric proxy without the need for accurate per-pixel depth. For example, a global

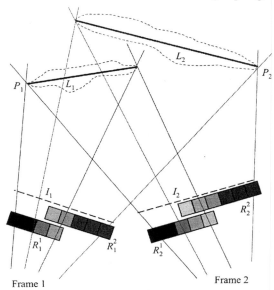

Fig. 16.2. A light field (with images I_1 and I_2) can be represented by a set of coherent layers (L_1 and L_2). A coherent layer is a collection of layered images in the light field. For instance, L_1 is represented by layered image R_1^1 (from I_1) and R_2^1 (from I_2). Each layered image has an alpha matte associated with its boundary. Part of the scene corresponding to each layer (e.g., L_1) is simply modeled as a plane (e.g., P_1).

planar surface (P_j) is used as the geometric proxy for each layer L_j in the example shown in Figure 16.2. To deal with complicated scenes and camera motions, a local planar surface P_j^i can also be used to model the layer in every image i of the light field.

A layer in the pop-up light field is considered as "coherent" if the layer can be rendered free of aliasing by using a simple planar geometric proxy (global or local). Anti-aliased rendering occurs at two levels when

1. the layer itself is rendered; and
2. the layer is rendered with its background layers.

Therefore, to satisfy the first requirement, the depth variation in each layer must be sufficiently small, as suggested in [33]. Moreover, the planar surface can be adjusted interactively to achieve the best rendering effect. This effect of moving the focal plane has been shown in [116, 33].

However, to meet the second requirement, accurate layer boundaries must be maintained across all the frames to construct the coherent layers. A natural approach to ensuring segmentation coherence across all frames is to propagate the segmented regions on one or more key frames to all the remaining frames [264, 347]. Sub-pixel precision segmentation may be obtained on the key frames by meticulously zooming on the images and tracing the boundaries. Propagation from key frames to other frames, however, causes inevitable under-segmentation or over-segmentation of a foreground layer. Typically over-segmentation of a foreground layer leads to the inclusion of background pixels, thus introducing ghosting along the occlusion boundaries in the rendered image. A possible example of foreground over-segmentation is exhibited in Figure 4(g) of [264] where black pixels on the front object's boundary can be observed. To alleviate the rendering artifacts caused by over-segmentation or under-segmentation of layers, the layer boundary needs to be refined using alpha matting [235].

Figure 16.2 illustrates coherent layers of a pop-up light field. All the pixels at each coherent layer have consistent depth values (to be exact, within a depth bound), but may have different fractional alpha values along the boundary. To produce fractional alpha mattes for all the regions in a coherent layer, a straightforward solution is to apply video matting [46]. The video matting problem is formulated as a maximum a posterior (MAP) estimation as in Bayesian matting [47],

$$
\begin{aligned}
&\arg \max_{F,B,\alpha} P(F,B,\alpha|C) \\
&= \arg \max_{F,B,\alpha} L(C|F,B,\alpha) + L(F) + L(B) + L(\alpha)
\end{aligned}
\tag{16.1}
$$

where $L(\cdot) = logP(\cdot)$ is log likelihood, C is the observed color for a pixel, and F, B and α are foreground color, background color and alpha value to be estimated, respectively. For color image, C, F and B are vectors in RGB space. In Bayesian matting and video matting, the log likelihood for the alpha $L(\alpha)$ is assumed constant so that $L(\alpha)$ is dropped from Equation (16.1).

In video matting, the optical flow is applied to the trimap (the map of foreground, background and uncertain region), but not to the output matte. The output foreground matte is produced by Bayesian matting on the current frame, based on the propagated trimap. Video matting works well if it is just a simple matter of replaying the foreground mattes against a different background. However, these foreground mattes may not have in-between frame coherence that is needed for generating novel views.

16.4.2 Coherence matting

An approach called *coherence matting* is used to construct the alpha mattes in a coherent layer that have in-between frame coherence. The workflow of this approach is similar to video matting and is illustrated in Figure 16.3. First, the user-specified boundaries are propagated across frames. Second, the uncertain region along the boundary is determined. Third, the under-segmented background regions from multiple images are combined to construct a sufficient background image. Fourth, the alpha matte for the foreground image (in the uncertain region) is estimated. The key to this approach is at the fourth step in Figure 16.3(d) where a coherent feathering function across the corresponding layer boundaries is introduced. Note that, for a given layer, a separate foreground matte is estimated independently for each frame in the light field, and the coherence across frames is maintained by foreground boundary consistency.

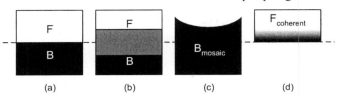

Fig. 16.3. Illustration of major steps in coherence matting. (a) The user specifies an approximate segmentation. (b) An uncertain region is added in between foreground and background. (c) A background mosaic is constructed from multiple undersegmented background images. (d) A coherent foreground layer is then constructed using coherent matting.

$L(B)$ in Equation (16.1) can be dropped, since the background can be explicitly estimated (see Section 16.5.4). By incorporating a coherence prior on the alpha channel $L(\alpha)$ across frames, coherence matting can be formulated as

$$L(F, B, \alpha | C) = L(C | F, B, \alpha) + L(F) + L(\alpha) \qquad (16.2)$$

where the log likelihood for the alpha $L(\alpha)$ is modelled as:

$$L(\alpha) = -(\alpha - \alpha_0)^2 / \sigma_a^2 \qquad (16.3)$$

where $\alpha_0 = f(d)$ is a feathering function of d and σ_a^2 is the standard deviation. d is the distance from the pixel to the layer boundary. The feathering function $f(d)$ define the α value for surrounding pixels of a boundary. In this chapter, the default feathering function $f(d) = (d/w) * 0.5 + 0.5$, where w is feathering width, as illustrated in Figure 16.4. It is often used to smooth a hard edge of boundary or selection in image composition.

Assume the observed color C and sampled foreground color F (from a set of neighboring foreground pixels) are all of Gaussian distribution:

$$L(C | F, B, \alpha) = -||C - \alpha F - (1 - \alpha)B||^2 / \sigma_C^2 \qquad (16.4)$$

$$L(F) = -(F - \overline{F})^T \Sigma_F^{-1} (F - \overline{F}) \qquad (16.5)$$

where σ_C is the standard deviation of the observed color C, $\overline{F}$ is the weighted average of foreground pixels around F and Σ_F is the weighted covariance matrix. Taking the partial derivatives of (16.2) with respect to F and α and forcing them equal to zero results in the following equations:

$$F = \frac{\Sigma_F^{-1} \overline{F} + C\alpha / \sigma_C^2 - B\alpha(1 - \alpha) / \sigma_C^2}{\Sigma_F^{-1} + I\alpha^2 / \sigma_C^2} \qquad (16.6)$$

$$\alpha = \frac{(C - B) \cdot (F - B) + \alpha_0 \cdot \sigma_C^2 / \sigma_a^2}{||F - B||^2 + \sigma_C^2 / \sigma_a^2} \qquad (16.7)$$

α and F are solved alternatively by using (16.6) and (16.7). Initially, α is set to α_0.

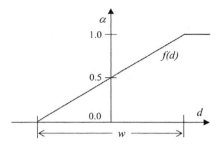

Fig. 16.4. The feathering function in coherence matting.

16.4.3 Rendering with coherent matting

Bayesian matting [47] and video matting [46] solve the matting from the equation

$$\alpha = \frac{(C - B) \cdot (F - B)}{||F - B||^2},$$

which works well in general but becomes unstable when $F \approx B$. In comparison, the coherence matting of Equation (16.7) can be solved more stably, because applying the coherence prior on α results in a non-zero denominator. The coherence prior

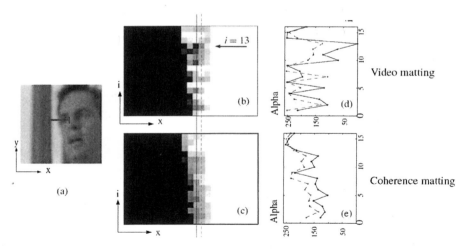

Fig. 16.5. Comparison between video matting and coherence matting. (a) is a small window on one frame in the Plaza data (Figure 16.11). (b) and (c) are two alpha epipolar plane images (α-EPI) corresponding to the red line in (a), using the algorithm of video matting and coherence matting respectively. (d) and (e) are the alpha curves of two adjacent columns, which are marked as blue and red lines in (b) and (c). (d) corresponds to video matting, and shows a large jump at $i = 13$, which causes an accidental transparency within the face. (e) corresponds to coherence matting, which provides a more reasonable result.

behaves similar to the smoothness constraint commonly used in visual reconstruction (e.g., shape from shading [109]).

The spatial inconsistency of the alpha matte from video matting can be observed in Figure 16.5. The plot of the alpha epipolar plane image (α-EPI) of a video matting result is shown. Similar to the conventional EPI [17], for a short segment of scanline from the Plaza sequence, the alpha values along this segment are stacked for all of the 16 frames ((b) and (c)). The alpha values along 2 lines (solid and dotted) in the α-EPI are plotted in (d) and (e). Each line represents the alpha values of the corresponding pixels across 16 frames. A close inspection of (b) around frame $i = 13$(video matting method), shows that the alpha value changes from about 126 to 0, then to 180, (the range of alpha is from 0 to 255) indicating a small part of the face accidentally becomes transparent.

The temporal incoherence of the alpha matte from video matting can be more problematic during rendering. The fluctuation of alpha values along both dotted and solid lines will generate incoherent alpha values and thus cause rendering artifacts as the viewpoint is changed (along axis i). Figure 16.5(e) shows the same solid and dotted lines with coherent matting results. Both lines have much less fluctuation between neighboring pixels, and appear temporally smoother than their counterparts in Figure 16.5(d).

16.5 Pop-up light field construction

An easy-to-use UI was developed to facilitate the construction of a pop-up light field. The user can easily specify, refine and propagate layer boundaries, and indicate rendering artifacts. More layers can be popped up and refined until the user is satisfied with the rendering quality.

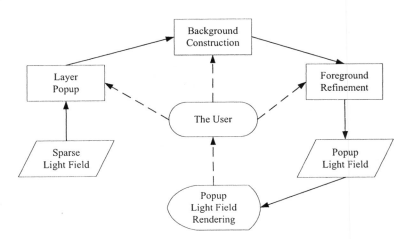

Fig. 16.6. Flowchart of pop-up light field construction UI.

16.5.1 UI operators

Figure 16.6 summarizes the operations in the pop-up light field construction UI. The key is that a human is in the loop. The user supplies the information needed for layer segmentation, background construction and foreground refinement. By visually inspecting the rendering image from the pop-up light field, the user also indicates where aliasing occurs and thus which layer needs to be further refined. The user input or feedback is automatically propagated across all the frames in the pop-up light field. The four steps of operations in the UI are summarized as follows.

1. *Layer pop-up.* This step segments layers and specifies their geometries. To start, the user selects a key frame in the input light field, specifies regions that need to be popped up, and assigns the layer's geometry by either a constant depth or a plane equation. This step results in a coarse segmentation represented by a polygon. The polygon region and geometry configuration can be automatically propagated across frames. Layers should be popped up in order of front to back. More details of layer pop-up are shown in section 16.5.3.

2. *Background construction.* This step obtains background mosaics that are needed to estimate the alpha mattes of foreground layers. Note that the background mosaic is useful only for the pixels around the foreground boundaries, i.e. in the uncertain region as shown in Figure 16.3. More details of background construction are discussed in section 16.5.4.

3. *Foreground refinement.* Based on the constructed background layers, this step refines the alpha matte of the foreground layer by applying the coherence matting algorithm described in section 16.4.2. Unlike layer pop-up in step 1, foreground refinement in this step should be performed in back-to-front order.

4. *Rendering feedback.* Any modification to the above steps will update the underlying pop-up light field data. The rendering window will be refreshed with the changes as well. By continuously changing the viewpoint the user can inspect for rendering artifacts. The user can mark any rendering artifacts such as ghosting areas by brushing directly on the rendering window. The corresponding frame and layer will then be selected for further refinement.

16.5.2 UI design

Figure 16.7 shows the appearance of the UI, including five workspaces where the user interacts with a frame and a layer in the pop-up light field. These workspaces and their functionalities are explained as follows.

Lower middle. The user chooses an active frame by clicking on the *Frame navigator*, shown at the lower middle in Figure 16.7. The active frame appears in the *Editing frame view*, shown at the upper left in Figure 16.7.

Upper left. In the *Editing frame view*, the user can create or select an active layer and edit its polygon region. This active layer is displayed in blue polygons with

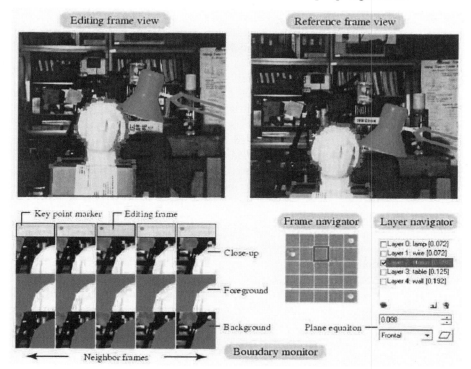

Fig. 16.7. The UI for Pop-up light field construction.

See color plate section near center of book.

crosses for each editable vertex. The information of the active layer is available in the *Layer navigator*, shown at the lower right of Figure 16.7.

Lower right. From the *Layer navigator*, the user can obtain the active layer's information. The user can select, add, or delete layers in the list. By selecting the layer in the check box, the user can turn on/off a layer's display in the *Editing frame view*, *Reference frame view* (shown at the upper right of Figure 16.7) and the rendering window . The plane equation of the active layer is displayed and can be modified through keyboard input. Layer equations can also be set through adjusting the rendering quality in the rendering window.

Upper right. The *Reference frame view* is used to display another frame in the light field. This workspace is useful for a number of operations where correspondences between the reference frame view and the editing frame view need to be considered, such as specifying plane equations.

Lower left. To fine tune the polygon location for the active layer, the *Boundary monitor* (lower left of Figure 16.7) shows close-up views of multiple frames in the light field. The first row shows the close-up around the moving vertex. The second and third rows show the foreground and background of the active layer composed with a fixed background selected by the user. For instance, using mono fuchsia color in Figure 16.7 as the background, it is easy for the user to observe over-segmentation or under-segmentation of the foreground across multiple frames simultaneously.

Not shown in the figure is the rendering window on which the user can render any novel view in real time and can inspect the rendering quality. The user can also specify the frontal plane's equation for an active layer by sliding the plane depth back and forth until the best rendering quality (i.e., minimum ghosting) is achieved. If the ghosting cannot be completely eliminated at the occlusion boundaries, the layer's polygon must be fine tuned. The user can brush on the ghosting regions, and the system can automatically select the affected frame and layer for modification. The affected layer is front-most and closest to the specified ghosting region.

To specify the slant plane equation for a layer, the user needs to select at least four pairs of corresponding points on the *Editing frame view* and the *Reference frame view*. The plane equation can be automatically computed and then used for rendering.

Also not shown in the above figure is a dialog box where the user can specify the feathering function. Specifying a feathering curve is useful for the coherence matting algorithm described in section 16.4.2.

16.5.3 Layer pop-up

To pop up a layer, the user needs to segment and specify the geometry of the layer for all frames in the light field. This section discusses the operations by which the user interacts with the system and the underlying algorithms.

16.5.3.1 Layer initialization

Polygons are used to represent layer boundaries, since the correspondence between polygons can be maintained well in all frames by the corresponding vertices. The user can specify the layer's boundary with a polygon (e.g., using the polygon lasso tool in Adobe Photoshop) and edit the polygon by dragging the vertices. The editing will be immediately reflected in the *Boundary monitor* window and in the rendering window (section 16.5.2).

First of all, the user needs to inspect the rendering window by changing the viewpoint and decide which region is going to be popped up (usually the front-most non-ghosting object). The user then selects a proper key frame to work with and draws a polygon on the frame.

Then, the user needs to specify the layer's geometry. For a frontal plane, the layer depth is the one that achieves the best rendering quality which can be observed on the rendering window by the user. For a slant plane, the user specifies at least four pairs of corresponding points on at least two frames to estimate the plane equation.

Once the layer geometry is decided, the polygon on the first key frame can be propagated to all other frames by back projecting its vertices, resulting in a coarse segmentation of the layer on all frames in the light field. All vertices on the key frame are marked as key points. At this stage, the layer has a global geometry which is shared across all the frames. Moreover, an accurate polygon boundary for layer initialization is not necessary. Because of occlusions and viewpoint changes, propagated polygon boundaries inevitably need to be refined.

16.5.3.2 Layer refinement

The following aspects need to be considered in layer refinement.

Boundary refinement in a key frame   All vertices on any frame can be added, deleted and moved. Once a vertex is modified, it is marked as a key point. The position of the modified vertex will be propagated across frames at once and the layer region will be updated in several UI workspaces. To adjust a vertex position, the user can observe how well foreground and background colors are separated in the *Boundary monitor* window, or how much the ghosting effect is removed in the rendering window.

Boundary propagation across multiple frames   For a specific vertex on the layer boundary, if it is marked as a key point on only one frame in the light field, the position of this vertex on any other frame is computed by back projecting the point based on the layer geometry. When a vertex is marked as key points on two or more frames by the user, however, the positions of this vertex on other frames must be computed by interpolating the back projections of these key points. The coordinate of non-key point is interpolated from key points using following schemes:

For the 1D image array, the two corresponding key points immediately to the left and right of each non-key point are selected. The coordinate of this non-key point is obtained by back-projecting the 3D point, via triangulating the two selected key points. For a 2D image array, the selection should proceed with care because interpolation is preferred over extrapolation. The back-projected point for the non-key points should change smoothly when the user edits the key point smoothly. The back-projected point for each non-key point is computed as follows: (1) compute the Delaunay triangulation of all key points. (2) for each non-key point in the interior of a triangle, run a 3-point back-projection algorithm by using three vertices of the triangle, e.g. non-key point b is interpolated from the key points A, B and D as illustrated in Figure 16.8; (3) for each non-key point P in the exterior of all triangles, select two key points P_0 and P_1 that maximize the angle $\angle P_0 P P_1$, e.g. non-key point a is interpolated from the key points A and D, c is interpolated from the key points A and C in Figure 16.8.

Note that not all vertices in a key frame are key points. Key points can also exist in non-key-frames. If a key point is marked later as a non-key-point, its position needs to be interpolated by its corresponding key points from other frames.

Coherence matting   It is difficult to accurately describe a layer boundary simply using polygons. It is hard for the user to manually adjust to sub pixel accuracy a boundary with subtle micro geometry. A pixel is often blended with colors from both

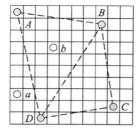

Fig. 16.8. Propagate information from key points to non-key points in 2D image arrary. The solid (yellow) points are key points and hollow (red) points are non-key points.

foreground and background due to the camera's point spread function. Therefore, the interface was designed to be tolerant and not require the user to specify very accurate sub-pixel boundary positions. Instead, an coherence matting algorithm is applied to further refine the layer boundary. Polygon editing (in a frame and across frames) and coherence matting can be alternatively performed with assistance from the user.

Local geometry   When the viewpoint changes significantly, a single planar geometry may not be sufficient to achieve anti-aliased rendering. The system uses a local geometry representation, with an example shown in Figure 16.13. This representation allows each key frame to have its own planar equation, and interpolate the planar equation for the layer on any intermediate non-key-frames.

16.5.4 Constructing background

The algorithm of coherence matting in section 16.4.2 assumes that the background for the uncertain regions (where matting is estimated) is known. A key observation is that because the uncertain regions are located around foreground boundaries, they can only appear on neighboring frames in the light field where these regions are disoccluded. The background reconstruction algorithm fills the disoccluded region using (warped) pixels from neighboring frames.

Once the foreground is popped up, the background image can be obtained by removing the foreground image as shown in Figure 16.9(a). Moreover, the background boundary is eroded by a few pixels (typically two pixels) before constructing the background mosaic because a possible under-segmentation of the foreground may leave some mixed foreground pixels on the background around boundaries.

An automatic algorithm is designed to construct the background, which warps the neighboring images to fill the holes using the background layer's geometry. This method works well if the the background is well approximated by plane, e.g. in Figure 16.1.

However, if the background contains objects with relatively large depth variation, the background layer would need to be further subdivided into sub-layers, each of which being represented as a plane. As shown in Figure 16.9(a), a background layer is segmented manually into four sub-layers using polygons. This time, the location

(a)   (b)

Fig. 16.9. (a) The background mosaic operator uses the polygon lasso operator to segment the layer into regions. (b) The resulting background mosaic fills in many missing pixels in (a). Although (b) still has many missing pixels, it is enough for coherence matting of the foreground.

of the polygon is not critical. Instead, the criterion here is to group the background boundaries into a better planar approximation.

The sub-layers are propagated from the key frame, where the user specifies the division, to all other frames using the existing background layer geometry. This propagation requires less accuracy as long as it covers the same group of boundaries. The relative motion of the sub-layer across frames is estimated hierarchically, starting from translation to affine, and from affine to perspective transform [291]. Only the pixels visible in both frames are used to estimate parameters. Figure 16.9(b) shows the resulting mosaic. Note that a hole-free mosaic is not required, as a few pixels surrounding the occlusion boundaries are adequate for coherence matting.

16.6 Real-time rendering of pop-up light field

An integral part of the UI is the real-time pop-up light field renderer which provides the user instant feedback on the rendering quality. The renderer was based on previous light field and Lumigraph rendering systems [22, 91, 116]. The pop-up light field rendering algorithm consists of three steps: (1) splitting a light field into layers, (2) rendering layers in back-to-front order, and (3) combining the layers.

16.6.1 Data structure

The data structure used in the rendering algorithm is shown below.

```
struct PopupLightField {
    Array<CameraParameter>  cameras;
    Array<Layer>  layers;
};
struct Layer {
    Array<Plane>  equations;
    Array<Image>  images;
};
struct Image {
    BoundingBox  box;
    Array2D<RGBA>  pixels;
};
```

The pop-up light field keeps the camera parameters associated with all the input frames. Each layer in the pop-up light field has corresponding layered images, one for each frame. Each layered image has a corresponding plane equation, so as to represent the local geometry. If global geometry is applied to a layer, all equations are the same for images in this layer.

Since these corresponding layered images vary their shapes in different views, they are stored as an array of images on each layer. Layers can be overlapping in the pop-up light field and each layered image is modified independently by mosaicing and coherent matting. Therefore it is necessary to keep both color and opacity of images for each layer separately. Each layered image is stored as an RGBA texture image of the foreground colors with its opacity map, and a bounding box as well. The opacity (alpha value) of the pixel is zero when this pixel is out of the foreground.

16.6.2 Layered rendering algorithm

The scene is rendered layer by layer using texture-mapped triangles in back-to-front order. Then the layers are sequentially combined by alpha blending. The rendering scheme is based on [22, 106] but extended to multiple layers. The pseudocode of the rendering algorithm is shown below.

```
ClearFrameBuffer()
T ←CreateRenderingPrimitives()
for all layers Layer from back to front do
        for all triangles △ ∈ T do
                SetupProjectiveTextureMapping(△)
                Render(△)
                BlendToFrameBuffer(△)
        end for
end for
```

After initializing a frame buffer, a set of triangular polygons are then generated, on which the original images are blended and drawn. The camera positions are first projected onto the image plane; these projected points together with the image plane's four corner points are subsequently triangulated.

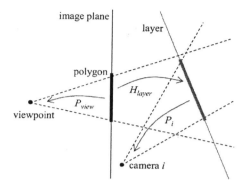

Fig. 16.10. Set-up of projective texture mapping

Then a triple of texture images $\{I_i\}_{i=1}^3$ are assigned to each triangle, which are blended across the triangle when rendering. The blending ratio $\{w_i^k\}_{k=1}^3 (0 \le w_i^k \le 1, \sum_{k=1}^3 w_i^k = 1)$ for three images are also assigned to each of the three vertices, and linearly interpolated across the triangle. The exact blending ratio based on ray angles is not necessarily distributed linearly on the screen. If the size of a triangle is not small enough with respect to the screen size, the triangle is iteratively subdivided into four triangles. On the vertex which is the projection of an I_i's camera, the blending ratio w_i^k is calculated using the following equation.

$$w_i^k = 1 \qquad \text{if camera } i \text{ is projected onto the } k\text{-th vertex}$$
$$= 0 \qquad \text{otherwise}$$

For the vertex which is not the projection of a camera, the weights are calculated using the angle between the ray through the camera and the ray through the vertex [22].

Then, each layer is rendered by blending texture images $\{I_i\}$ using blending ratios $\{w_i^k\}$. At the point other than the vertices on the triangle, the blending ratios $\{\tilde{v}_i\}$ are calculated by interpolating $\{w_i^k\}_{k=1}^3$. Using $\{I_i\}$ and $\{\tilde{v}_i\}$, the pixels on the triangle are drawn in the color $\sum_{i=1}^3 \tilde{v}_i I_i$.

The texture images are mapped onto each triangle projectively as illustrated in Figure 16.10. Let P_{view} be the projection matrix for the rendering camera (to produce the novel view), P_i be the projection matrix for the camera corresponding to I_i, and H_{layer} be a planar homography from the triangle to the layer plane. Then, the texture image I_i is mapped onto the triangle using a projection matrix $P_i H_{layer}$.

16.6.3 Hardware implementation

Light field rendering can be accelerated using graphics hardware. Although several hardware-accelerated light field rendering approaches have been proposed [22, 91,

Table 16.1. Rendering performance (FPS = frames per second).

Data	Resolution (w × h × #cameras)	Original LF (MB)	Pop-up LF (MB [#layers])	FPS
Tsukuba	384 × 288 × 25	8.3	19.4 [5]	62.5
Plaza	640 × 486 × 16	14.9	38.3 [16]	58.8
Pokemon	512 × 384 × 81	47.8	117.8 [5]	31.3
Fur	1136 × 852 × 23	66.8	140.4 [5]	46.9
Statuette	1136 × 852 × 43	124.8	161.2 [4]	37.1

106, 116], they cannot be used directly for pop-up light field rendering. In these previous approaches, texture images are blended by multi-pass rendering of triangles and alpha-blending them in the frame buffer. In a layered rendering algorithm, layers must be alpha-blended using the alpha values assigned in texture images, which means each layer must be rendered onto the frame buffer in a single pass.

One straightforward way would be to copy the frame buffer into memory and composite them after rendering each layer. Unfortunately, this is too slow for interactive use. Instead, a single pass rendering method is used; it involves multitexture mapping and programmable texture blending, which is available on modern graphics hardware.

In order to blend all textures on a single triangle, we first bind three different textures assigned to each triangle, then assign three blending ratios $\{w_1, w_2, w_3\}$ as the primary color { R,G,B } on each vertex. The primary color is smoothly interpolated on the triangle. Hence the interpolated blending ratios $\{\bar{v}_i\}$ are obtained simply by referring to the primary color at an arbitrary point on the triangle. Then the texture images on the triangle can be blended using the blending equation programmed in the pixel shader in graphics hardware.

The layers can be composed simply by alpha-blending each triangle on the frame buffer when it is rendered because the triangles are arranged without overlap in a layer and each triangle is drawn in a single pass.

The rendering system has been implemented using OpenGL and its extensions for multi-texturing and per-pixel shading, and tested on a PC (CPU 660 MHz, memory 768 MB) equipped with an NVidia GeForce4 or ATI Radeon9700 graphics card with 128MB of graphics memory. The performance of rendering is shown in Table 16.1.

16.7 Experimental results

This section shows the result of constructing several pop-up light fields from real scenes. The "Tsukuba" data set and the "Plaza" sequence are courtesy of Prof. Ohta of University of Tsukuba, and Dayton Taylor, respectively. "Pokemon" data is captured by a computer-controlled vertical X-Y table in a lab. Data sets of "Statuette" (with unstructured camera motion) and "Furry toys" (with the camera moving along a line) are captured by a Canon G2 Digital Camera.

Fig. 16.11. Result of pop-up light field rendering of the *Plaza* sequence rendered from a novel viewpoint (in the position midway between the 11th and 12th frames). The input consists of only 16 images. Sixteen layers are used to model the pop-up light field.

one focal plane in the front   one focal plane at the back   5 layers are popped up

Fig. 16.12. Results on *Pokemon* 9×9: comparison of conventional light field rendering and pop-up light field rendering.

See color plate section near center of book.

As shown in Table 16.1, rendering of all the pop-up light fields can be done in real-time (with a frame rate greater than 30 fps). Table 16.2 summarizes the amount of work required to construct these pop-up light fields. For most scenes in the experiments, it takes a couple of hours for a graphics graduate student to interactively model the pop-up light field. It, however, took the student 5 hours to construct the pop-up light field from the Plaza sequence where 16 layers are segmented.

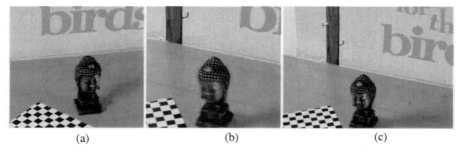

(a)   (b)   (c)

Fig. 16.13. Results on sparse images taken with unstructured camera positions. (a) The global planar surface is set as a frontal-parallel plane in this view, (b) rendering result from another view with the global plane, (c) rendering result from the same view of (b) with local geometry.

See color plate section near center of book.

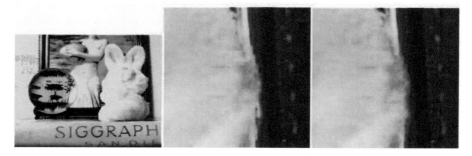

Fig. 16.14. Comparison of results on *furry rabbit*, with video matting (middle image) and with coherence matting (right image). The alpha matte from coherent matting is smoother than that from video matting in the rendering image.

See color plate section near center of book.

Tsukuba. Some rendering results of the *Tsukuba* 5×5 light field are shown in Figure 16.1. It is demonstrated that with 4 layers, anti-aliased rendering can be achieved. The same rendering quality can be achieved with 7 layers if the light field is down-sampled to 3×3.

Pokemon. Figure 16.12 again demonstrates the progressive improvement of visual quality when more layers are popped up. With 5 layers, anti-aliased rendering of

Table 16.2. User interaction statistics.

Data	Tsukuba	Plaza	Pokemon	Fur	Statuette
# frame	25	16	81	23	43
# layer	4	16	5	5	4
points/frame	129	379	90	167	47
key points/frame	6.3	62.4	8.6	16.1	12.3
Time (hours)	≈ 0.5	≈ 4	≈ 1	≈ 1	≈ 1.5

pop-up light fields (*Pokemon* 9×9) is achieved. The four layers that model the three toys and the background use frontal-parallel planes while the table plane is slanted.

Statuette. For complicated scenes, instead of using a global planar surface defined in the world coordinate system, local geometry should be used. Figure 16.13 shows the rendering result from a sequence of 42 images taken with unstructured camera motion. If a global planar surface is set as a frontal-parallel plane in the frame (Figure 16.13(a)), rendering at a very different viewpoint will have noticeable artifacts, as shown in Figure 16.13(b). Figure 16.13(c) shows a good rendering result using view-dependent geometry. Specifically, the plane orientations are changed for different views.

Furry rabbit. A sparse light field of a furry toy rabbit (23 images with the camera path along a line) is used to show the efficacy of coherence matting. Figure 16.14 compares the results with video matting and with coherence matting. The zoomed up views of the left ear demonstrate that coherence matting obtains a more consistent matte than video matting.

Plaza. Figure 16.11 shows an aliasing-free novel view rendered using the pop-up light field constructed from the *Plaza* sequence, which is a collection of only 16 images. The sequence was captured by a series of "time-frozen cameras" arranged along a line or curve. Because the scene is very complex, stereo reconstruction is very difficult. Note that nearly perfect matting is achieved for the floating papers in the air. The boundaries for the foreground characters are visually acceptable, made possible mainly by the coherent layers produced by coherence matting.

16.8 Discussion

In a way, coherence matting is similar to prefiltering the alpha channel in the foreground layer. It was suggested in [160, 33] that prefiltering can be used to reduce aliasing, at the expense of lower rendering resolution. With coherence matting, prefiltering the entire light field is unnecessary. It is adequate to prefilter only occlusion boundaries.

Moreover, background and foreground layers are handled differently in a pop-up light field. Because we have multiple images, we can construct a complete background mosaic from under-segmented background layers. For the foreground image, the pixels around the boundary are "prefiltered" (by coherence matting) before rendering. The rendering artifact is that the boundary of the foreground image is slightly blurred.

The pop-up light field is based on sparse light field input. It is unable to handle specular highlights and other significant appearance changes. Moreover, coherent matting does not work well for semi-transparent surfaces and long hairs because the prior $L(\alpha)$ used in the formulation is approximately modeled as a point spread function.

16.9 Concluding remarks

The system described in this chapter was inspired by the real-time 3D model acquisition system of [248], in which the user specifies areas that need to be modeled depending on the current merge model from multiple scans. Though the goal and methodology are very different, the key concept in the pop-up light field system is similar: the user is in the modeling loop and specifies, through a real-time pop-up light field renderer, where aliasing occurs and how the scene should be further modeled.

Another motivation stems from the difficulty of recovering accurate per-pixel depth using stereo or other vision techniques. The pop-up light field is an image-based modeling technique that does not rely on accurate 3D depth/surface reconstruction. Rather, it is based on accurate layer extraction/segmentation in the light field images. In a way, it trades the difficult correspondence problem in 3D reconstruction for another equally difficult segmentation problem. However, it is much easier for a user to specify accurate contours in images than accurate depth for each pixel.

17

Feature-Based Light Field Morphing

In the previous three chapters, we described systems that renders static scenes in their own unique ways. One system focuses on the reduction of data while maintaining the quality of the viewing experience (Chapter 14), another involves customizing viewer experience for large scenes (Chapter 15), and the third requires some user interaction to overcome the difficult segmentation process (Chapter 16). However, all these systems result in visualization of the captured scene—they do not show how captured data can be edited to create *new* content.

How does one edit a light field? In this chapter, we describe a feature-based technique for morphing 3D objects represented by light fields. This technique enables morphing of image-based objects whose geometry and surface properties are too difficult to model with traditional vision and graphics techniques. Light field morphing is not based on 3D reconstruction; instead it relies on *ray correspondence*, i.e., the correspondence between rays of the source and target light fields. We address two main issues in light field morphing: feature specification and visibility changes.

An intuitive and easy-to-use user interface (UI) was developed for feature specification. The key to this UI is *feature polygons*, which are intuitively specified as 3D polygons and are used as a control mechanism for ray correspondence in the abstract 4D ray space. *Ray-space warping* is used to handle visibility changes due to object shape changes. It is capable of filling arbitrarily large holes caused by object shape changes; these holes are usually too large to be properly handled by traditional image warping. The light field method can deal with non-Lambertian surfaces, including specular surfaces (with dense light fields). This capability allows convincing 3D morphing effects to be generated.

17.1 The morphing problem

Metamorphosis, or morphing, is a popular technique for visual effects. When used effectively, morphing can give a compelling illusion that an object is smoothly transforming into another. Following the success of image morphing [12, 321], graphics researchers have developed a variety of techniques for morphing 3D objects [152,

Fig. 17.1. Light field morphing: A 3D morphing sequence from a furry toy cat (real object) to the Stanford bunny (synthetic object).

89]. These techniques are designed for geometry-based objects, i.e., objects whose geometry and surface properties are known, either explicitly as for boundary-based techniques (e.g., [144, 62, 153]) or implicitly as for volume-based techniques (e.g., [112, 157, 49]).

In this chapter, we describe a feature-based morphing technique for 3D objects represented by light fields [160] or Lumigraphs [91]. Unlike traditional graphics rendering, light field rendering generates novel views directly from images; no knowledge about object geometry or surface properties is assumed [160]. Light field morphing thus enables morphing between image-based objects, whose geometry and surface properties, including surface reflectance, hypertexture, and subsurface scattering [67], may be unknown or difficult to model with traditional graphics techniques.

The light field morphing problem can be stated as follows: Given the source and target light fields L_0 and L_1 representing objects O_0 and O_1, construct a set of intermediate light fields $\{L_\alpha \mid 0 < \alpha < 1\}$ that smoothly transforms L_0 into L_1, with each L_α representing a plausible object O_α having the essential features of O_0 and O_1. The intermediate light field L_α is called a light field morph, or simply a *morph*.

A naive approach to light field morphing is to apply image morphing to individual images in the source and target light fields and assemble the light field morphs from the intermediate images of image morphing. Unfortunately, this approach will fail for a fundamental reason: light field morphing is a 3D morphing and image morphing is not. This difference manifests itself when a hidden part of the morphing object becomes visible because of object shape change, as image morphing will produce "ghosting" that betrays a compelling 3D morphing.

The plenoptic editing proposed by Seitz and Kutulakos [262] represents another approach to image-based 3D morphing. They first recover a 3D voxel model from the image data and then apply traditional 3D warping to the recovered model. The visibility issues can be resolved with the recovered geometry, but there are problems, including the Lambertian surface assumption needed for voxel carving [262] and the difficulties with recovering detailed geometry. Most of the problems are related to the fundamental difficulties of recovering surface geometry from images [72].

Light field morphing is an image-based 3D morphing technique that is not based on 3D surface reconstruction. The basis of light field morphing is *ray correspondence*, i.e., the correspondence between rays of the source and target light fields [160]. The role of ray correspondence in light field morphing is the similar to that of

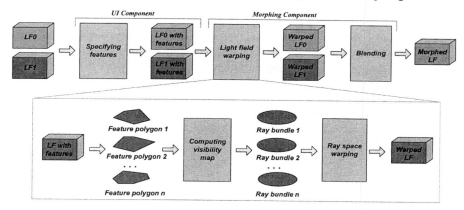

Fig. 17.2. Overview of light field morphing. The overall pipeline is illustrated in the upper part, whereas the warping of a light field is detailed in the lower part.

vertex correspondence in geometry-based 3D morphing [152, 153]. Like vertex correspondence (e.g., see [62, 95]), ray correspondence is controlled by user-specified feature elements.

A key issue in light field morphing is thus the construction of a user interface (UI) for specifying feature elements. Since there is no intrinsic solution to a morphing problem, user interaction is essential to the success of any morphing system [157, 321, 152]. For light field morphing, the main challenge is the design of intuitive feature elements for an abstract 4D ray space [160]. To address this challenge, *feature polygons* are introduced as the central feature elements for light field morphing. As 3D polygons, feature polygons are intuitive to specify. More importantly, feature polygons partition a light field into groups of rays. The rays associated with a feature polygon P constitute a *ray bundle*, and the ray correspondence of this ray bundle is controlled by the control primitives of the feature polygon P. Note that feature polygons do not make a rough geometry of the underlying object; they are needed only at places where visibility changes (due to object shape change).

Another key issue in light field morphing, and more generally in image-based 3D morphing, is visibility change. Two types of visibility change exist. The first is due to viewpoint changes. In light field morphing, this type of visibility change is automatically taken care of by the input light fields. The second type of visibility change is that caused by object shape changes, which must be handled. For a given view, a hole is created when a hidden surface patch in the source light field L_0 becomes visible in the target light field L_1 due to object shape change. This type of hole may be arbitrarily large and thus cannot be dealt with properly by traditional image warping methods (e.g., [40, 260]). This problem is handled using a technique called *ray-space warping*, which is inspired by Beier and Neely's image warping [12]. With ray-space warping, a hole can be filled by approximating an occluded ray with the "nearest visible ray." Not surprisingly, ray-space warping requires visibility processing and the key to visibility processing is the *global visibility map*, which associates each light field ray with a feature polygon.

Ray-space warping produces accurate results under the popular Lambertian surface assumption [40, 260]. For non-Lambertian surfaces, ray-space warping tries to minimize the errors by using the "nearest visible rays." Unlike plenoptic editing [262], non-Lambertian surfaces, including specular surfaces, can be handled.

Light field morphing is easy to use and flexible. Feature specification usually takes about 20 to 30 minutes and sparse light fields can be used as input to save storage and computation. When the input light fields are very sparse (e.g., 2 to 3 images per light field), light field morphing is termed key-frame morphing to emphasize its similarity with image morphing. Key-frame morphing may be regarded as a generalization of view morphing [260] because key-frame morphing allows the user to add more input images as needed to eliminate the holes caused by visibility changes. Note that although view morphing can generate morphing sequences that appear strikingly 3D, it is not a general scheme for image-based 3D morphing because the viewpoint is restricted to move along a prescribed line.

In this chapter, we show results for a few applications of light field morphing, including generating 3D morphs for interactive viewing, creating animation sequences of a 3D morphing observed by a camera moving along an arbitrary path in 3D, key-frame morphing, and transferring textures from one 3D object to another. In addition, we show how an animation sequence of a 3D morphing can be efficiently computed without fully evaluating all the morphs. The techniques we present can be used as visualization tools for illustration/education purposes [13], in the entertainment industry, and for warping/sculpting image-based objects [159, 262].

The rest of the chapter is organized as follows. In Section 17.2, we give an overview of the light field morphing system. Section 17.3 describes the specification of feature elements, in particular feature polygons, and visibility processing. Section 17.4 presents the warping algorithms for warping a light field and for generating 3D animation sequences. Experimental results are reported in Section 17.5, followed by discussion and concluding remarks in Sections 17.6 and 17.7.

17.2 Overview

As shown in Figure 17.2, the light field morphing system has two main components. The first is a UI for specifying feature element pairs through side-by-side interactive displays of the source and target light fields. Three types of feature elements are used: feature lines, feature polygons, and background edges. The second component is a morphing unit that automatically computes the morph L_α for a given α through the following steps. First, the feature elements of L_α are obtained by linearly interpolating those of L_0 and L_1. Second, L_0 and L_1 are warped to $\hat{L}_0$ and $\hat{L}_1$ respectively for feature alignment. Finally, L_α is obtained by linearly interpolating the warped light fields $\hat{L}_0$ and $\hat{L}_1$. Of these steps, both the first and last steps are simple; the warping step is the main part of the second component.

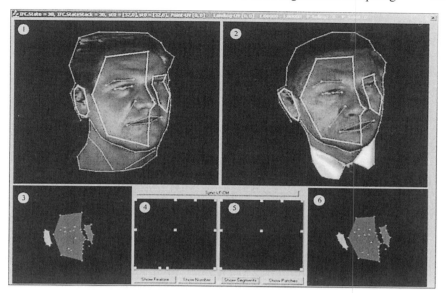

Fig. 17.3. The user interface for feature specification. On the top, windows (1) and (2) are interactive renderings of the source and target light fields. Three pairs of feature polygons are drawn using wireframe rendering (white lines) on top of the source and target objects. The background edges are drawn as yellow polylines. On the bottom, windows (3) and (6) are interactive renderings of the global visibility maps, showing the visibility of the feature polygons using color-coded polygons. Windows (4) and (5) display the (s, t)-planes of the two light fields, with each yellow dot representing a key view used for specifying background edges.

The two most important operations in light field morphing are feature specification and visibility processing. The key to feature specification is the feature polygons, which are 3D (non-planar) polygons approximating surface patches of 3D objects. The key to visibility processing is the global visibility map, which associates each ray of a light field L with a feature polygon of L.

The critical roles of feature polygons and global visibility maps in the warping of a light field L are illustrated in the lower part of Figure 17.2. The global visibility map of L can be computed from the user-specified feature polygons of L. The global visibility map partitions L into ray bundles such that each feature polygon P is associated with a ray bundle $R(P)$. Light field warping is then performed by warping one ray bundle at a time using ray-space warping, with the ray correspondence of a ray bundle $R(P)$ determined by P's control primitives.

Feature polygons are only needed where visibility changes. Rays not in any ray bundle are called background rays, which can be easily treated by image warping because there is no visibility change involved.

Following the Lumigraph [91] convention (Chapter 2), the (u, v)-plane is the image (focal) plane and the (s, t)-plane the camera plane. For a given light field

L, L can be thought of as either a collection of images $\{L_{(s,t)}\}$ or a set of rays $\{L(u, v, s, t)\}$. An image $L_{(s_0,t_0)}$ is also called a view of the light field L. In $L_{(s_0,t_0)}$, the pixel at position (u_0, v_0) is denoted as $L_{(s_0,t_0)}(u_0, v_0)$, which is equivalent to ray $L(u_0, v_0, s_0, t_0)$.

17.3 Features and visibility

In feature-based morphing [89], the corresponding features of the source and target objects are identified by a pair of feature elements. In this section, we show how to specify such feature element pairs when the source and target objects are described by light fields. We also describe visibility processing using feature polygons.

17.3.1 Feature specification

The user specifies feature element pairs using the UI shown in Figure 17.3. Three types of feature elements are used: feature lines, feature polygons, and background edges.

Feature lines. A feature line is a 3D line segment connecting two points called its vertices, which are also called *feature points*. The purpose of a feature line is to approximate a curve on the surface of a 3D object. The user specifies a feature line E by identifying the pixel locations of its vertices. Once E is specified, the system displays E on top of the interactive rendering of the light field.

To determine the 3D position of a vertex $\mathbf{v}$, *geometry-guided manual correspondence* is used: the user manually identifies projections $p_1(\mathbf{v})$ and $p_2(\mathbf{v})$ of $\mathbf{v}$ in two different views $L_{(s_1,t_1)}$ and $L_{(s_2,t_2)}$ under the guidance of epipolar geometry [72]. After the user specifies $p_1(\mathbf{v})$ in view $L_{(s_1,t_1)}$, the epipolar line of $p_1(\mathbf{v})$ is drawn in view $L_{(s_2,t_2)}$ as a guide for specifying $p_2(\mathbf{v})$ since $p_2(\mathbf{v})$ must be on the epipolar line of $p_1(\mathbf{v})$. Because the camera parameters of both views are known, calculating $\mathbf{v}$ from $p_1(\mathbf{v})$ and $p_2(\mathbf{v})$ is straightforward.

Feature polygons. A feature polygon P is a 3D polygon defined by n feature lines $\{E^1, ..., E^n\}$, which are called the edges of P. P has *control primitives* $\{E^1, ..., E^{n+k}\}$ which includes both the edges of P and *supplementary feature lines* $\{E^{n+1}, ..., E^{n+k}\}$ for additional control inside the feature polygon. The purpose of a feature polygon is to approximate a surface patch of a 3D object. In general, P is allowed to be non-planar so that it can approximate a large surface patch as long as the surface patch is relatively flat.

To specify a feature polygon, the user draws a series of connected feature lines (two consecutive lines sharing a vertex) in counterclockwise order in the interactive display of a light field. A technical difficulty in the specification process is that, because light field rendering does not perform visible surface computation, all vertices are visible in every view. Fortunately, the user can easily distinguish vertices on visible surfaces from those on hidden surfaces, for two reasons. First, there are relatively few vertices and a vertex on the visible surface can be identified by the landmark it labels. More importantly, the interactive display gives different motion parallax to visible vertices in the front and invisible ones in the back.

To ensure that the patches are well approximated by feature polygons, the geometry of the patches is restricted. More specifically, for a surface patch S approximated by a feature polygon P, S is required to have no self-occlusion and be relatively flat. S is split if either requirement is not met. By requiring S to have no self-occlusion, self-occlusion in P can be avoided if it is a sufficiently close approximation of S. For such a P, we only have to check occlusion caused by other feature polygons during visibility processing. Note that whether S satisfies the two conditions is solely judged within the viewing range of L. For example, consider any one of the faces in Figure 17.3. The surface patch approximated by a feature polygon has no self-occlusion for the viewing range of the light field shown. However, when the viewpoint moves beyond the viewing range of this light field, e.g., to the side of the face, the nose will cause self-occlusion within the surface patch.

Background edges.   Background edges are introduced to control rays that do not belong to any feature polygons. These rays exist for two reasons. First, feature polygons only roughly approximate surface patches of a 3D object. In each light field view, rays near the object silhouette may not be covered by the projection of any feature polygons. Second, parts of the object surface may not be affected by the visibility change caused by object shape change. There is no need to specify feature polygons for the corresponding rays.

For rays that do not belong to any feature polygons, they are controlled with background edges, which are 2D image edges specified by the user. Background edges play the same role as the feature edges in image morphing [12]. A series of connected background edges form a background polyline. As shown in Figure 17.3, a background polyline is manually specified in a few key views and interpolated into other views by linear interpolation.

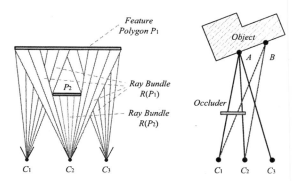

Fig. 17.4. Left: Example of ray bundles in a light field with three cameras. The rays associated with a feature polygon P_1 across all views of the light field constitute ray bundle $R(P_1)$. In this example, $R(P_1)$ includes all pink rays but not the green rays. Right: Example of nearest visible rays in a light field with three cameras. Occluded ray $C_1 B$ (pink) is replaced by $C_2 B$ (pink), while $C_1 A$ is replaced by $C_3 A$.

17.3.2 Global visibility map

After specifying all feature elements of a light field L, the global visibility map (or visibility map for short) of L can be defined as follows:

Definition 1 *The global visibility map of a light field L with feature polygons $\{P_1, ..., P_{n_p}\}$ is a mapping $V : L \rightarrow N$ from the ray space L to the set of integers N such that*

$$V(u, v, s, t) = \begin{cases} i & \text{if ray } L(u, v, s, t) \text{ belongs to } P_i \\ -1 & \text{otherwise} \end{cases}$$

Intuitively, V may be regarded as a light field of false colors, with $V(u, v, s, t)$ indicating the id of the feature polygon visible at ray $L(u, v, s, t)$. Figure 17.3 shows examples of visibility maps.

Visibility computation. The visibility map V is computed based on the vertex geometry of feature polygons[1] as well as the fact that feature polygons have no self-occlusion by construction. The main calculation is that of the visibility of a set of relatively flat but non-planar polygons. This is a calculation that can be done efficiently using OpenGL.

Consider rendering a non-planar polygon P_i into a view $L_{(s,t)}$. A problem with this rendering is that the projection of P_i into the view $L_{(s,t)}$ may be a concave polygon, which OpenGL cannot display correctly. One solution to this problem is a two-pass rendering method using the stencil buffer. This method works for feature polygons since they have no self-occlusion as we mentioned earlier. Alternatively, visibility map computation can be simplified by restricting feature polygons to be triangles without supplementary feature lines. However, the user is required to draw many feature polygons, which makes feature specification unnecessarily tedious.

Ray bundles. Based on the visibility map V, the rays of L can be grouped according to their associated feature polygons. A group so obtained is called a ray bundle, denoted as $R(P_i)$ where P_i is the associated feature polygon. As we shall see in Section 17.4, $R(P_i)$ can be warped using ray-space warping with the control primitives of P_i (see the ray-space warping equation (17.1) in Section 17.4.1). The ray correspondence of $R(P_i)$ is thus completely determined by the control primitives of P_i. Rays that do not belong to any ray bundle are called background rays. Background rays are controlled by the background edges. Ray bundles have been used in the context of global illumination [292].

17.4 Warping

As mentioned, for each $0 < \alpha < 1$, the light field morph L_α is obtained by blending two light fields $\hat{L}_0$ and $\hat{L}_1$, which are warped from L_0 and L_1 for feature alignment.

---

[1] Using feature polygons to handle occlusion is related to layered representations in image-based rendering (e.g., [264]).

In this section, we discuss the warping from L_0 to $\hat{L}_0$ since the warping from L_1 to $\hat{L}_1$ is essentially the same. We also describe an efficient warping algorithm for animation sequences of 3D morphing. The warping from L_0 to $\hat{L}_0$ takes the following steps: (a) calculate feature polygons and background edges of $\hat{L}_0$, (b) build the visibility map of $\hat{L}_0$, (c) compute ray bundles of the warped light field $\hat{L}_0$, and (d) treat background rays.

17.4.1 Basic ray-space warping

Because the rays of a light field L are grouped ray bundles, the basic operation of light field warping is to warp a ray bundle $R(P_i)$. For simplicity, let us assume that L has only an n-sided feature polygon P_i, whose feature lines are $\{E^1, ..., E^{n+k}\}$ before warping and $\{\hat{E}^1, ..., \hat{E}^{n+k}\}$ afterwards.

The basic ray-space warping regards the warped light field $\hat{L}$ as a 4D ray space and directly computes color values of individual rays:

$$\hat{L}(u, v, s, t) = L(u', v', s', t'),$$

where

$$(u', v')^T = \mathbf{f}(u, v, E^1_{(s',t')}, ..., E^{n+k}_{(s',t')}, \hat{E}^1_{(s,t)}, ..., \hat{E}^{n+k}_{(s,t)}) \qquad (17.1)$$

and (s', t') are free variables in the (s, t)-plane. The vector function $\mathbf{f}()$ is the Beier-Neely field warping function [12]. For a given point (u, v) in view $\hat{L}_{(s,t)}$, $\mathbf{f}()$ finds the preimage (u', v') in view $L_{(s',t')}$ based on the correspondence between the feature lines $E^1_{(s',t')}, ..., E^{n+k}_{(s',t')}$ in $L_{(s',t')}$ and $\hat{E}^1_{(s,t)}, ..., \hat{E}^{n+k}_{(s,t)}$ in $\hat{L}_{(s,t)}$.

For each ray $\hat{L}(u, v, s, t)$, the basic ray-space warping provides a set of rays $\{L(u', v', s', t')\}$ whose colors can be assigned to $\hat{L}(u, v, s, t)$. Possible values of (s', t') include (s, t), in which case ray-space warping yields the same result as image warping [12].

17.4.2 Light field warping

To warp the light field L_0 to $\hat{L}_0$, we apply the basic warping methods described above to feature polygons of L_0. The warping takes four steps. First, feature polygons and background edges of $\hat{L}_0$ are calculated. The vertices of feature lines in $\hat{L}_0$ are linearly interpolated from their counterparts of L_0 and L_1. Figure 17.5 (top row) shows an example of feature interpolation. For $i = 0, 1$, let $\{v^i_1, ..., v^i_n\}$ be the vertices of feature lines in L_i. The vertices of feature lines in $\hat{L}_0$ are $\{\hat{v}_1, ..., \hat{v}_n\}$, where

$$\hat{v}_k = (1 - \alpha)v^0_k + \alpha v^1_k, \quad k = 1, ..., n.$$

The connections between the vertices are the same in $\hat{L}_0$ and L_0. Thus we can easily obtain the feature polygons of $\hat{L}_0$ as well as their control primitives.

Second, the visibility map of $\hat{L}_0$ is built; this gives us information about the visibility changes caused by object shape change. Using the edge geometry of feature

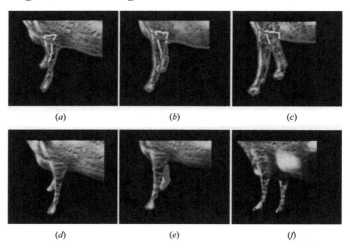

Fig. 17.5. Issues in light field warping. Top row shows interpolation of feature polygons, viewed from $(s,t) = (3,0)$. (a) Light field L_0 with feature lines. (b) Light field $L_{0.5}$ with feature lines. (c) Light field L_1 with feature lines. Bottom row shows a hole caused by object shape change. (d) Light field L_0 viewed from $(s,t) = (3,0)$. (e) Warped light field $\hat{L}_0$ viewed from $(s,t) = (3,0)$. The area highlighted in green is a hole corresponding to the occluded part of a feature polygon in (d). (f) Light field L_0 viewed from $(s,t) = (32,32)$. The feature polygon occluded at view $(s,t) = (3,0)$ is now fully visible.

See color plate section near center of book.

polygons of $\hat{L}_0$, we can perform the visibility calculation of these polygons, with non-planar polygons rendered by the view-dependent triangulation as before. The result of this visibility calculation is the visibility map of $\hat{L}_0$.

Third, the warped ray bundles of light field $\hat{L}_0 = \{\hat{L}_{0\,(s,t)}\}$ are computed view-by-view. Consider processing ray bundle $R(\hat{P}_0)$ in view $\hat{L}_{0\,(s,t)}$ for feature polygon $\hat{P}_0$ that corresponds to feature polygon P_0 in L_0. $\hat{L}_0(u,v,s,t)$ is evaluated in three steps:

1. Visibility testing. The visibility map of L_0 is checked to see whether P_0 is visible at ray $L_0(u',v',s,t)$ determined by the ray-space warping equation (17.1) with $(s',t') = (s,t)$.

2. Pixel mapping. If P_0 is visible at ray $L_0(u',v',s,t)$, we let $\hat{L}_0(u,v,s,t) = L_0(u',v',s,t)$.

3. Ray-space warping. Otherwise, $\hat{L}_{0\,(s,t)}(u,v)$ is in a hole, and ray-space warping is used to fill the hole. Figure 17.5 (top row) shows an example of a hole. The basic ray-space warping described earlier provides a set of values $\{L_0(u',v',s',t')\}$ parameterized by free variable (s',t'). Using the visibility map of L_0, a search is performed for the "nearest visible ray" $L_0(u',v',s',t')$ such that P_0 is visible at ray $L_0(u',v',s',t')$ determined by the ray-space warping equation (17.1) and (s',t') is as close to (s,t) as possible. This search starts from the immediate neighbors of (s,t)

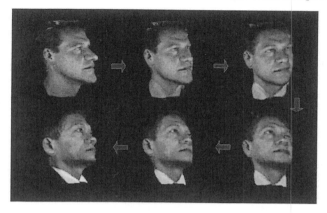

Fig. 17.6. 3D facial morphing.

in the (s, t)-plane and propagates outwards, accepting the first valid (s', t'). Note that the search will never fail because P_0 by construction is fully visible in at least one view of L_0. Once (s', t') is found, we set

$$\hat{L}(u, v, s, t) = L_0(u', v', s', t')$$

according to the ray-space warping equation (17.1). Figure 17.4 illustrates the "nearest visible ray."

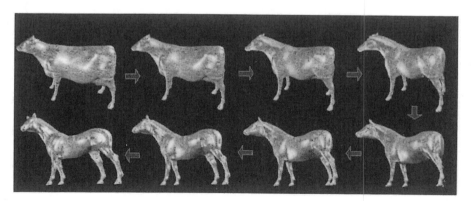

Fig. 17.7. A morphing example with large occlusions and specular surfaces.

In the last step, background rays, which correspond to pixels not covered by the projection of any feature polygon, are processed. Image warping is applied to these pixels, using the background edges and (projected) feature polygon edges as control primitives.

The idea behind choosing the "nearest visible ray" is the following. For $\hat{L}_0(u, v, s, t)$, the basic ray-space warping provides a set of values $\{L_0(u', v', s', t')\}$. Under the

Lambertian surface assumption, all rays are equally valid. However, the Lambertian surface assumption only approximately holds despite its widespread use in image-based rendering [40, 260]. By choosing the visible ray nearest to ray $L_0(u', v', s, t)$ when P_0 is occluded at the latter, we are trying to minimize the error caused by the Lambertian surface assumption.

Note that for the "nearest visible ray," we choose a visible ray $L_0(u', v', s', t')$ with (s', t') as close to (s, t) as possible. This is the measure of "closeness" used in [160]. A more natural measure is the angle deviation in [22]. Unfortunately, calculation of angle deviation requires a good estimation of the depth at pixel $L_{0\ (s,t)}(u', v')$. The estimated depth from the associated feature polygon may not be accurate enough.

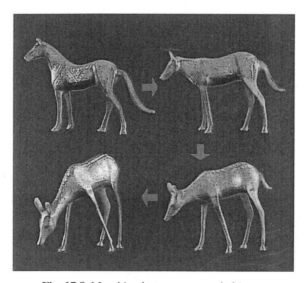

Fig. 17.8. Morphing between two real objects.

17.4.3 Warping for animation

The system can produce animation sequences that allow the user to observe a morphing process from a camera moving along an arbitrary path in 3D. In particular, the camera does not have to be inside the (s, t)-plane. One way to compute such a 3D morphing sequence is to first compute a sequence of light field morphs $M = \{L_0, L_{1/n}, ..., L_{(n-1)/n}, L_1\}$ and then create the 3D morphing sequence by rendering the light field morphs in M. Unfortunately, the CPU/storage costs for computing M can be very high. We now describe a method for generating a 3D morphing sequence without fully evaluating the sequence M.

Suppose we are given α and we want to compute the image I_α in the morphing sequence. From the known camera path and α, we can find the camera position $\mathbf{v}_\alpha$.

The image I_α is a blend of two images $\hat{I}_0$ and $\hat{I}_1$, where $\hat{I}_0$ is warped from L_0 and $\hat{I}_1$ is warped from L_1. The image $\hat{I}_0$ is warped from L_0 by first calculating, for each pixel (x_α, y_α) in the image $\hat{I}_0$, its corresponding ray $(u_\alpha, v_\alpha, s_\alpha, t_\alpha)$ and then applying ray-space warping. The image $\hat{I}_1$ is warped from L_1 the same way.

17.5 Results

The light field morphing algorithm was implemented on a Pentium III 667 MHz PC. In this section, we show some results and a few applications of light field morphing.

17.5.1 3D morphs and animations

A 3D morph L_α represents a plausible object having the essential features of both the source and target objects. L_α can be interactively displayed through light field rendering [160]. An animation sequence of the 3D morphing process from a camera moving along an arbitrary path in 3D can also be generated.

Figure 17.6 shows a 3D facial morphing between an Asian male and a Caucasian male. Both light fields are 33×33 in the (s, t)-plane and 256×256 in the (u, v)-plane. These light fields are rendered in 3D Studio Max from two Cyberscan models, each having about 90k triangles; the models are not used for morphing. Feature specification took about 20 minutes, and it involved the following: 12 pairs of feature polygons and 9 pairs of supplementary feature lines with 41 pairs of feature points, and 8 background polylines with 53 background edges. On the average, a background polyline was specified in 8 out of the 1089 views of each light field and interpolated into other views. With unoptimized implementation, the two global visibility maps took 37 seconds each, whereas light field warping and blending took 15 seconds and 0.5 seconds respectively per image.

Figure 17.7 provides an example with large occlusion and specular surfaces. The light fields are acquired the same way at the same resolution as in the 3D facial morphing example. Feature specification took about 30 minutes. Here, 50 pairs of feature polygons and 1 pair of supplementary feature lines with 126 pairs of feature points were specified. Four background polylines (made of 50 background edges) were also specified. On average, a background polyline was specified in 8 out of the 1089 views of each light field and interpolated into other views.

Figure 17.11 shows the morphing between a real bronze statue and the famous statue of Egyptian queen Nefertiti. The surface of the antique bronze statue shows complicated material property. This is very difficult to model with a textured geometry model, although some nice progress has been made in this area [67].

To acquire the light fields of the bronze statue, outside-looking-in CMs [267] were used; CMs are a slightly different parameterization of the light field. Three hundred images of the bronze statue were acquired at a resolution of 360×360. The CMs of the Nefertiti statue was rendered at the same resolution in 3D Studio Max from a textured model (model not used for morphing). Feature specification took about 20 minutes. As for morphing time, the global visibility map took 15 seconds per map. Warping and blending took 7 seconds per image.

Figure 17.8 shows the morphing between two real bronze statues: a deer and a horse. The light fields were acquired at the same resolution as in the Nefertiti example. Feature specification took about 25 minutes. The global visibility map took 25 seconds per map. Warping and blending took 5 seconds per image.

Fig. 17.9. A frame (second from the left) enlarged from the 3D morphing sequence shown in Figure 17.1.

17.5.2 Key-frame morphing

As mentioned, key-frame morphing is light field morphing with very sparse light fields. Figure 17.1 and Figure 17.9 show the result of key-frame morphing between a real furry toy cat and the Stanford bunny. Note that the fur on the cat surface is very difficult to model with textured geometry models.

Three photographs of the toy cat were taken using a calibrated camera. Three images of the bunny were rendered using the same camera parameters as the real photographs. Feature specification took about 7 minutes. The global visibility map took 4 seconds to compute for each object. Warping and blending each image of the 3D morphing sequence took 24 seconds.

The number of key frames needed depends on both the visibility complexity of the source and target objects and the presence of non-lambertian surfaces. As expected, the quality of key-frame morphing improves as more input images are used. In this regard, key-frame morphing is more flexible than view morphing [260]. This flexibility is particularly important when, e.g., there are many visibility changes due to object shape change. In such a case, the nearest visible ray will be frequently needed to fill the holes, and it is desirable to have the nearest visible rays to be actually nearer for highly non-Lambertian surfaces.

17.5.3 Plenoptic texture transfer

Given the source and target objects O_0 and O_1 represented by light fields L_0 and L_1, the texture of O_0 is transferred to O_1 by constructing a light field L_{01} as follows. First, the feature elements of L_{01} are obtained as those of L_1. Second, L_0 is warped to $\hat{L}_0$ for feature alignment. Finally, L_{01} is obtained as the warped light fields $\hat{L}_0$. Intuitively, a morph is created using the feature elements of L_1 and the radiance of L_0. Unlike 2D texture transfer (e.g., [71]), plenoptic texture transfer is a 3D effect. Figure 17.10 shows the result of plenoptic texture transfer from the furry cat toy in Figure 17.1 onto the Stanford bunny. Note that for plenoptic texture transfer to work well, the two objects should be similar to avoid texture distortions.

17.6 Discussion

In light field morphing, it is easy to handle complex surface properties. Geometry-based 3D morphing, for example, will have difficulties with the furry cat example in Figure 17.1. On the other hand, the lack of geometry causes problems in light field morphing. An example is the view point restriction imposed by the input light fields. Geometry-based 3D morphing can be incorporated into light field morphing by using image-based visual hulls [190] as rough geometry to morph two light fields. However, the visual hull geometry cannot replace feature polygons because visual hull geometry is obtained from the silhouette and thus cannot handle visibility changes not on the object silhouette.

Light field morphing can be regarded as a generalization of image morphing (an image is a 1×1 light field) and as such can suffer the ghost problem in image morphing for poorly-specified feature lines [12]. Fortunately, the usual fixes in image morphing also work for light field morphing [12].

17.7 Concluding remarks

This chapter describes an algorithm for morphing 3D objects represented by light fields. The principal advantage of light field morphing is the ability to morph between image-based objects whose geometry and surface properties may be too difficult to model with traditional vision and graphics techniques. Light field morphing is based on ray correspondence, not surface reconstruction. The morphing algorithm is feature-based. An intuitive and easy-to-use UI is used for specifying feature polygons for controlling the ray correspondence between two light fields. The visibility changes due to object shape changes can be effectively handled by ray-space warping. Finally, it is worth noting that light field morphing is a flexible morphing scheme. The user can perform 3D morphing by starting with a few input images and adding more input images as necessary to improve the quality of 3D morphing sequences.

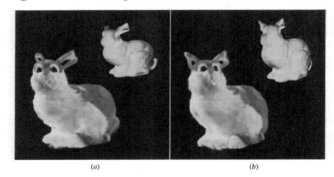

(a)   (b)

Fig. 17.10. Plenoptic texture transfer. The fur of the furry cat toy in Figure 17.1 is transferred onto the surface of the Stanford bunny. The images show the bunny before and after the texture transfer from different viewpoints.

See color plate section near center of book.

Fig. 17.11. A morphing example involving a surface of complicated material property (the antique bronze statue).

See color plate section near center of book.

References

1. M. D. Adams and F. Kossentini. Reversible integer-to-integer wavelet transforms for image compression: Performance evaluation and analysis. *IEEE Transactions on Image Processing*, 9:1010–1024, June 2000.
2. E. H. Adelson and J. R. Bergen. The plenoptic function and elements of early vision. *Computational Models of Visual Processing*, pages 3–20, 1991.
3. A. Agarwala, C. Zheng, C. Pal, M. Agrawala, M. Cohen, B. Curless, D. Salesin, and R. Szeliski. Panoramic video textures. *Proceedings of SIGGRAPH (ACM Transactions on Graphics)*, pages 821–827, August 2005.
4. D. G. Aliaga and I. Carlbom. Plenoptic stitching: A scalable method for reconstructing 3D interactive walkthroughs. *Computer Graphics (SIGGRAPH)*, pages 443–450, August 2001.
5. M. Antonini, M. Barland, P. Mathieu, and I. Daubechies. Image coding using wavelet transform. *IEEE Transactions on Image Processing*, 1:205–220, April 1992.
6. S. Avidan and A. Shashua. Novel view synthesis in tensor space. In *IEEE Conference on Computer Vision and Pattern Recognition*, pages 1034–1040, San Juan, Puerto Rico, June 1997.
7. H. Aydinoglu and M.H. Hayes. Stereo image coding: A projection approach. In *IEEE International Conference on Image Processing*, volume 8(4), pages 506–516, 1998.
8. F. Ayres, Jr. *Theory and Problems of Projective Geometry*. McGraw-Hill Inc., 1967.
9. S. Baker, R. Szeliski, and P. Anandan. A layered approach to stereo reconstruction. In *IEEE Conference on Computer Vision and Pattern Recognition*, pages 434–441, Santa Barbara, June 1998.
10. J. Baldwin, A. Basu, and H. Zhang. Panoramic video with predictive windows for telepresence applications. In *IEEE International Conference on Robotics and Automation*, volume 3, pages 1922–1927, Detroit, May 1999.
11. M. Bass, editor. *Handbook of Optics*. McGraw-Hill, New York, 1995.
12. T. Beier and S. Neely. Feature-based image metamorphosis. In *Computer Graphics (SIGGRAPH)*, Annual Conference Series, pages 35–42, Chicago, IL, July 1992.
13. B. P. Bergeron. Morphing as a means of generating variability in visual medical teaching materials. *Computers in Biology and Medicine*, 24:11–18, January 1994.
14. A. Berman, A. Dadourian, and P. Vlahos. Method for removing from an image the background surrounding a selected object. In *U.S. Patent 6,134,346*, 2000.
15. V. Bhaskaran and K. Konstantinides. *Image and Video Compression Standards*. Kluwer Academic Publishers, 2nd edition, 1997.

16. M. Bierling. Displacement estimation by hierarchical block matching. *SPIE Visual Communication and Image Processing*, 1001:942–951, 1988.

17. R. C. Bolles, H. H. Baker, and D. H. Marimont. Epipolar-plane image analysis: An approach to determining structure from motion. *International Journal of Computer Vision*, 1:7–55, 1987.

18. F. Bossen and T. Ebrahimi. A simple and efficient binary shape coding technique based on bitmap representation. In *IEEE International Conference on Acoustics, Speech, and Signal Processing*, volume 4, pages 3129–3132, Munich, Germany, April 1997.

19. M. Botsch, A. Hornung, M. Zwicker, and L. Kobbelt. High-quality surface splatting on today's GPUs. In *Proceedings of the Eurographics Symposium on Point-Based Graphics*, pages 17–24, 2005.

20. T. Boult. Remote reality demonstration. In *IEEE Conference on Computer Vision and Pattern Recognition*, pages 966–967, Santa Barbara, CA, June 1998.

21. N. Brady and F. Bossen. Shape compression of moving objects using context-based arithmetic encoding. *Signal Processing: Image Communications*, 15(7-8):601–617, May 2000.

22. C. Buehler, M. Bosse, L. McMillan, S. Gortler, and M. Cohen. Unstructured Lumigraph rendering. In *Computer Graphics (SIGGRAPH)*, pages 425–432, Los Angeles, CA, August 2001.

23. C. S. Burrus, R. A. Gopinath, and H. Guo. *Introduction to Wavelets and Wavelet Transforms: A primer*. Prentice-Hall, Englewood Cliffs, NJ, 1997.

24. E. Camahort. 4D light-field modeling and rendering. Technical Report TR01-52, The University of Texas at Austin, May 2001.

25. E. Camahort, A. Lerios, and D. Fussell. Uniformly sampled light fields. In *9th Eurographics Workshop on Rendering*, pages 117–130, Vienna, Austria, June/July 1998.

26. F. M. Candocia. Simultaneous homographic and comparametric alignment of multiple exposure-adjusted pictures of the same scene. *IEEE Transactions on Image Processing*, 12(12):1485–1494, December 2003.

27. D. Capel and A. Zisserman. Super-resolution from multiple views using learnt image models. In *Conf. on Computer Vision and Pattern Recognition*, volume 2, pages 627–634, Kauai, HI, December 2001.

28. R. L. Carceroni and K. N. Kutulakos. Multi-view scene capture by surfel sampling: From video streams to non-rigid 3D motion, shape and reflectance. In *International Conference on Computer Vision*, volume II, pages 60–67, 2001.

29. J. Carranza, C. Theobolt, M. A. Magnor, and H.-P. Seidel. Free-viewpoint video of human actors. *Proceedings of SIGGRAPH (ACM Transactions on Graphics)*, pages 569–577, July 2003.

30. CCITT Recommendation H.261, Geneva. *Video Coding for Audiovisual Services at px64 kbits/s*, March 1994.

31. J.-X. Chai, S. B. Kang, and H.-Y. Shum. Rendering with non-uniform approximate concentric mosaics. In *3D Structure from Multiple Images of Large-Scale Environments (SMILE)*, pages 94–108, Dublin, Ireland, July 2000.

32. J.-X. Chai and H.-Y. Shum. Parallel projections for stereo reconstruction. In *IEEE Conference on Computer Vision and Pattern Recognition*, volume 2, pages 493–500, Hilton Head Island, SC, June 2000.

33. J.-X. Chai, X. Tong, S.-C. Chan, and H.-Y. Shum. Plenoptic sampling. *Computer Graphics (SIGGRAPH)*, pages 307–318, July 2000.

34. S.-C. Chan, C. W. Kok, and S. W. Chau. Codebook generation and search algorithm for vector quantization using arbitrary hyperplanes. In *IEEE International Symposium on Circuits and Systems*, volume 3, pages 1885–1888, 1993.

35. S.-C. Chan, K. T. Ng, Z. F. Gan, K. L Chan, and H.-Y. Shum. The compression of simplified dynamic light fields. In *IEEE International Conference on Acoustics, Speech, and Signal Processing*, volume 3, pages 653–656, Hong Kong, April 2003.

36. S.-C. Chan, K. T. Ng, Z. F. Gan, K. L. Chan, and H.-Y. Shum. The plenoptic videos: Capturing, rendering and compression. In *IEEE International Symposium on Circuits and Systems*, volume 3, pages 905–908, May 2004.

37. S.-C. Chan, K. T. Ng, Z. F. Gan, K. L. Chan, and H.-Y. Shum. The plenoptic video. *IEEE Transactions on Circuits and Systems for Video Technology*, 15(12):1650–1659, December 2005.

38. T. F. Chan and L. A. Vese. Active contours without edges. *IEEE Transactions on Image Processing*, 10(2):266–277, February 2001.

39. C. Chang, G. Bishop, and A. Lastra. LDI tree: A hierarchical representation for image-based rendering. *Computer Graphics (SIGGRAPH)*, pages 291–298, August 1999.

40. S. Chen and L. Williams. View interpolation for image synthesis. *Computer Graphics (SIGGRAPH)*, pages 279–288, August 1993.

41. S. E. Chen. QuickTime VR – An image-based approach to virtual environment navigation. *Computer Graphics (SIGGRAPH)*, pages 29–38, August 1995.

42. W.-C. Chen, J.-Y. Bouguet, M. H. Chu, and R. Grzeszczuk. Light field mapping: Efficient representation and hardware rendering of surface light fields. *Proceedings of SIGGRAPH (ACM Transactions on Graphics)*, pages 447–456, July 2002.

43. G. Cheung, S. Baker, J. Hodgins, and T. Kanade. Markerless human motion transfer. In *International Symposium on 3D Data Processing Visualization and Transmission*, pages 373–378, Thessaloniki, Greece, September 2004.

44. G. Cheung, S. Baker, and T. Kanade. Visual hull alignment and refinement across time: A 3D reconstruction algorithm combining shape-from-silhouette with stereo. In *IEEE Conference on Computer Vision and Pattern Recognition*, volume 2, pages 375–382, Madison, WI, June 2003.

45. T. Chiang and Y. Q. Zhang. A new rate control scheme using quadratic rate-distortion modeling. *IEEE Transactions on Circuits and Systems for Video Technology*, 7(1):246–250, February 1997.

46. Y.-Y. Chuang, A. Agarwala, B. Curless, D. H. Salesin, and R. Szeliski. Video matting of complex scenes. In *Proceedings of SIGGRAPH (ACM Transactions on Graphics)*, pages 243–248, 2002.

47. Y.-Y. Chuang, B. Curless, D. H. Salesin, and R. Szeliski. A Bayesian approach to digital matting. In *IEEE Conference on Computer Vision and Pattern Recognition*, volume 2, pages 264–271, 2001.

48. L. Coconu and H.-C. Hege. Hardware-accelerated point-based rendering of complex scenes. In *Eurographics Workshop on Rendering*, pages 43–52, Aire-la-Ville, Switzerland, 2002.

49. D. Cohen-Or, A. Solomovici, and A. Levin. Three-dimensional distance field metamorphosis. *ACM Transactions on Graphics*, 17(2):116–141, April 1998.

50. R. T. Collins. A space-sweep approach to true multi-image matching. In *IEEE Conference on Computer Vision and Pattern Recognition*, pages 358–363, San Francisco, June 1996.

51. D. Comaniciu and P. Meer. Mean shift: A robust approach toward feature space analysis. *IEEE Transactions on Pattern Analysis and Machine Intelligence*, 24(5):603–619, May 2002.

52. J. H. Conway and N. J. A. Sloane. Fast quantizing and decoding algorithms for lattice quantizers and codes. *IEEE Transactions on Information Theory*, 28:227–232, March 1982.

53. J. H. Conway and N. J. A. Sloane. A fast encoding method for lattice codes and quantizers. *IEEE Transactions on Information Theory*, 29:820–824, November 1983.

54. T. M. Cover and J. A. Thomas. *Elements of Information Theory*. John Wiley and Sons, New York, 1991.

55. H. S. M. Coxeter. *Introduction to Geometry*. John Wiley and Sons, 1969.

56. A. Criminisi, P. Perez, and K. Toyama. Object removal by exemplar-based inpainting. In *EEE Conference on Computer Vision and Pattern Recognition*, volume 2, pages 721–728, Madison, WI, June 2003.

57. A. Criminisi, I. Reid, and A. Zisserman. Single view metrology. In *International Conference on Computer Vision*, pages 434–441, Kerkyra, Greece, September 1999.

58. B. Curless and M. Levoy. A volumetric method for building complex models from range images. *Computer Graphics (SIGGRAPH)*, pages 303–312, August 1996.

59. J. Davis. Mosaics of scenes with moving objects. In *IEEE Conference on Computer Vision and Pattern Recognition*, pages 354–360, Santa Barbara, CA, June 1998.

60. P. Debevec, Y. Yu, and G. Borshukov. Efficient view-dependent image-based rendering with projective texture-mapping. In *Eurographics Workshop on Rendering*, pages 105–116, 1998.

61. P. E. Debevec, C. J. Taylor, and J. Malik. Modeling and rendering architecture from photographs: A hybrid geometry- and image-based approach. *Computer Graphics (SIGGRAPH)*, pages 11–20, August 1996.

62. D. DeCarlo and D. Metaxas. The integration of optical flow and deformable models with applications to human face shape and motion estimation. In *IEEE Conference on Computer Vision and Pattern Recognition*, pages 231–237, San Francisco, CA, June 1996.

63. X. Decoret, F. Durand, F. X. Sillion, and J. Dorsey. Billboard clouds for extreme model simplification. *Proceedings of SIGGRAPH (ACM Transactions on Graphics)*, pages 689–696, July 2003.

64. X. Decoret, F. Sillion, G. Schaufler, and J. Dorsey. Multi-layered impostors for accelerated rendering. *Computer Graphics Forum*, 18(3):61–73, September 1999.

65. M Deering. Geometry compression. In *Computer Graphics (SIGGRAPH)*, Annual Conference Series, pages 13–20, August 1995.

66. Microsoft Document. Creating compressed textures. DirectX SDK documentation, MSDN Library, October 2000.

67. Julie Dorsey and Pat Hanrahan. Modeling and rendering of metallic patinas. *Computer Graphics (SIGGRAPH)*, pages 387–396, August 1996.

68. J. Duan and J. Li. Compression of the layered depth image. In *IEEE Data Compression Conference*, pages 331–340, March 2001.

69. P. Dutré, P. Bekaert, and K. Bala. *Advanced Global Illumination*. AK Peters, Natick, MA, 2003.

70. B. Romeny (Ed.). *Geometry-Driven Diffusion in Computer Vision*. Kluwer Academic Publishers, Netherlands, 1994.

71. A. A. Efros and W. T. Freeman. Image quilting for texture synthesis and transfer. *Computer Graphics (SIGGRAPH)*, pages 341–346, August 2001.

72. O. Faugeras. *Three-dimensional Computer Vision: A Geometric Viewpoint*. MIT Press, Cambridge, MA, 1993.

73. O. Faugeras, L. Robert, S. Laveau, G. Csurka, C. Zeller, C. Gauclin, and I. Zoghlami. 3-D reconstruction of urban scenes from image sequences. *Computer Vision and Image Understanding*, 69(3):292–309, March 1998.

74. M. A. Fischler and R. C. Bolles. Random Sample Consensus: A paradigm for model fitting with applications to image analysis and automated cartography. *Communications of the ACM*, 24(6):381–395, June 1981.

75. A. Fitzgibbon, Y. Wexler, and A. Zisserman. Image-based rendering using image-based priors. In *International Conference on Computer Vision*, volume 2, pages 1176–1183, 2003.

76. M. Flierl and B. Girod. *Video Coding with Superimposed Motion-Compensated Signals: Applications to H.264 and Beyond*, volume 760 of *Engineering and Computer Science*. Kluwer Academic Publishers, Boston, MA, 2004.

77. J. Foote and D. Kimber. Flycam: Practical panoramic video and automatic camera control. In *IEEE International Conference on Multimedia and Expo.*, volume 3, pages 1419–1422, New York, July 2000.

78. L. Gall and A. Tabatabai. Subband coding of digital images using symmetric short kernel filters and arithmetic coding techniques. In *IEEE International Conference on Acoustics, Speech, and Signal Processing*, pages 761–765, New York, NY, 1988.

79. Z. F. Gan, S.-C. Chan, K. T. Ng, K. L Chan, and H.-Y. Shum. An object-based approach to plenoptic videos. In *IEEE International Symposium on Circuits and Systems*, pages 3435–3438, Kobe, Japan, May 2005.

80. Z. F. Gan, S.-C. Chan, K. T. Ng, and H.-Y. Shum. On the rendering and post-processing of simplified dynamic light fields with depth information. In *IEEE International Conference on Acoustics, Speech, and Signal Processing*, volume 3, pages 321–324, Montreal, Canada, May 2004.

81. Z. F. Gan, S.-C. Chan, K. T. Ng, and H.-Y. Shum. Object tracking for a class of dynamic image-based representations. In *SPIE Conference on Visual Communication and Image Processing*, volume 5960-134, Beijing, China, July 2005.

82. A. Gersho and R. M. Gray. *Vector quantization and signal compression*. Kluwer Academic Press, 1992.

83. C. Geyer and K. Daniilidis. Omnidirectional video. *The Visual Computer*, 19(6):405–416, October 2003.

84. B. Girod. Motion compensation: Visual aspects, accuracy, and fundamental limits. In *Motion Analysis and Image Sequence Processing*. Kluwer, 1995.

85. B. Girod, P. Eisert, M. Magnor, E. Steinback, and T. Wiegand. 3-D image models and compression–synthetic hybrid or natural fit? In *IEEE International Conference on Image Processing*, pages 525–529, Kobe, Japan, October 1999.

86. B. Girod, F. Harung, and U. Horn. Subband image coding. In *Subband and wavelet transforms: Design and applications*. Kluwer Academic Publishers, Boston, MA, 1995.

87. B. Goldlücke, M. Magnor, and B. Wilburn. Hardware-accelerated dynamic light field rendering. In *Workshop on Vision, Modeling, and Visualization*, pages 455–462, Erlangen, Germany, November 2002.

88. D. B. Goldman and J.-H. Chen. Vignette and exposure calibration and compensation. In *International Conference on Computer Vision*, pages 899–906, Beijing, China, October 2005.

89. J. Gomes, B. Costa, L. Darsa, and L. Velho. *Warping and Morphing of Graphics Objects*. Morgan Kaufmann, 1998.

90. C. Gomila. The H.264/MPEG-4 AVC video coding standard. *Short tutorial, EURASIP News Letter*, 15:19–34, 2004.

91. S. J. Gortler, R. Grzeszczuk, R. Szeliski, and M. F. Cohen. The Lumigraph. In *Computer Graphics (SIGGRAPH)*, pages 43–54, New Orleans, August 1996.

92. C. Granheit, A. Smolic, and T. Wiegand. Efficient representation and interactive streaming of high-resolution panoramic views. In *IEEE International Conference on Image Processing*, volume 3, pages 209–212, Rochester, NY, Sept. 2002.

93. N. Greene. Environment mapping and other applications of world projections. *IEEE Computer Graphics and Applications*, 6(11):21–29, November 1986.

94. N. Greene and P. S. Heckbert. Creating raster Omnimax images from multiple perspective views using the Elliptical Weighted Average filter. *IEEE Computer Graphics and Applications*, 6(6):21–27, June 1986.

95. A. Gregory, A. State, M. C. Lin, D. Manocha, and M. A. Livingston. Interactive surface decomposition for polyhedral morphing. *The Visual Computer*, 15(9):453–470, 1999.

96. G. Guennebaud and M. Paulin. Efficient screen space approach for hardware accelerated surfel rendering. In *Workshop on Vision, Modeling, and Visualization*, pages 1–10, 2003.

97. R. Gupta and R.I. Hartley. Linear pushbroom cameras. *IEEE Transactions on Pattern Analysis and Machine Intelligence*, 19(9):963–975, September 1997.

98. O. Hall-Holt and S. Rusinkiewicz. Stripe boundary codes for real-time structured-light range scanning of moving objects. In *International Conference on Computer Vision*, volume II, pages 359–366, 2001.

99. M. Halle. Holographic stereograms as discrete imaging systems. In *Practical Holography VIII (SPIE)*, volume 2176, pages 73–84, 1994.

100. M. Halle. Multiple viewpoint rendering. *Computer Graphics (SIGGRAPH)*, pages 243–254, July 1998.

101. R. Hartley and A. Zisserman. *Multiple View Geometry in Computer Vision (2nd Edition)*. Cambridge University Press, 2004.

102. S. W. Hasinoff, S. B. Kang, and R. Szeliski. Boundary matting for view synthesis. In *IEEE Workshop on Image and Video Registration*, Washington, DC, July 2004.

103. Z. He and S. K. Mitra. Optimum bit allocation and accurate rate control for video coding via ρ-domain source modeling. *IEEE Transactions on Circuits and Systems for Video Technology*, 12(10):840–849, October 2002.

104. P. S. Heckbert. Survey of texture mapping. *IEEE Computer Graphics and Applications*, 11(6):56–67, November 1986.

105. W. Heidrich, H. Lensch, M. F. Cohen, and H.-P. Seidel. Light field techniques for reflections and refractions. In *Eurographics Rendering Workshop*, pages 195–375, June 1999.

106. B. Heigl, R. Koch, M. Pollefeys, J. Denzler, and L. Van Gool. Plenoptic modeling and rendering from image sequences taken by hand-held camera. In *DAGM*, pages 94–101, 1999.

107. V. Hlaváč, A. Leonardis, and T. Werner. Automatic selection of reference views for image-based scene representations. In *European Conference on Computer Vision*, pages 526–535, 1996.

108. H. Hoppe. Progressive meshes. In *Computer Graphics (SIGGRAPH)*, pages 99–108, New Orleans, LA, August 1996.

109. B. K. P. Horn and M. J. Brooks, editors. *Shape From Shading*. MIT Press, Cambridge, MA, 1989.

110. Y. Horry, K. Anjyo, and K. Arai. Tour into the picture: Using a spidery mesh interface to make animation from a single image. In *Computer Graphics (SIGGRAPH)*, Annual Conference Series, pages 225–232, August 1997.

111. D. A. Huffman. A method for the construction of minimum redundancy codes. In *Institute of Radio Engineers*, volume 40, pages 1098–1101, September 1952.

112. J. F. Hughes. Scheduled Fourier volume morphing. *Computer Graphics (SIGGRAPH)*, 26(2):43–46, July 1992.

113. I. Ihm, S. Park, and R. Lee. Rendering of spherical light fields. In *Pacific Graphics*, pages 59–68, Seoul, Korea, October 1997.

114. iMove Inc. http://www.imoveinc.com., 2001.

115. M. Irani and S. Peleg. Improving resolution by image registration. *Graphical Models and Image Processing*, 53(3):231–239, May 1991.

116. A. Isaksen, L. McMillan, and S. Gortler. Dynamically reparameterized light fields. *Computer Graphics (SIGGRAPH)*, pages 297–306, July 2000.

117. ISO-IEC 14496-2:2001, Pattaya, Thailand. *MPEG-4, Coding of Audio Visual Objects - Part 2: Visual*, January 2001.

118. ISO-IEC, Int. Standard DIS 10918. *Digital Compression and Coding of Continuous-Tone Still Images*, 1994.

119. ISO/IEC 11172-2. *Information technology - Coding of moving pictures and associated audio for digital storage media at up to about 1.5 Mbit/s: Part 2 Video*, August 1993.

120. ISO/IEC 13818-2, Pattaya, Thailand. *MPEG-2, Information technology - Generic coding of moving pictures and associated audio information*, 2nd edition, 2000.

121. ISO/IEC JTC1/SC29 WG1 (JPEG/JBIG). *Lossless and near lossless coding of continuous tone still images (JPEG-LS)*, July 1997.

122. ISO/IEC JUTC1/SC 29/WG1 N1359. *Information technology - coded representation of picture and audio information - lossy/lossless coding of bi-level images*, July 1999.

123. ISO/IEC15444-1:2004. *Information technology – JPEG 2000 image coding system – Part 1: Core coding system*, 2004.

124. ISO/IECJTC1/SC19/ WG11 N3908, Pisa. *MPEG-4 video verification model v18.0: Coding of Moving Pictures and Audio*, January 2001.

125. ITU-T, Pattaya, Thailand. *H.264, Draft ITU-T Recommendation and Final Draft International Standard*, 2003.

126. ITU-T Recommendation. *H.263 Video Coding for Low Bit Rate Communication*, November 2000.

127. A. K. Jain. *Fundamentals of Digital Image Processing*. Prentice-Hall, 1989.

128. L. R. Jain and A. K. Jain. Displacement measurement and its application in interframe image coding. *IEEE Transactions on Communications*, 29:1799–1808, December 1981.

129. N. S. Jayant and P. Noll. *Digital Coding of Waveforms*. Prentice-Hall, Englewood Cliffs, NJ, 1984.

130. H. W. Jensen. *Realistic Image Synthesis Using Photon Mapping*. A K Peters Ltd., 2001.

131. S. Jeschke, M. Wimmer, and H. Schumann. Layered environment-map impostors for arbitrary scenes. In *Proceedings of Graphics Interface*, pages 1–8, May 2002.

132. *JPEG 2000 Home Page*. http://www.jpeg.org/jpeg2000/.

133. T. Kanade, P. W. Rander, and P. J. Narayanan. Virtualized Reality: Constructing virtual worlds from real scenes. *IEEE MultiMedia Magazine*, 1(1):34–47, Jan-March 1997.

134. S. B. Kang. A survey of image-based rendering techniques. In *Videometrics VI (SPIE International Symposium on Electronic Imaging: Science and Technology)*, volume 3641, pages 2–16, San Jose, CA, January 1999.

135. S. B. Kang. Catadioptric self-calibration. In *IEEE Conference on Computer Vision and Pattern Recognition*, volume 1, pages 201–207, Hilton Head Island, USA, June 2000.

136. S. B. Kang. Radial distortion snakes. In *IAPR Workshop on Machine Vision Applications*, pages 603–606, Tokyo, Japan, Nov. 2000.

137. S. B. Kang, S. M. Seitz, and P.-P. Sloan. Visual tunnel analysis for visibility prediction and camera planning. In *IEEE Conference on Computer Vision and Pattern Recognition*, volume 2, pages 195–202, Hilton Head Island, SC, June 2000.

138. S. B. Kang and R. Szeliski. 3-D scene data recovery using omnidirectional multibaseline stereo. *IEEE Conference on Computer Vision and Pattern Recognition*, pages 364–370, June 1996.

139. S. B. Kang and R. Szeliski. Extracting view-dependent depth maps from a collection of images. *International Journal of Computer Vision*, 58(2):139–163, July 2004.

140. S. B. Kang, R. Szeliski, and J. Chai. Handling occlusions in dense multi-view stereo. In *IEEE Conference on Computer Vision and Pattern Recognition*, volume I, pages 103–110, Kauai, HI, December 2001.

141. A. Katayama, K. Tanaka, T. Oshino, and H. Tamura. A viewpoint dependent stereoscopic display using interpolation of multi-viewpoint images. In S. Fisher, J. Merritt, and B. Bolas, editors, *Stereoscopic Displays and Virtual Reality Systems II (SPIE)*, volume 2409, pages 11–20, 1995.

142. A. K. Katsaggelos, L. P. Kondi, F. W. Meier, J. Ostermann, and G. M. Schuster. MPEG-4 and rate distortion based shape coding techniques. In *Proceedings of the IEEE, Special Issue on Multimedia Signal Processing, Part 2*, volume 86, pages 1126–1154, June 1998.

143. Q. Ke and T. Kanade. A subspace approach to layer extraction. In *IEEE Conference on Computer Vision and Pattern Recognition*, volume I, pages 255–262, Hawaii, December 2001.

144. J. R. Kent, W. E. Carlson, and R. E. Parent. Shape transformation for polyhedral objects. *Computer Graphics (SIGGRAPH)*, 26(2):47–54, July 1992.

145. L. Kobbelt and M. Botsch. A survey of point-based techniques in computer graphics. *Computers and Graphics*, 28(6):801–814, 2004.

146. T. Koga, K. Iinuma, A. Hirano, Y. Iijima, and T. Ishiguro. Motion-compensated interframe coding for video conferencing. In *National Telecommunication Conference*, pages C9.6.1–5, New Orleans, LA, November/December 1981.

147. V. Kolmogorov and R. Zabih. Multi-camera scene reconstruction via graph cuts. In *European Conference on Computer Vision*, pages 82–97, 2002.

148. R. Krishnamurthy, B. B. Chai, H. Tao, and S. Sethuraman. Compression and transmission of depth maps for image based rendering. In *IEEE International Conference on Image Processing*, volume 3, pages 828–831, Thessaloniki, Greece, October 2001.

149. Y. Kunita, M. Inami, T. Maeda, and S. Tachi. Real-time rendering system of moving objects. In *IEEE Workshop on Multi-View Modeling and Analysis of Visual Scenes*, pages 81–88, June 1999.

150. P. Lalonde and A. Fournier. Interactive rendering of wavelet projected light fields. In *Graphics Interface*, pages 107–114, Kingston, Canada, June 1999.

151. S. Laveau and O. D. Faugeras. 3-D scene representation as a collection of images. In *International Conference on Pattern Recognition*, volume A, pages 689–691, Jerusalem, Israel, October 1994.

152. F. Lazarus and A. Verroust. Three-dimensional metamorphosis: A survey. *The Visual Computer*, 14(8-9):373–389, 1998.

153. A. Lee, D. Dobkin, W. Sweldens, and P. Schröder. Multiresolution mesh morphing. In *Computer Graphics (SIGGRAPH)*, Annual Conference Series, pages 343–350, August 1999.

154. H.-J. Lee, T. Chiang, and Y.-Q. Zhang. Scalable rate control for MPEG-4 video. *IEEE Transactions on Circuits and Systems for Video Technology*, 10(6):878–894, September 2000.

155. J. Lengyel. The convergence of graphics and vision. *IEEE Computer*, 31(7):46–53, 1998.

156. J. Lengyel and J. Snyder. Rendering with coherent layers. In *Computer Graphics (SIGGRAPH)*, pages 233–242, August 1997.

157. A. Lerios, C. D. Garfinkle, and M. Levoy. Feature-based volume metamorphosis. In *Computer Graphics (SIGGRAPH)*, pages 449–456, August 1995.

158. H. W. Leung and T. Chen. Compression with mosaic prediction for image-based rendering applications. In *IEEE International Conference on Multimedia and Expo*, New York, NY, July 2000.

159. M. Levoy. Expanding the horizons of image-based modeling and rendering. In *Image-Based Rendering: Really New or Deja Vu (SIGGRAPH Panel)*, 1997.

160. M. Levoy and P. Hanrahan. Light field rendering. *Computer Graphics (SIGGRAPH)*, pages 31–42, August 1996.

161. M. Levoy and T. Whitted. The use of points as a display primitive. Technical report, UNC Technical Report 85-022, University of North Carolina, Chapel Hill, NC, 1985.

162. M. Lhuillier and L. Quan. Image interpolation by joint view triangulation. In *IEEE Conference on Computer Vision and Pattern Recognition*, volume 2, pages 139–145, Fort Collins, CO, June 1999.

163. M. Lhuillier and L. Quan. Image-based rendering by joint view triangulation. *IEEE Transactions on Circuits and Systems for Video Technology*, 13(11):1051–1063, November 2003.

164. J. Li, H.-Y. Shum, and Y. Q. Zhang. On the compression of image based rendering scene. In *IEEE International Conference on Image Processing*, volume 2, pages 21–24, Sept. 2000.

165. Y. Li, J. Sun, C. K. Tang, and H. Y. Shum. Lazy snapping. In *Proceedings of SIGGRAPH (ACM Transactions on Graphics)*, pages 303–308, August 2004.

166. S. Lin, Y. Li, S. B. Kang, X. Tong, and H.-Y. Shum. Simultaneous separation and depth recovery of specular reflections. In *European Conference on Computer Vision*, volume 3, pages 210–224, Copenhagen, Denmark, May/June 2002.

167. Z. Lin and H.-Y. Shum. On the number of samples needed in light field rendering with constant-depth assumption. In *IEEE Conference on Computer Vision and Pattern Recognition*, pages 588–579, Hilton Head Island, SC, June 2000.

168. Z. Lin and H.-Y. Shum. A geometric analysis of light field rendering. *International Journal of Computer Vision*, 58(2):121–138, July 2004.

169. Y. Linde, A. Buzo, and R. M. Gray. An algorithm for vector quantizer design. *IEEE Transactions on Communications*, 28:84–95, January 1980.

170. A. Lippman. Movie maps: An application of the optical videodisc to computer graphics. *Computer Graphics (SIGGRAPH)*, 14(3):32–43, July 1980.

171. D. Lischinski and A. Rappoport. Image-based rendering for non-diffuse synthetic scenes. In *Eurographics Rendering Workshop*, pages 301–314, June 1998.

172. B. Liu and A. Zaccarin. New fast algorithm for estimation of block motion vectors. *IEEE Transactions on Circuits and Systems for Video Technology*, 3:148–157, 1993.

173. H. Lohscheller and U. Franke. Color picture coding-algorithm optimization and technical realization. *Frequenz*, 41:127–135, 1987.

174. M. E. Lukacs. Predictive coding of multi-viewpoint image sets. In *IEEE International Conference on Acoustics, Speech, and Signal Processing*, pages 521–524, 1986.

175. L. Luo, Y. Wu, J. Li, and Y.-Q. Zhang. Compression of concentric mosaic scenery with alignment and 3d wavelet transform. In *SPIE Image and Video Communications and Processing*, San Jose, CA, January 2000.

176. M. Magnor. *Video-based Rendering*. A K Peters, 2005.

177. M. Magnor and B. Girod. Adaptive block-based light field coding. In *3rd International Workshop on Synthetic and Natural Hybrid Coding and Three-Dimensional Imaging*, pages 140–143, Santorini, Greece, Sept. 1999.

178. M. Magnor and B. Girod. Hierarchical coding of light fields with disparity maps. In *IEEE International Conference on Image Processing*, Kobe, Japan, October 1999.

179. M. Magnor and B. Girod. Data compression for light-field rendering. *IEEE Transactions on Circuits and Systems for Video Technology*, 10(3):338–343, Apr. 2000.

180. M. Magnor and B. Girod. Model-aided coding of multi-viewpoint image data. In *IEEE International Conference on Image Processing*, volume 2, pages 919–922, Vancouver, Canada, September 2000.

181. M. Magnor and B. Girod. Model-based coding of multi-viewpoint imagery. In *SPIE Visual Communication and Image Processing*, volume 4067(2), pages 14–22, Perth, Australia, June 2000.

182. M. Magnor, P. Ramanathan, and B. Girod. Multi-view coding for image-based rendering using 3-D scene geometry. *IEEE Transactions on Circuits and Systems for Video Technology*, 13(11):1092–1106, November 2003.

183. J. Mairal and R. Keriven. A GPU implementation of variational stereo. Technical Report Research Report 05-13, CERTIS, November 2005.

184. S. Mallat. A theory of multiresolution signal decomposition: the wavelet representation. *Proc. of IEEE Data Compression Conference*, 11:523–541, March 2000.

185. H. S. Malvar. *Signal Processing with Lapped Transforms*. Artech House, Boston, MA, 1992.

186. S. Mann. Pencigraphy with AGC: Joint parameter estimation in both domain and range of functions in same orbit of the projective-Wyckoff group. In *International Conference on Image Processing*, volume 3, pages 193–196, Los Alamitos, CA, 1996.

187. S. Mann and R. W. Picard. Virtual bellows: Constructing high-quality images from video. In *International Conference on Image Processing*, volume I, pages 363–367, Austin, TX, November 1994.

188. A. R. Mansouri. Region tracking via level set PDEs without motion computation. *IEEE Transactions on Pattern Analysis and Machine Intelligence*, 24(7):947–961, July 2002.

189. W. Mark, L. McMillan, and G. Bishop. Post-rendering 3D warping. In *Symposium on I3D Graphics*, pages 7–16, April 1997.

190. W. Matusik, C. Buehler, R. Raskar, S. Gortler, and L. McMillan. Image-based visual hulls. *Computer Graphics (SIGGRAPH)*, pages 369–374, July 2000.

191. W. Matusik, H. Pfister, P. Ngan, P. Beardsley, R. Ziegler, and L. McMillan. Image-based 3D photography using opacity hulls. *Proceedings of SIGGRAPH (ACM Transactions on Graphics)*, pages 427–437, July 2002.

192. L. McMillan. An image-based approach to three-dimensional computer graphics. Technical report, Ph.D. Dissertation, UNC Computer Science TR97-013, 1999.

193. L. McMillan and G. Bishop. Head-tracked stereoscopic display using image warping. In *Stereoscopic Displays and Virtual Reality Systems II (SPIE)*, pages 21–30, February 1995.

194. L. McMillan and G. Bishop. Plenoptic modeling: An image-based rendering system. *Computer Graphics (SIGGRAPH)*, pages 39–46, August 1995.

195. G. Medioni, M. S. Lee, and C. K. Tang. *A Computational Framework for Feature Extraction and Segmentation*. Elseviers Science, Amsterdam, 2000.

196. G. Miller, S. Rubin, and D. Ponceleon. Lazy decompression of surface light fields for precomputed global illumination. *Eurographics Rendering Workshop*, pages 281–292, October 1998.

197. Y. Mishima. Soft edge chroma-key generation based upon hexoctahedral color space. In *U.S. Patent 5,355,174*, 1993.

198. J. L. Mitchell. *MPEG video compression standard*. Chapman and Hall, 1997.

199. J. L. Mitchell and W. B. Pennebaker. Optimal hardware and software arithmetic coding procedures for the q-coder binary arithmetic coder. *IBM Journal of Research and Development*, 32(6):727–736, November 1988.

200. S. Moezzi, A. Katkere, D. Y. Kuramura, and R. Jain. Reality modeling and visualization from multiple video sequences. *IEEE Computer Graphics and Applications*, 16(6):58–63, November 1996.

201. B. Mok, M. Chen, J. Dorsey, and F. Durand. Image-based modeling and photo editing. In *Computer Graphics (SIGGRAPH)*, pages 433–442, 2001.

202. Moonlight cordless Ltd. http://www.moonlight.co.il, 2000.

203. *MPEG Home Page*. http://www.chiariglione.org/mpeg/.

204. T. Naemura, M. Kaneko, and H. Harashima. Compression and representation of 3-D images. *IEICE Transactions on Information and Systems*, E82-D(3):558–567, 1999.

205. T. Naemura, J. Tago, and H. Harashima. Real-time video-based modeling and rendering of 3D scenes. *IEEE Computer Graphics and Applications*, 22(2):66–73, Mar. 2002.

206. V. S. Nalwa. A true omnidirectional viewer. Technical report, Bell Laboratories, Holmdel, NJ, February 1996.

207. S. K. Nayar. Catadioptric omnidirectional camera. In *IEEE Conference on Computer Vision and Pattern Recognition*, pages 482–488, San Juan, Puerto Rico, June 1997.

208. A. N. Netravali and B. G. Haskell. *Digital Pictures, Representation, Compression, and Standards*. Plenum Press, 2nd edition, 1995.

209. K. T. Ng, S.-C. Chan, and H.-Y. Shum. Scalable coding and progressive transmission of concentric mosaic using nonlinear filter banks. In *IEEE International Conference on Image Processing*, volume 2, pages 113–116, Thessaloniki, Greece, October 2001.

210. K. T. Ng, S.-C. Chan, and H.-Y. Shum. The data compression and transmission aspects of panoramic videos. *IEEE Transactions on Circuits and Systems for Video Technology*, 15(1):82–95, Jan. 2005.

211. K. T. Ng, S.-C. Chan, H.-Y. Shum, and S. B. Kang. On the data compression and transmission aspects of panoramic video. In *IEEE International Conference on Image Processing*, volume 2, pages 105–108, Thessaloniki, Greece, October 2001.

212. R. Ng. Fourier slice photography. *Proceedings of SIGGRAPH (ACM Transactions on Graphics)*, 24(3):735–744, July 2005.

213. J. R. Ohm. Encoding and reconstruction of multiview video objects: Looking at data compression in the context of the MPEG-4 multimedia standard. *IEEE Signal Processing Magazine*, 16(3):47–54, May 1999.

214. J. R. Ohm. Stereo/multiview encoding using the MPEG family of standards. In *Electronic Imaging*, San Diego, CA, January 1999.

215. M. Okutomi and T. Kanade. A multiple baseline stereo. *IEEE Transactions on Pattern Analysis and Machine Intelligence*, 15(4):353–363, April 1993.

216. M. M. Oliveira, G. Bishop, and D. McAllister. Relief texture mapping. In *Computer Graphics (SIGGRAPH)*, pages 359–368, New Orleans, LA, July 2000.

217. OMNI Scientific Entertainment Magazine (in Japanese). The wizard of the toric camera, April 1986.

218. R. Ooi, T. Hamamoto, T. Naemura, and K. Aizawa. Pixel independent random access image sensor for real time image-based rendering system. In *IEEE International Conference on Image Processing*, volume 2, pages 193–196, October 2001.

219. M. T. Orchard and G. J. Sullivan. Overlapped block motion compensation: An estimation-theoretic approach. *IEEE Transactions on Image Processing*, 3:693–699, September 1994.

220. A. Ortega and K. Ramchandran. Rate-distortion methods for image and video compression. *IEEE Signal Processing Magazine*, 15:23–50, November 1998.

221. S. J. Osher and R. P. Fedkiw. *Level Set Methods and Dynamic Implicit Surfaces*. Springer Verlag, 2002.
222. S. J. Osher and J. A. Sethian. Fronts propagation with curvature dependent speed: Algorithms based on Hamilton-Jacobi formulations. *Journal of Computational Physics*, 79(1):12–49, 1988.
223. J. D. Owens, D. Luebke, N. Govindaraju, M. Harris, J. Kruger, A. E. Lefohn, and T. J. Purcell. A survey of general-purpose computation on graphics hardware. In *Eurographics, State of the Art Reports*, pages 21–51, August 2005.
224. N. Paragios and R. Deriche. Geodesic active contours and levels sets for detection and tracking of moving objects. *IEEE Transactions on Pattern Analysis and Machine Intelligence*, 22(3):266–280, March 2000.
225. S. Peleg and M. Ben-Ezra. Stereo panorama with a single camera. In *IEEE Conference on Computer Vision and Pattern Recognition*, pages 395–401, Fort Collins, CO, June 1999.
226. S. Peleg and J. Herman. Panoramic mosaics by manifold projection. In *IEEE Conference on Computer Vision and Pattern Recognition*, pages 338–343, San Juan, Puerto Rico, June 1997.
227. S. Peleg, B. Rousso, A. Rav-Acha, and A. Zomet. Mosaicing on adaptive manifolds. *IEEE Transactions on Pattern Analysis and Machine Intelligence*, pages 1144–1154, October 2000.
228. W. B. Pennebaker and J. L. Mitchell. *JPEG Still Image Data Compression Standard*. Van Nostrand Reinhold, New York, 1993.
229. P. Perona and J. Malik. Scale-space and edge detection using anisotropic diffusion. *IEEE Transactions on Pattern Analysis and Machine Intelligence*, 12(7):629–639, 1990.
230. I. Peter and W. Strasser. The wavelet stream: Interactive multi resolution light field rendering. *Eurographics Rendering Workshop*, pages 262–273, June 2001.
231. H. Pfister, M. Zwicker, J. van Baar, and M. Gross. Surfels: Surface elements as rendering primitives. In *Computer Graphics (SIGGRAPH)*, pages 335–342, July 2000.
232. M. Pharr and G. Humphreys. *Physically Based Rendering*. Morgan Kaufmann, 2004.
233. F. Policarpo, M. M. Oliveira, and J. L. D. Comba. Real-time relief mapping on arbitrary polygonal surfaces. In *ACM SIGGRAPH Symposium on Interactive 3D Graphics and Games*, April 2005.
234. D. Porquet, J.-M. Dischler, and D. Ghazanfarpour. Real-time high quality view-dependent texture mapping using per-pixel visibility. In *International Conference on Computer Graphics and Interactive Techniques in Australasia and Southeast Asia (Graphite)*, November/December 2005.
235. T. Porter and T. Duff. Compositing digital images. In *Computer Graphics (SIGGRAPH)*, pages 253–259, July 1984.
236. W. K. Pratt. *Digital Image Processing*. John Wiley and Sons, New York, NY, 2nd edition, 1991.
237. K. Pulli, M. Cohen, T. Duchamp, H. Hoppe, J. McDonald, L. Shapiro, and W. Stuetzle. View-based rendering: Visualizing real objects from scanned range and color data. In *Eurographics Workshop on Rendering*, St. Etienne, France, June 1997.
238. A. Puri, R. V. Kollarits, and B. G. Haskell. Basics of stereoscopic video, new compression results with MPEG-2 and a proposal for MPEG-4. *Signal Processing: Image Comm.*, 10:201–234, 1997.
239. P. Rademacher. View-dependent geometry. In *Computer Graphics (SIGGRAPH)*, pages 439–446, Los Angeles, CA, August 1999.
240. P. Rademacher and G. Bishop. Multiple-center-of-projection images. In *Computer Graphics (SIGGRAPH)*, pages 199–206, Orlando, FL, July 1998.

241. M. Ramasubramanian, S. Pattanaik, and D. Greenberg. A perceptually based physical error metric for realistic image synthesis. In *Computer Graphics (SIGGRAPH)*, pages 73–82, Los Angeles, CA, August 1999.

242. K. R. Rao and P. Yip. *The Discrete Cosine Transform*. Academic Press, New York, NY, 1990.

243. A. Rav-Acha, Y. Pritch, D. Lischinski, and S. Peleg. Dynamosaicing: Video mosaics with non-chronological time. In *IEEE Conference on Computer Vision and Pattern Recognition*, pages 58–65, San Diego, CA, June 2005.

244. M. J. P. Regan, G. S. P. Miller, S. M. Rubin, and C. Kogelnik. A real-time low-latency hardware light-field renderer. In *Computer Graphics (SIGGRAPH)*, pages 287–290, Los Angeles, CA, August 1999.

245. L. Ren, H. Pfister, and M. Zwicker. Object space EWA surface splatting: A hardware accelerated approach to high quality point rendering. *Eurographics, Computer Graphics Forum*, 21(3):461–470, 2002.

246. J. Ribas-Corbera and S. Lei. Rate control in DCT video coding for low-delay communications. *IEEE Transactions on Circuits and Systems for Video Technology*, 9(1):172–185, February 1999.

247. I. E. Richardson and J. E. Richardson. *H.264 and MPEG-4 Video Compression: Video Coding for Next Generation Multimedia*. John Wiley and Sons, New York, NY, 2003.

248. S. Rusinkiewicz, O. Hall-Holt, and M. Levoy. Real-time 3D model acquisition. In *Proceedings of SIGGRAPH (ACM Transactions on Graphics)*, pages 438–446, San Antonio, TX, 2002.

249. M. A. Ruzon and C. Tomasi. Alpha estimation in natural images. In *IEEE Conference on Computer Vision and Pattern Recognition*, volume 1, pages 18–25, Hilton Head Island, SC, June 2000.

250. A. Said and W. A. Pearlman. A new, fast, and efficient image codec based on set partitioning in hierarchical trees. *IEEE Transactions on Circuits and Systems for Video Technology*, 6(3):243–250, June 1996.

251. M. Sainz and R. Pajarola. Point-based rendering techniques. *Computers and Graphics*, 28(6):869–879, 2004.

252. G. Sapiro. *Geometric Partial Differential Equations and Image Analysis*. Cambridge University Press, Cambridge, England, 2001.

253. D. Scharstein. Stereo vision for view synthesis. In *IEEE Conference on Computer Vision and Pattern Recognition*, pages 852–857, San Francisco, CA, June 1996.

254. D. Scharstein and R. Szeliski. A taxonomy and evaluation of dense two-frame stereo correspondence algorithms. *International Journal of Computer Vision*, 47(1):7–42, May 2002.

255. G. Schaufler. Per-object image warping with layered impostors. In *Eurographics Workshop on Rendering*, pages 145–156, June/July 1998.

256. H. Schirmacher, W. Heidrich, and H. Seidel. Adaptive acquisition of Lumigraphs from synthetic scenes. In *Eurographics*, pages 151–159, Milan, Italy, September 1999.

257. H. Schirmacher, W. Heidrich, and H.-P. Seidel. High-quality interactive Lumigraph rendering through warping. In *Graphics Interface*, pages 87–94, Montreal, Canada, May 2000.

258. H. Schirmacher, L. Ming, and H.-P. Seidel. On-the-fly processing of generalized Lumigraphs. *Eurographics, Computer Graphics Forum*, 20(3):165–173, 2001.

259. A. Schödl, R. Szeliski, D. H. Salesin, and I. Essa. Video textures. In *Computer Graphics (SIGGRAPH)*, pages 489–498, New Orleans, LA, July 2000.

260. S. M. Seitz and C. M. Dyer. View morphing. In *Computer Graphics (SIGGRAPH)*, pages 21–30, New Orleans, LA, August 1996.

261. S. M. Seitz and C. M. Dyer. Photorealistic scene reconstruction by voxel coloring. In *IEEE Conference on Computer Vision and Pattern Recognition*, pages 1067–1073, 1997.

262. S. M. Seitz and K. N. Kutulakos. Plenoptic image editing. In *International Conference on Computer Vision*, pages 17–24, 1998.

263. J. A. Sethian. *Level Set Methods: Evolving Interfaces in Geometry, Fluid Mechanics, Computer Vision and Materials Sciences*. Cambridge University Press, Cambridge, England, 1996.

264. J. Shade, S. Gortler, L.-W. He, and R. Szeliski. Layered depth images. In *Computer Graphics (SIGGRAPH)*, pages 231–242, Orlando, July 1998.

265. J. Shade, D. Lischinski, D. Salesin, T. DeRose, and J. Snyder. Hierarchical image caching for accelerated walkthroughs of complex environments. In *Computer Graphics (SIGGRAPH)*, pages 75–82, New Orleans, LA, August 1996.

266. Y. Shoham and A. Gersho. Efficient bit allocation for an arbitrary set of quantizers. *IEEE Transactions on Acoustic, Speech, and Signal Processing*, 36:1445–1453, September 1988.

267. H.-Y. Shum and L.-W. He. Rendering with Concentric Mosaics. In *Computer Graphics (SIGGRAPH)*, pages 299–306, Los Angeles, August 1999.

268. H.-Y. Shum, K. T. Ng, and S.-C. Chan. Virtual reality using the Concentric Mosaic: construction, rendering and data compression. In *IEEE International Conference on Image Processing*, volume 3, pages 644–647, September 2000.

269. H.-Y. Shum, K. T. Ng, and S.-C. Chan. A virtual reality system using the Concentric Mosaic: Construction, rendering, and data compression. *IEEE Transactions on Multimedia*, 7(1):85–95, February 2005.

270. H.-Y. Shum, J. Sun, S. Yamazaki, Y. Li, and C. K. Tang. Pop-up light field: An interactive image-based modeling and rendering system. *ACM Transactions on Graphics*, 23(2):143–162, April 2004.

271. H.-Y. Shum and R. Szeliski. Construction and refinement of panoramic mosaics with global and local alignment. In *International Conference on Computer Vision*, pages 953–958, Bombay, India, January 1998.

272. H.-Y. Shum and R. Szeliski. Stereo reconstruction from multiperspective panoramas. In *International Conference on Computer Vision*, pages 14–21, 1999.

273. H.-Y. Shum, L. Wang, and J.-X. Chai. Rendering by manifold hopping. In *SIGGRAPH 2001 technical sketch*, page 253, August 2001.

274. H.-Y. Shum, L. Wang, J.-X. Chai, and X. Tong. Rendering by manifold hopping. *International Journal of Computer Vision*, 50(2):185–201, 2002.

275. T. Sikora, S. Bauer, and B. Makai. Efficiency of shape-adaptive transforms for coding of arbitrarily shaped image segments. *IEEE Transactions on Circuits and Systems for Video Technology*, 5(3):254–258, June 1995.

276. Sleeping Train Productions, Inc. The work of director Michel Gondry, 2003.

277. P. P. Sloan, M. F. Cohen, and S. J. Gortler. Time critical Lumigraph rendering. In *Symposium on Interactive 3D Graphics*, pages 17–23, Providence, RI, April 1997.

278. N. J. A. Sloane. Tables of sphere packings and spherical codes. *IEEE Transactions on Information Theory*, 27:327–338, May 1981.

279. A. R. Smith and J. F. Blinn. Blue screen matting. In *Computer Graphics (SIGGRAPH)*, pages 259–268, New Orleans, LA, August 1996.

280. A. Smolić and H. Kimata. AHG on 3DAV Coding. ISO/IEC JTC1/SC29/WG11 MPEG03/M9635, Trondheim, Norway, July 2003.

281. R. Srinivasan and K.R. Rao. Predictive coding based on efficient motion estimation. *IEEE Transactions on Communications*, 33:1011–1014, September 1985.

282. J. Starck and A. Hilton. Model-based multiple view reconstruction of people. *International Conference on Computer Vision*, pages 915–922, October 2003.

283. C. Stiller and J. Konrad. Estimating motion in image sequences: A tutorial on modeling and computation of 2D motion. *IEEE Signal Processing Magazine*, 16:70–91, July 1999.

284. G. Strang and T. Q. Nguyen. *Wavelets and Filter Banks*. Wellesley-Cambridge Press, Wellesley, MA, 1996.

285. M. G. Strintzis and S. Malasiotis. Object-based coding of stereoscopic and 3D image sequences: A review. *IEEE Signal Processing Magazine*, 16(3):14–28, May 1999.

286. G. J. Sullivan and T. Wiegand. Rate-distortion optimization for video compression. *IEEE Signal Processing Magazine*, 16:74–90, November 1998.

287. R. Swaminathan, S. B. Kang, R. Szeliski, A. Criminisi, and S. K. Nayar. On the motion and appearance of specularities in image sequences. In *European Conference on Computer Vision*, volume 1, pages 508–523, Copenhagen, Denmark, May/June 2002.

288. R. Szeliski. Video mosaics for virtual environments. *IEEE Computer Graphics and Applications*, pages 22–30, March 1996.

289. R. Szeliski, S. Avidan, and P. Anandan. Layer extraction from multiple images containing reflections and transparency. In *IEEE Conference on Computer Vision and Pattern Recognition*, pages 246–253, Hilton Head Island, NC, June 2000.

290. R. Szeliski and M. Cohen. Sprites with depth–Fast rendering techniques for sprites with depth offsets. Technical report, Microsoft Research Vision Technology Group, Technical Note No. 5, 1998.

291. R. Szeliski and H.-Y. Shum. Creating full view panoramic image mosaics and environment maps. *Computer Graphics (SIGGRAPH)*, pages 251–258, August 1997.

292. L. Szirmay-Kalos and W. Purgathofer. Global ray-bundle tracing with hardware acceleration. *Eurographics Rendering Workshop*, pages 247–258, June 1998.

293. H. Tao, H.S. Sawhney, and R. Kumar. A global matching framework for stereo computation. In *International Conference on Computer Vision*, volume 1, pages 532–539, 2001.

294. D. S. Taubman and M. W. Marcellin. *JPEG2000: Image Compression Fundamentals, Standards, and Practice*. Kluwer Academic Publishers, 2001.

295. A. M. Tekalp. *Digital Video Processing*. Prentice Hall, 1995.

296. S. Teller. Toward urban model acquisition from geo-located images. In *Pacific Graphics*, pages 45–51, Singapore, October 1998.

297. X. Tong, J.-X. Chai, and H.-Y. Shum. Layered Lumigraph with LOD control. *Journal of Visualization and Computer Animation*, 13(4):249–261, 2002.

298. X. Tong and R. M. Gray. Coding of multi-view images for immersive viewing. In *IEEE International Conference on Acoustics, Speech, and Signal Processing*, volume 4, pages 1879–1882, June 2000.

299. X. Tong, H.-Y. Shum, S. B. Kang, T. Feng, and R. Szeliski. Locally reparameterized light fields. In *SIGGRAPH (Sketch)*, page 221, Los Angeles, CA, August 2001.

300. R. Y. Tsai. A versatile camera calibration technique for high-accuracy 3D machine vision metrology using off-the-shelf TV cameras and lenses. *IEEE Journal of Robotics and Automation*, RA-3(4):323–344, 1987.

301. Y. Tsin, S. B. Kang, and R. Szeliski. Stereo matching with reflections and translucency. In *IEEE Conference on Computer Vision and Pattern Recognition*, volume 1, pages 702–709, Madison, WI, June 2003.

302. M. Uyttendaele, A. Criminisi, S. B. Kang, S. Winder, R. Hartley, and R. Szeliski. High-quality image-based interactive exploration of real-world environments. *IEEE Computer Graphics and Applications*, 24(3):52–63, May/June 2004.

303. M. Uyttendaele, A. Eden, and R. Szeliski. Eliminating ghosting and exposure artifacts in image mosaics. In *IEEE Conference on Computer Vision and Pattern Recognition*, volume 2, pages 509–516, December 2001.

304. P. P. Vaidyanathan. *Multirate Systems and Filter Banks*. Prentice Hall, Englewood Cliffs, NJ, 1993.

305. V. Vaish, B. Wilburn, N. Joshi, and M. Levoy. Using plane + parallax for calibrating dense camera arrays. In *IEEE Conference on Computer Vision and Pattern Recognition*, volume 1, pages 2–9, Washington, DC, July 2004.

306. S. Vedula, S. Baker, and T. Kanade. Image-based spatio-temporal modeling and view interpolation of dynamic events. *ACM Transaction on Graphics*, 24(2):240–261, April 2005.

307. S. Vedula, S. Baker, P. Rander, R. Collins, and T. Kanade. Three-dimensional scene flow. In *International Conference on Computer Vision*, volume 2, pages 722–729, Kerkyra, Greece, September 1999.

308. S. Vedula, S. Baker, P. Rander, R. Collins, and T. Kanade. Three-dimensional scene flow. *IEEE Transactions on Pattern Analysis and Machine Intelligence*, 27(3):475–480, March 2005.

309. S. Vedula, S. Baker, S. Seitz, and T. Kanade. Shape and motion carving in 6D. In *IEEE Conference on Computer Vision and Pattern Recognition*, volume 2, pages 592–598, Hilton Head Island, NC, June 2000.

310. A. Vetro, H. Sun, and Y. Wang. MPEG-4 rate control for multiple video objects. *IEEE Transactions on Circuits and Systems for Video Technology*, 9(1):186–199, February 1999.

311. M. Vetterli and J. Kovacevic. *Wavelets and Subband Coding*. Prentice Hall, Englewood Cliffs, NJ, 1995.

312. B. Wallace. Merging and transformation of raster images for cartoon animation. In *Computer Graphics (SIGGRAPH)*, pages 253–262, Dallas, TX, 1981.

313. J. Wang and E. Adelson. Layered representation for motion analysis. In *IEEE Conference on Computer Vision and Pattern Recognition*, pages 361–366, New York, June 1993.

314. Y. Wang, J. Ostermann, and Y. Q. Zhang. *Video processing and communications*. Prentice Hall, 2002.

315. Y. Wexler, A. W. Fitzgibbon, and A. Zisserman. Bayesian estimation of layers from multiple images. In *European Conference on Computer Vision*, volume 3, pages 487–501, 2002.

316. T. Whitted. Overview of IBR: Software and hardware issues. In *International Conference on Image Processing*, volume 2, page 14, Vancouver, Canada, September 2000.

317. B. Wilburn, M. Smulski, H. H. K. Lee, and M. Horowitz. The light field video camera. In *SPIE Electonic Imaging: Media Processors*, volume 4674, pages 29–36, 2002.

318. L. Williams. Pyramidal parametrics. In *Computer Graphics (SIGGRAPH)*, pages 1–11, Detroit, MI, July 1983.

319. P. A. Wintz. Transform picture coding. *Proceedings of the IEEE*, 60(7):809–820, July 1972.

320. J. Woetzel and R. Koch. Real-time multi-stereo depth estimation on gpu with approximative discontinuity handling. In *1st European Conference on Visual Media Production (CVMP)*, London, UK, March 2004.

321. G. Wolberg. Image morphing: A survey. *The Visual Computer*, 14(8-9):360–372, 1998.

322. T. Wong, P. Heng, S. Or, and W. Ng. Image-based rendering with controllable illumination. In *Eurographics Workshop on Rendering*, pages 13–22, St. Etienne, France, June 1997.

323. D. N. Wood, D. I. Azuma, K. Aldinger, B. Curless, T. Duchamp, D. H. Salesin, and W. Stuetzle. Surface light fields for 3D photography. In *Computer Graphics (SIGGRAPH)*, Annual Conference Series, pages 287–296, New Orleans, LA, July 2000.

324. D. N. Wood, A. Finkelstein, J. F. Hughes, C. E. Thayer, and D. H. Salesin. Multiperspective panoramas for cel animation. In *Computer Graphics (SIGGRAPH)*, pages 243–250, Los Angeles, CA, August 1997.

325. O. Woodford and A. Fitzgibbon. Fast image-based rendering using hierarchical image-based priors. In *British Machine Vision Conference*, Oxford, UK, September 2005.

326. M. Wu and H.-Y. Shum. Real-time stereo rendering of concentric mosaics with linear interpolation. In *IEEE/SPIE Visual Communications and Image Processing*, pages 23–30, Perth, Australia, June 2000.

327. Q. Wu, K.-T. Ng, S.-C. Chan, and H.-Y. Shum. An object-based compression system for a class of dynamic image-based representations. In *IEEE International Conference on Image Processing*, volume 3, pages 405–408, Genova, Italy, September 2005.

328. Y. Wu, L. Luo, J. Li, and Y. Q. Zhang. Rendering of 3D wavelet compressed concentric mosaic scenery with progressive inverse wavelet synthesis (PIWS). In *SPIE Visual Communication and Image Processing*, volume 4067(4), Perth, Australia, June 2000.

329. Y. Wu, C. Zhang, J. Li, and J. Xu. Smart rebinning for compression of the concentric mosaics. In *ACM International Conference Multimedia*, pages 201–209, Los Angeles, CA, October 2000.

330. Y. Xiong and K. Turkowski. Creating image-based VR using a self-calibrating fisheye lens. In *IEEE Conference on Computer Vision and Pattern Recognition*, pages 237–243, San Juan, Puerto Rico, June 1997.

331. Z. Xiong, O. G. Guleryuz, and M. T. Orchard. A DCT-based embedded image coder. *IEEE Signal Processing Letters*, 3(11):1278–1288, 1996.

332. G. Xu and Z. Zhang. *Epipolar Geometry in Stereo, Motion and Object Recognition: A Unified Approach*. Kluwer Academic Publishers, 1996.

333. Y. Yamada, S. Tazaki, and R. M. Gray. Asymptotic performance of block quantizers with difference distortion measures. *IEEE Transactions on Information Theory*, 26(1):6–14, January 1980.

334. H. Yamaguchi, Y. Tatehira, K. Akiyama, and Y. Kobayashi. Stereo-scopic images disparity for predictive coding. In *IEEE International Conference on Acoustics, Speech, and Signal Processing*, pages 1976–1979, May 1989.

335. J. C. Yang, M. Everett, C. Buehler, and L. McMillan. A real-time distributed light field camera. In *Eurographics Workshop on Rendering*, pages 77–85, Pisa, Italy, June 2002.

336. R. Yang, G. Welch, and G. Bishop. Real-time consensus-based scene reconstruction using commodity graphics hardware. In *Pacific Graphics*, pages 225–234, Beijing, China, 2002.

337. A. Yilmaz, X. Li, and M. Shah. Object contour tracking using level sets. In *Asian Conference on Computer Vision*, Jeju Island, Korea, 2004.

338. C. Zhang and T. Chen. Spectral analysis for sampling image-based rendering data. *IEEE Transactions Circuit System Video Technology*, 13(11):1038–1050, November 2003.

339. C. Zhang and T. Chen. A survey on image-based rendering - Representation, sampling and compression. *EURASIP Signal Processing: Image Communication*, 19(1):1–28, January 2004.

340. C. Zhang and J. Li. Compression and rendering of concentric mosaics with reference block codec (rbc). In *SPIE Visual Communication and Image Processing*, volume 4067(5), Perth, Australia, June 2000.

341. C. Zhang and J. Li. Compression of Lumigraph with multiple reference frame (MRF) prediction and just-in-time rendering. In *IEEE Data Compression Conference*, pages 254–263, Snowbird, UT, March 2000.

342. C. Zhang and J. Li. Interactive browsing of 3D environment over the Internet. In *SPIE Visual Communication and Image Processing*, volume 4310(51), pages 509–520, San Jose, CA, January 2001.

343. L. Zhang, B. Curless, and S. M. Seitz. Spacetime stereo: Shape recovery for dynamic scenes. In *IEEE Conference on Computer Vision and Pattern Recognition*, pages 367–374, Madison, WI, June 2003.

344. Y. Zhang and C. Kambhamettu. On 3D scene flow and structure estimation. In *IEEE Conference on Computer Vision and Pattern Recognition*, volume II, pages 778–785, 2001.

345. Z. Zhang. Image-based geometrically-correct photorealistic scene/object modeling (IBPhM): A review. In *Asian Conference on Computer Vision*, pages 340–349, Hong Kong, China, January 1998.

346. Z. Zhang. A flexible new technique for camera calibration. *IEEE Transactions on Pattern Analysis and Machine Intelligence*, 22(11):1330–1334, November 2000.

347. Z. Zhang, L. Wang, B. Guo, and H.-Y. Shum. Feature-based light field morphing. *Proceedings of SIGGRAPH (ACM Transactions on Graphics)*, pages 457–464, July 2002.

348. J. Y. Zheng and S. Tsuji. Panoramic representation for route recognition by a mobile robot. *International Journal of Computer Vision*, 9:55–76, 1992.

349. C. L. Zitnick, S. B. Kang, M. Uyttendaele, S. Winder, and R. Szeliski. High-quality video view interpolation using a layered representation. *Proceedings of SIGGRAPH (ACM Transactions on Graphics)*, pages 600–608, August 2004.

350. J. Ziv. On universal data compression - an intuitive overview. *J. Vis. Comm. Image Rep.*, 5(4):317–321, December 1994.

351. J. Ziv and A. Lempl. A universal algorithm for sequential data compression. *IEEE Transactions on Information Theory*, 23(3):337–343, May 1977.

352. A. Zomet, D. Feldman, S. Peleg, and D. Weinshall. Mosaicing new views: The crossed-slits projection. *IEEE Transactions on Pattern Analysis and Machine Intelligence*, 25(6):741–754, June 2003.

353. D. Zorin and A. Barr. Correction of geometric perceptual distortions in pictures. In *Computer Graphics (SIGGRAPH)*, pages 257–264, Los Angeles, CA, August 1995.

Index

DATE DUE

GAYLORD No. 2333   PRINTED IN U.S.A.